DETERMINATE STRUCTURES

STATICS, STRENGTH, ANALYSIS, DESIGN

TECHNICAL MANUAL

BY

SAMUEL E. FRENCH, PH.D., P.E.

PROFESSOR OF ENGINEERING
UNIVERSITY OF TENNESSEE AT MARTIN

Delmar Publishers

I(T)P   An International Thomson Publishing Company

Albany • Bonn • Boston • Cincinnati • Detroit • London • Madrid • Melbourne
Mexico City •. New York • Pacific Grove • Paris • San Francisco • Singapore • Tokyo
Toronto • Washington

Notice to Reader

Publisher does not warrant or guarantee any of the products described herein or perform any independent analysis in connection with any of the product information contained herein. Publisher does not assume, and expressly disclaims, any obligation to obtain and include information other than that provided to it by the manufacturer.

The reader is expressly warned to consider and adopt all safety precautions that might be indicated by the activities herein and to avoid all potential hazards. By following the instructions contained herein, the reader willingly assumes all risks in connection with such instructions.

The publisher makes no representation or warranties of any kind, including but not limited to, the warranties of fitness for particular purpose or merchantability, nor are any such representations implied with respect to the material set forth herein, and the publisher takes no responsibility with respect to such material. The publisher shall not be liable for any special, consequential, or exemplary damages resulting, in whole or in part, from the readers' use of, or reliance upon, this material.

Delmar Staff

Publisher: Susan Simpfenderfer
Acquisitions Editor: Paul Shepardson
Developmental Editor: Jeanne Mesick
Project Editor: Thomas Smith
Production Coordinator: Karen Smith

COPYRIGHT © 1996
By Delmar Publishers
a division of International Thomson Publishing Inc.
The ITP logo is a trademark under license

Printed in the United States of America

For more information, contact:

Delmar Publishers
3 Columbia Circle, Box 15015
Albany, New York, 12212-5015

International Thomson Publishing Europe
Berkshire House 168-173
High Holborn
London WC1 V 7AA
England

Thomas Nelson Australia
102 Dodds Street
South Melbourne, 3205
Victoria, Australia

Nelson Canada
1120 Birchmount Road
Scarborough, Ontario
Canada M1K 5G4

International Thomson Editores
Campos Eliseos 385, Piso 7
Col Polanco
11560 Mexico D F Mexico

International Thomson Publishing GmbH
Königswinterer Strasse 418
53227 Bonn
Germany

International Thomson Publishing Asia
221 Henderson Road
#05 - 10 Henderson Building
Singapore 0315

International Thomson Publishing - Japan
Hirakawacho Kyowa Building, 3F
2-2-1 Hirakawacho
Chiyoda-ku, Tokyo 102
Japan

Delmar Publishers' Online Services

To access Delmar on the World Wide Web, point your browser to:
http://www.delmar.com/delmar.html
To access through Gopher: gopher://gopher.delmar.com
(Delmar Online is part of "thomson.com", an Internet site with information on more than 30 publishers of the International Thomson Publishing organization.)
For information on our products and services:
email: info@delmar.com
or call 800-347-7707

2 3 4 5 6 7 8 9 10 XXX 02 01 00 99 98 97
Library of Congress Catalog Card Number: 95-25175
ISBN: 0-8273-7645-6   CIP

Contents

Preface

The contents of this part of the textbook are published separately primarily as a means to improve manageability. It is obvious that the unusually large scope of this textbook requires a textbook of rather unwieldy proportions; it became apparent very soon that the book would have to be published in two volumes. Several options were considered as a means to provide a more convenient size for the book.

Of the several possibilities, the option that was eventually adopted was to separate only the outside problems and the design tables as a second volume, leaving the entire text as a single volume. While such a division admittedly produces a rather large text, it has the advantage that the text remains one volume, a very desirable feature.

There are other benefits, however, that affected the decision to separate the design tables in this way. It should be recognized that the design tables will almost certainly be the most-thumbed, most-used and most marked-up part of the book. Having the tables in a separate publication allows that part of the book to be used as a handbook (or manual) of relatively small size, easily handled and readily accessible. The prospect of having the design tables as a hard-to-find appendix at the end of a thick volume of text is considered to be a rather unattractive option.

Placing the outside problems in a separate volume is somewhat unusual but it, too, has advantages. When design tables are needed to solve the outside problems, the tables are readily accessible without having to search through another volume. Further, the clutter of placing the many outside problems within the text produces a real distraction in the continuity of the text. And in addition, when a new edition is prepared (well, one can hope), the outside problems can be edited, extended and revised with little disturbance to the main text.

The means adopted here for making the size of the book more manageable has been found to be an acceptable compromise. In his own work, the author uses the design tables as an independent volume and one suspects that those who use the textbook for any length of time will also use it the same way.

Samuel E. French
Martin, Tennesse, 1995

Delmar Publishers' Online Services
To access Delmar on the World Wide Web, point your browser to:
http://www.delmar.com/delmar.html
To access through Gopher: gopher://gopher.delmar.com
(Delmar Online is part of "thomson.com", an Internet site with information on more than 30 publishers of the International Thomson Publishing organization.)
For information on our products and services:
email: info@delmar.com
or call 800-347-7707

PART 1
OUTSIDE PROBLEMS

CHAPTER 1
Statics

The following problems relate to the subject matter presented in Section 1.5.

Problems 1.1 through 1.8 Solve for the magnitude and direction of the resultant force using the parallelogram law. Include a sketch showing your solution and the magnitude and direction of the resultant.

Problem 1.1

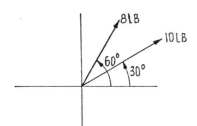

Problem 1.2

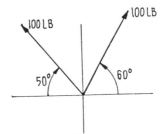

Problem 1.3

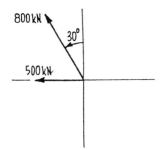

Problem 1.4

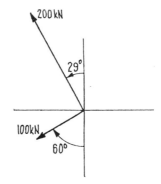

Problem 1.5

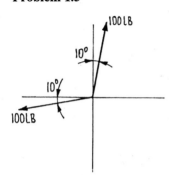

Problem 1.6

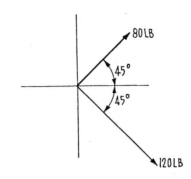

Problem 1.7

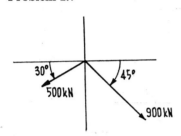

Problem 1.8

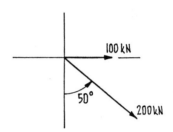

Problem 1.9 through 1.16: Solve problems 1.1 through 1.8 respectively, using the head-to-tail technique. Incude a sketch of your solution, giving the magnitudes and directions of the forces and of the resultants.

Problem 1.17 through 1.24: Solve for the single resultant force of the following force systems. Include a sketch showing your solution, giving the magnitudes and directions of the forces and of the resultant.

Problem 1.17

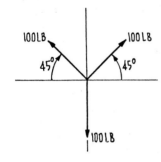

Problem 1.18

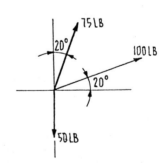

Problem 1.19

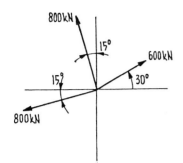

Problem 1.20

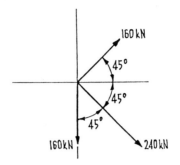

Problem 1.21

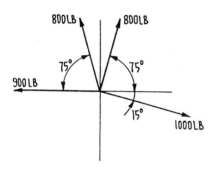

Problem 1.22

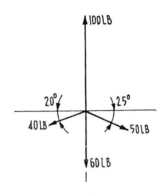

Problem 1.23

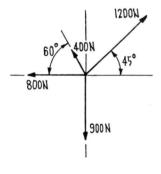

Problem 1.24

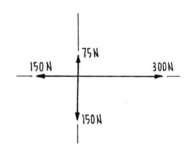

CHAPTER 2

Statics

The following problems relate to the subject matter developed in sections 2.1 to 2.3.

Problem 2.1 through 2.4: Resolve each of the given forces into its *x* and *y* components. Determine directions of the components visually and show your results on a sketch of the force system.

Problem 2.1

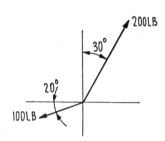

Problem 2.2

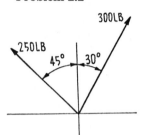

Problem 2.3

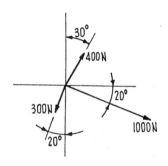

Problem 2.4

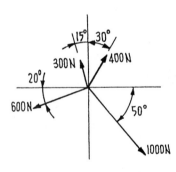

Problem 2.5 through 2.8: Recompute the *x* and *y* components of the forces given in Problems 2.1 through 2.4 respectively, using only angles measured counterclockwise from the positive *x* and *y* axes. Compare your results (with their plus and minus signs) to the solutions obtained in Problems 2.1 through 2.4 where directions were determined visually.

The following problems relate to the subject matter developed in Section 2.4.

Problems 2.9 through 2.12: Determine the single reaction force produced by the given systems of forces. Show your results on a sketch of the system.

Problem 2.9

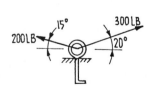

Problem 2.10

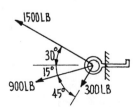

Problem 2.11

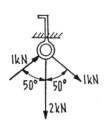

Problem 2.12

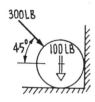

Problems 2.13 through 2.16 Determine the reactions at the supporting surfaces.

Problem 2.13

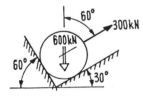

Problem 2.14

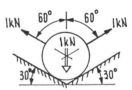

Problem 2.15

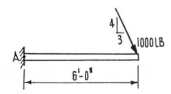

Problem 2.16

The following problems relate to the subject matter developed in Section 2.5.

Problems 2.17 through 2.24: Determine the moment produced by the given forces about an axis of rotation at Point A. Show the resultant moment in its correct direction on a sketch of the system.

Problem 2.17

Problem 2.18

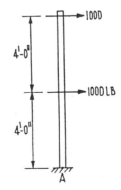

Problem 2.19

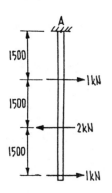

Problem 2.20

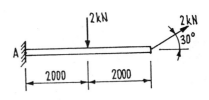

Problem 2.21

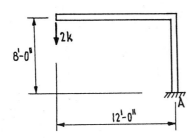

Problem 2.22

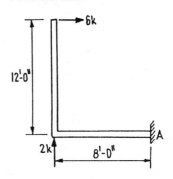

Problem 2.23

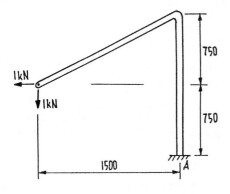

Problem 2.24

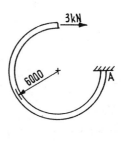

The following problems relate to the subject matter developed in Section 2.6.

Problems 2.25 through 2.28: Determine the force in each leg of the given systems.

Problem 2.25

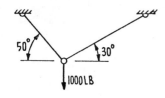

Problem 2.26

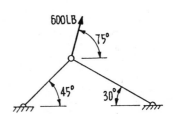

Problem 2.27

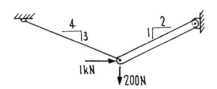

Problem 2.28

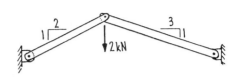

Problems 2.29 through 2.32: Determine the force *P* that will produce equilibrium in the indicated configurations.

Problem 2.29

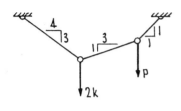

Problem 2.30

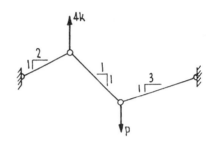

Problem 2.31

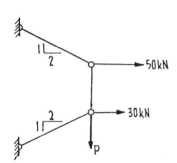

Problem 2.32

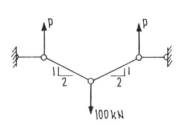

Problem 2.33 through 2.38: Determine the forces at the supports of the given structures. Show your results on a centerline sketch of the structure.

Problem 2.33

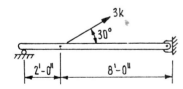

Problem 2.34

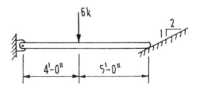

Problem 2.35

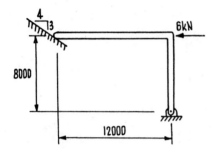

Problem 2.36

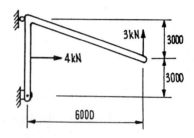

Problem 2.37

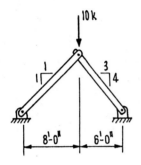

Problem 2.38

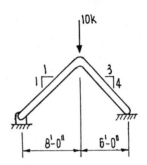

CHAPTER 3

Statics

The following problems relate to the subject matter developed in Sections 3.1 and 3.2.

Problems 3.1 through 3.8: Determine the magnitude and location of the resultant force produced by the given systems of parallel forces. Show your results on a sketch of the system.

Problem 3.1

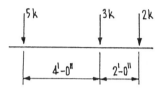

Problem 3.2

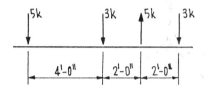

Problem 3.3

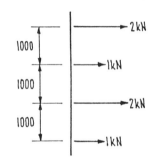

Problem 3.4

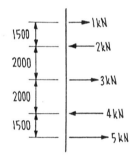

Problem 3.5

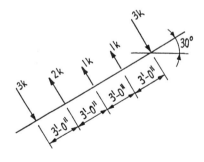

Problem 3.6

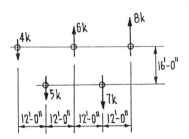

Problem 3.7   **Problem 3.8**

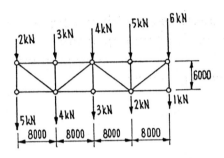   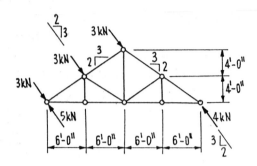

The following problems relate to subject matter developed in Section 3.3.

Problems 3.9 through 3.12: Determine the cable force T for the given arrangements.

Problem 3.9   **Problem 3.10**

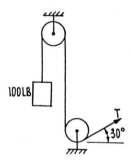   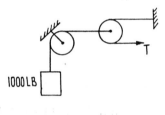

Problem 3.11   **Problem 3.12**

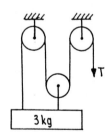   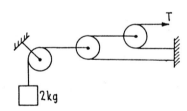

The following problems relate to the subject matter developed in Section 3.4 and 3.5.

Problems 3.13 through 3.24: Determine the location of the centroid of the given plane figures. (Use Table A-1 of the Appendix if needed). Show your results on a sketch of the figure, drawn reasonably to scale.

Problem 3.13

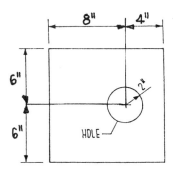

Problem 3.14

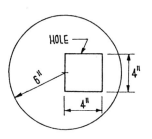

Problem 3.15

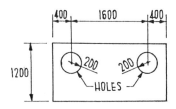

Problem 3.16

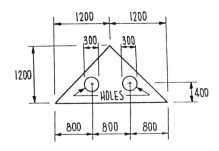

Problem 3.17

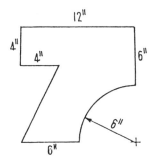

Problem 3.18

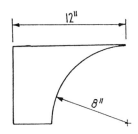

Problem 3.19

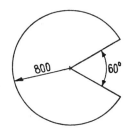

Problem 3.20

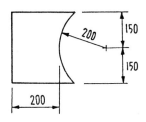

Problem 3.21

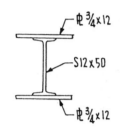

Problem 3.23

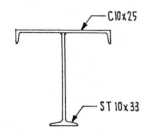

Problem 3.22

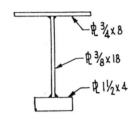

Problem 3.24

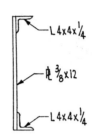

The following problems relate to the subject matter developed in Section 3.6.

Problems 3.25 through 3.28: Determine the forces at the supports of the given systems.

Problem 3.25

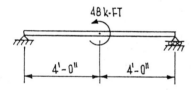

Problem 3.26

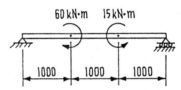

Problem 3.27

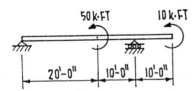

Problem 3.28

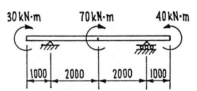

The following problems relate to the subject matter developed in Section 3.7.

Problems 3.29 through 3.32: Resolve the given force at Point *A* into a force and a moment at *B*.

Problem 3.29

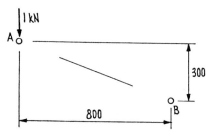

Problem 3.30

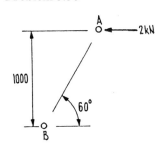

Problem 3.31

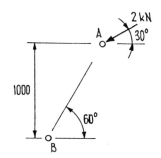

Problem 3.32

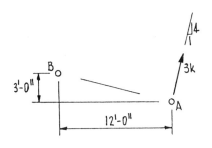

Problems 3.33 through 3.36: Commbine the given force and moment into a single force at a distance d from the given position. Determine the value of the distance d. Give your results in a sketch, showing both the initial conditions and final conditions.

Problem 3.33

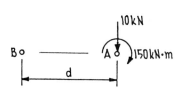

Problem 3.34

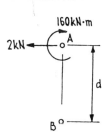

Problem 3.35

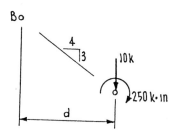

Problem 3.36

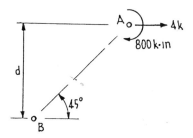

CHAPTER 4

Statics

The following problems relate to the subject matter developed in Sections 4.1 and 4.2.

Problems 4.1 through 4.8: Determine the magnitude, direction and location of the resultant of the applied forces.

Problem 4.1

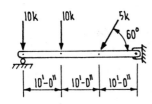

Problem 4.2

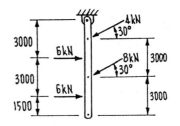

Problem 4.3

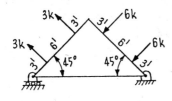

Problem 4.4

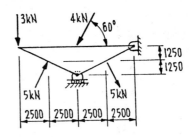

Problem 4.5

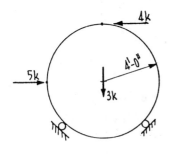

Problem 4.6

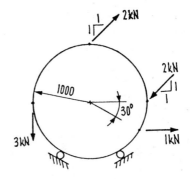

Problem 4.7

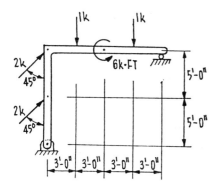

Problem 4.8

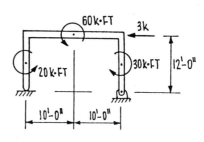

The following problems relate to the subject matter developed in Section 4.3.

Problems 4.9 through 4.24: Determine the reactions to the given systems of forces and moments.

Problem 4.9
Beam dead weight = 1600 lbs.

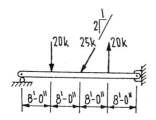

Problem 4.10
Beam dead weight = 10kN

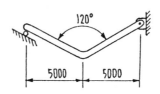

Problem 4.11
Beam dead weight = 960 lbs.

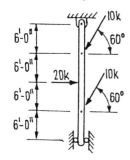

Problem 4.12
Beam dead weight = 8kN

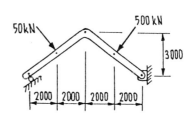

Problem 4.13
Dead weight of frame = 1500 lbs.

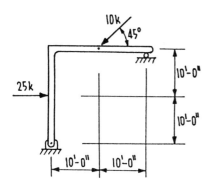

Problem 4.14
Dead weight of frame = 12kN

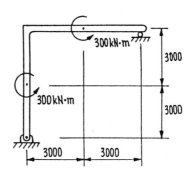

Problem 4.15
Dead weight of frame = 1620 lbs.

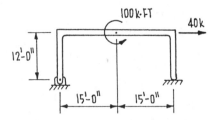

Problem 4.16
Dead weight of frame = 12kN

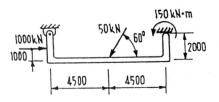

Problem 4.17
Each cylinder is 1 ft long
Each cylinder is 2 ft diameter
Each cylinder weighs 100 lb.

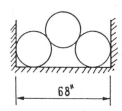

Problem 4.18
Each cylinder is 1000 mm long
Each cylinder is 500 mm diameter
Each cylinder weighs 1kN

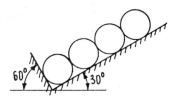

Problem 4.19
Ignore dead weights

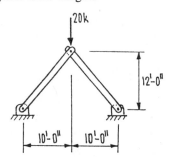

Problem 4.20
Ignore dead weights

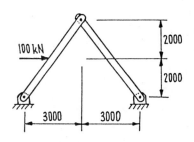

Problem 4.21
Ignore dead weights

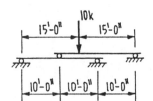

Problem 4.22
Ignore dead weights

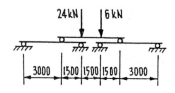

Problem 4.23
Ignore dead weights

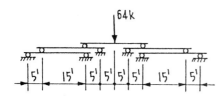

Problem 4.24
Ignore dead weights

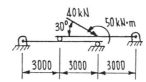

The following problems relate to the subject matter developed in Section 4.4.

Problems 4.25 through 4.28: Determine the force P that will cause motion to begin.

Problem 4.25
$\mu_s = 0.20, \mu_k = 0.15$

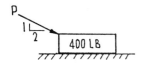

Problem 4.26
$\mu_s = 0.30, \mu_k = 0.24$

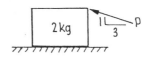

Problem 4.27
$\mu_s = 0.18, \mu_k = 0.14$

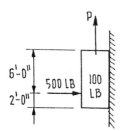

Problem 4.28
$\mu_s = 0.18, \mu_k = 0.12$

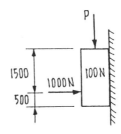

Problems 4.29 through 4.32: For Problems 4.25 through 4.28 respectively, determine the force P that will
sustain motion after it has started.
Problems 4.33 through 4.36: Determine whether the body will tip or slide when motion impends.

Problem 4.33
$\mu_s = 0.18$

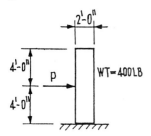

Problem 4.34
$\mu_s = 0.50$

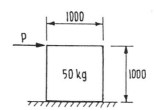

Problem 4.35

$\mu_s = 0.24$

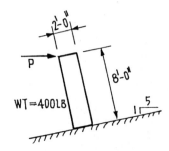

Problem 4.36

$\mu_s = 0.22$

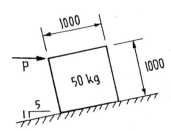

Problems 4.37 through 4.40: Determine the force P that will cause motion to begin, whether by motion of the bodies or by sliding at the supports.

Problem 4.37

$\mu_s = 0.30$

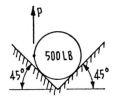

Problem 4.38

$\mu_s = 0.25$

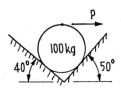

Problem 4.39

$\mu_s = 0.20$

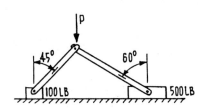

Problem 4.40

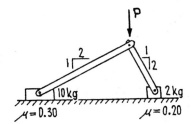

CHAPTER 5

Statics

The following problems relate to the subject matter developed in Section 5.2.

Problems 5.1 through 5.8: Determine the rectangular components of each force or moment.

Problem 5.1

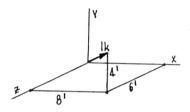

Problem 5.2

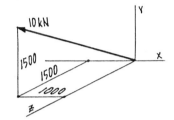

Problem 5.3

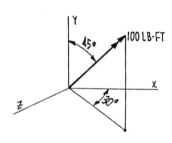

Problem 5.4

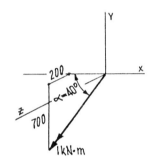

Problems 5.5 through 5.6: Determine the projection of the given force (or moment) on the alternative line of action *OA* (or axis *OA*)

Problem 5.5

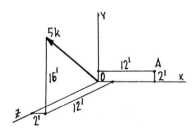

Problem 5.6

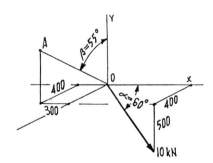

Problem 5.7

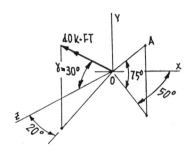

Problem 5.8

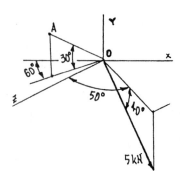

The following problems relate to the subject matter developed in Section 5.3.

Problems 5.9 through 5.20: Determine the resultant of the given systems. Show your result on a sketch, giving the magnitude, location, and direction angles of the resultant.

Problem 5.9

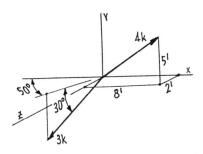

Problem 5.10

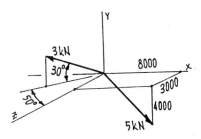

Problem 5.11

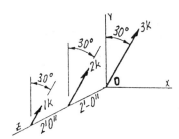

Problem 5.12

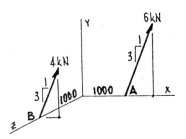

Problem 5.13

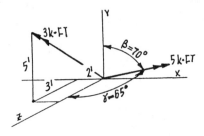

Problem 5.14

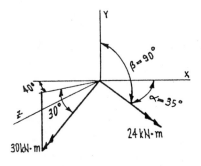

Problem 5.15

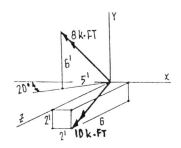

Problem 5.16

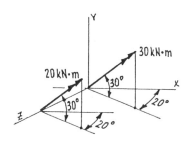

Problem 5.17

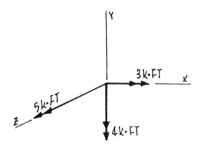

Problem 5.18

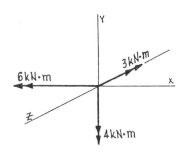

Problem 5.19

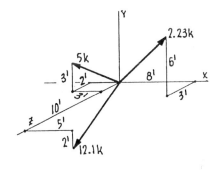

Problem 5.20

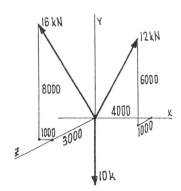

The following problems relate to the subject matter developed in Sections 5.4 and 5.5.

Problems 5.21 through 5.28: Determine the forces in the supporting members for the given systems of forces.

Problem 5.21

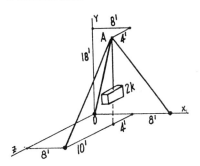

Problem 5.22

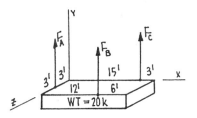

Problem 5.23

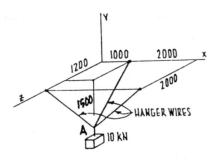

Problem 5.24

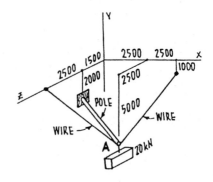

Problem 5.25

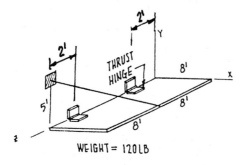

Problem 5.26

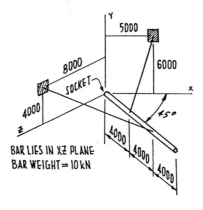

Problem 5.27

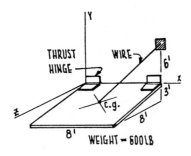

Problem 5.28

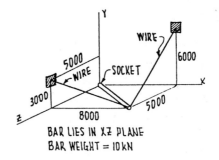

CHAPTER 6

Statics

The following problems relate to the subject matter developed in Section 6.4.

Problems 6.1 through 6.8: Determine the reactions at the supports for the given trusses.

Problem 6.1

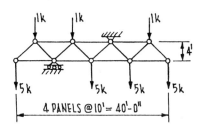

Problem 6.2

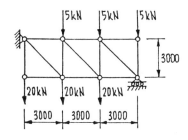

Problem 6.3

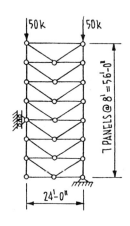

Problem 6.4

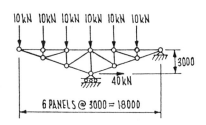

Problem 6.5

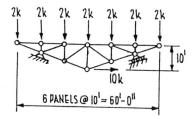

Problem 6.6

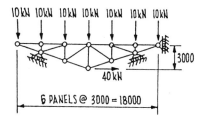

Problem 6.7

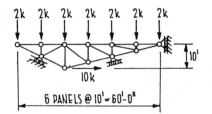

Problem 6.8

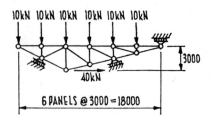

The following problems relate to the subject matter developed in Section 6.5.

Problems 6.9 through 6.16: Determine the force in each truss member using the method of joints. Show your results on a summary sketch; indicate whether each force is tension or compression.

Problem 6.9

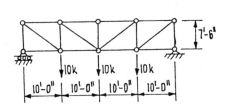

Problem 6.10

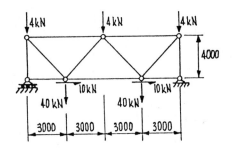

Problem 6.11

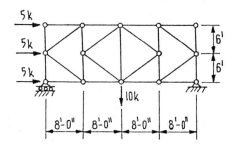

Problem 6.12

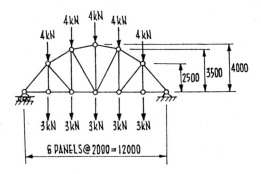

Problem 6.13

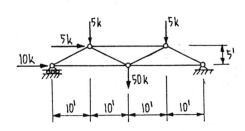

Problem 6.14

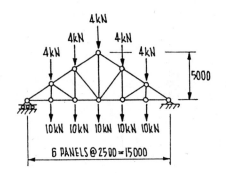

Problem 6.15   **Problem 6.16**

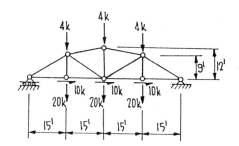

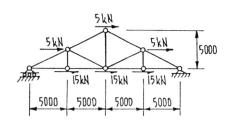

Problems 6.17 through 6.24: Except for problem 6.19, determine the force in each truss member of problems 6.9 through 6.16, respectively, using the method of sections. Show your results on a summary sketch; indicate whether each force is in tension or compression. For problem 6.19, examine the truss of problem 6.11 for a solution by the method of sections and state your conclusions.

CHAPTER 7

Statics

The following problems relate to the subject matter developed in Chapter 7.

Problems 7.1 through 7.24: Determine the reactions at the supports of the given structures. Ignore dead weight of the members.

Problem 7.1

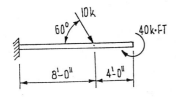

Problem 7.2

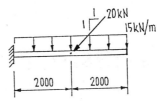

Problem 7.3

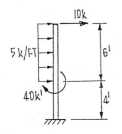

Problem 7.4

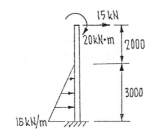

Problem 7.5

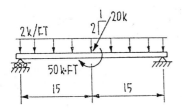

Problem 7.6

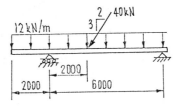

Problem 7.7

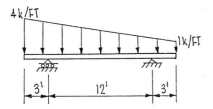

Problem 7.8

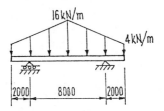

Problem 7.9

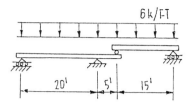

Problem 7.10

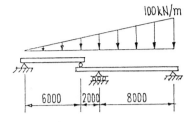

Problem 7.11

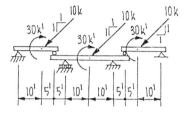

Problem 7.12

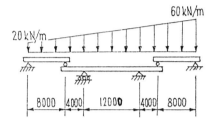

Problem 7.13

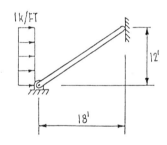

Problem 7.14

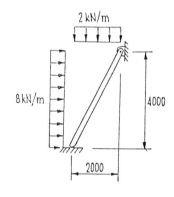

Problem 7.15

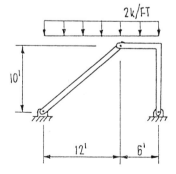

Problem 7.16

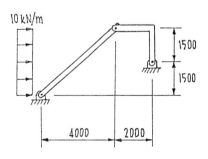

Problem 7.17

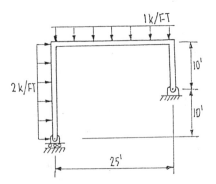

Problem 7.18

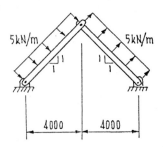

Problem 7.19

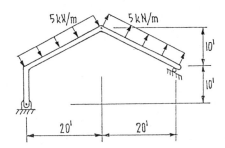

Problem 7.20

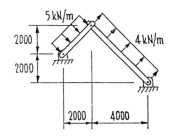

Problem 7.21

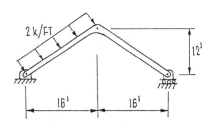

Problem 7.22

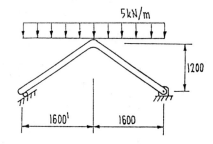

Problem 7.23

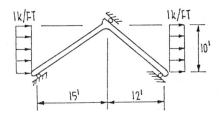

Problem 7.24

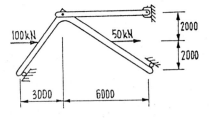

CHAPTER **8**

Strength of Materials

The following problems relate to the subject matter developed in Section 8.3.

Problem 8.1 A solid steel bar 1 inch in diameter and 18'-0" long, is subjected to an axial tension of 16,000 lb. Determine the axial direct stress in the material.

Problem 8.2 A structural steel tube 4 × 4 × 1/4, 12'-0" long, is used as a column carrying an axial load of 40 kips. Determine the axial direct stress in the material.

Problem 8.3 A truss member carrying a tensile load of 72 kips is built as a double angle steel strut, 2 - 5 × 3 × 1/4 angles, long legs back to back, 3/8 in. between angles. Length of the member is 15'-4". Determine the axial direct stress in the material.

Problem 8.4 A timber compression member 12'-0" long carrying 30 kips of load is built of 2 - 4 × 10 members spaced 1 1/2 in. apart. Determine the axial direct stress in the wood. Axial modulus of elasticity is 1,500,000 psi.

Problem 8.5 For the member of Problem 8.1, determine the unit strain and the total elongation under the given load.

Problem 8.6 For the member of Problem 8.2, determine the unit strain and the total shortening under the given load.

Problem 8.7 For the member of Problem 8.3, determine the unit strain and the total elongation under the given load.

Problem 8.8 For the member of Problem 8.4, determine the unit strain and the total shortening under the given load.

Problem 8.9 For the steel bar of Problem 8.1, determine the change in diameter of the bar due to the load.

Problem 8.10 For one of the 5 × 3 × 1/4 in. angles of Problem 8.3, determine the change in dimension of both the 5 in. leg and the 3 in. leg due to the axial load.

Problem 8.11 An aluminum bar 3/8 in. diameter and 18 in. long deforms 0.0601 in. under a tensile load of 4000 lbs. Determine the modulus of elasticity of the material.

Problem 8.12 A test sample of coldworked copper, 10 mm in diameter and 100 mm long, elongates 0.0255 mm under an axial load of 3 kN. Determine the modulus of elasticity of the material.

Problem 8.13 A bar 1 in. in diameter and 6'-0" long elongates 0.0506 in. under an axial load of 16,000 lb. The bar is made from steel having an ultimate strength of 58,000 lb/in^2. For the same size and load, what will be the elongation of a steel bar having an ultimate strength of 78,000 psi?

Problem 8.14 Determine the total elongation of the steel bar under the load.

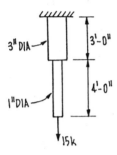

Problem 8.15 Determine the total elongation of the aluminum bar under the load. $E_{ALUMINUM}$ = 70,000 N/mm^2

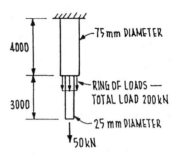

Problem 8.16 Determine the angle of rotation of the bar that will occur when the load is applied. $E_{ALUMINUM}$ =10,000,000 psi.

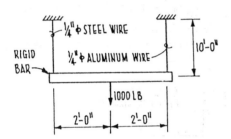

Problem 8.17 Determine the total elongation of the bar of Problem 8.14 if the upper segment is made of steel and the bottom segment is aluminum.

Problem 8.18 Determine the vertical displacement of the Point A when the load is applied.

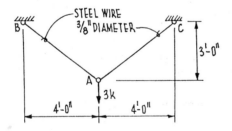

Problem 8-19 Determine the vertical displacement of the Point A when the load is applied. Ignore deflections in the heavy steel strut.

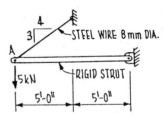

Problem 8.20 Determine the total displacement of the Point A when the weight is attached. Ignore deformations in and around the pulley wheels.

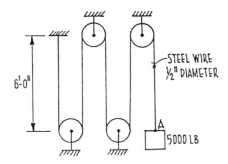

Problem 8.21 Determine the displacement of the Point A when the weight is added. Ignore deformations in and around the pulley wheels.

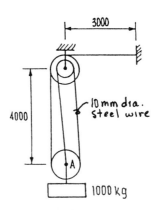

The following problems relate to the subject matter developed in Section 8.4.

Problem 8.22 For a load of 70k, determine the bearing stress:

    a) on the concrete

    b) on the soil

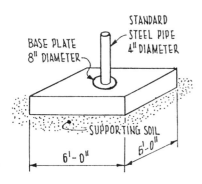

Problem 8.23 For an end reaction of 7 kips, determine the bearing stress on the wood beam. (See Table T-1 for actual sizes of wood beams).

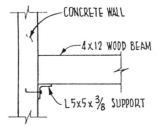

Problem 8.24 A crate measuring 2000 × 2000 × 2000 rests on soil having a general slope of 20° from horizontal. The crate weighs 90 kN. Determine the bearing stress on the soil.

Problem 8.25 Determine the bearing area of a pneumatic tire on a pavement under a wheel load of 4kN and tire pressure of $0.2 N/mm^2$.

The following problems relate to subject matter developed in Section 8.5

Problem 8.26 Determine the direct shear on the bolts in the connection.

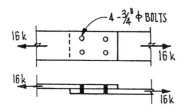

Problem 8.27 Determine the direct shear stress on the bolts in the connection.

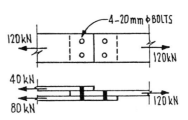

Problem 8.28 Determine the shear stress in the 3/8 in. fillet welds.

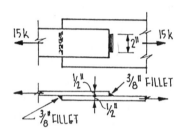

Problem 8.29 Determine the shear stress in the 10 mm fillet welds.

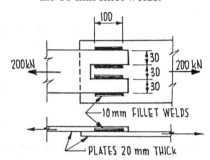

Problem 8.30 Determine the location of the 12k load for the stresses in both 3/16 in. welds to be equal.

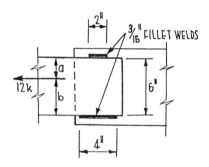

Problem 8.31 Determine the magnitude and the location of the Load P for the stress in all three 6 mm fillet welds to be 100 N/mm^2.

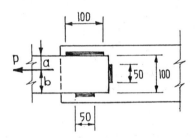

Problem 8.32 Determine the load P that will produce a shear stress of 5000 psi in the aluminum shear panel.

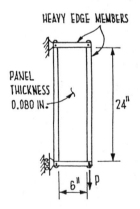

Problem 8.33 Determine the required thicknesses of the aluminum shear panel to sustain P = 60 kN with an allowable shear stress of 40 N/mm^2.

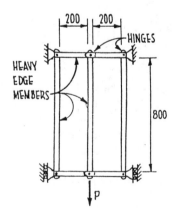

Problem 8.34 Determine the detrusion angle γ in the aluminum panel of Problem 8.32. For aluminum, E = 10,000,000 psi.

Problem 8.35 Determine the detrusion angle γ in the aluminum panel of Problem 8.33. For aluminum, E = 67,000 N/mm^2.

CHAPTER 9

Strength of Materials

The following problems relate to the subject matter developed in Section 9.1.

Problem 9.1 Determine the torque on the shaft *AB*.

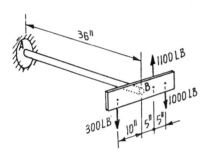

Problem 9.2 Determine the torque on the shaft *AB*.

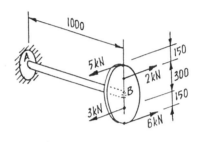

Problem 9.3 Determine the torque between points A and B, between points B and E, and between points C and D. All wheels are 10 in. in diameter. All forces in lbs.

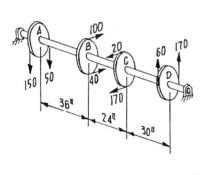

Problem 9.4 Determine the torque between points A and B, between B and C, and D.

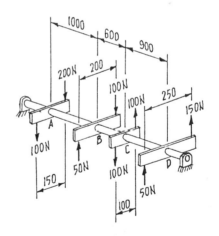

The following problems relate to the subject matter developed in Section 9.2.

Problem 9.5 If the shaft of Problem 9.1 is solid and is 3/4 in. diameter, find the maximum torsional shear stress in the shaft.

Problem 9.6 If the shaft of Problem 9.2 is solid and is 80 mm diameter, find the maximum torsional shear stress in the shaft.

Problem 9.7 If the shaft of Problem 9.3 is solid and is 1/2 in. diameter, find the maximum torsional shear stress in the shaft.

Problem 9.8 If the shaft of Problem 9.4 is solid and is 25 mm diameter, find the maximum torsional shear stresses in the shaft.

Problem 9.9 If the shaft of Problem 9.1 is hollow, 1 in. o.d. and 7/8 in. i.d., find the torsional shear stress at the inside face and outside face of the hollow shaft.

Problem 9.10 If the shaft of Problem 9.2 is hollow, 100 mm o.d. and 80 mm i.d., find the torsional shear stress at the inside face and outside face of the hollow shaft.

Problem 9.11 If the shaft of Problem 9.3 is hollow, 5/8 in. o.d. and 1/2 in. i.d., find the torsional shear stress at the inside face and outside face of the hollow shaft.

Problem 9.12 If the shaft of Problem 9.4 is hollow, 30 mm o.d. and 25 mm i.d., find the torsional stress at the inside face and outside face of the hollow shaft.

The following problems relate to the subject matter developed in Section 9.3.

Problem 9.13 If the shaft of Problem 9.1 is solid steel and is 7/8 in. diameter, determine the angular rotation between points A and B due to torsion.

Problem 9.14 If the shaft of Problem 9.2 is solid aluminum and is 75 mm diameter, determine the angular rotation between points A and B due to torsion. $G = 26, 500 \text{ N/mm}^2$

Problem 9.15 If the shaft of Problem 9.3 is steel and is hollow, 3/4 in. o.d. and 5/8 in. i.d., determine the angular rotation between points A and B, between points B and C and between C and D.

Problem 9.16 If the shaft of Problem 9.4 is aluminum and is hollow, 35 mm o.d. and 30 mm i.d., find the angular rotation between points A and B, between points B and C and between points C and D.

The following problems relate to the subject matter developed in Section 9.4.

Problem 9.17 Find the shear at sections *a-a* and *b-b* and the moment at sections *c-c* and *d-d* of the given beam. Include algebraic sign.

Problem 9.18 Find the shear at sections *a-a* and *b-b* and the moment at sections *c-c* (midspan) and *d-d* of the given beam. Include algebraic sign.

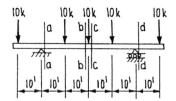

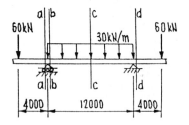

Problem 9.19 Find the shear at sections *a-a* and *b-b* and the moment at sections *c-c* (midspan) and *d-d* of the given beam. Include algebraic signs.

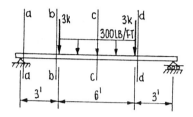

Problem 9.20 Find the shear and moment at the support. Include algebraic signs. Include algebraic signs.

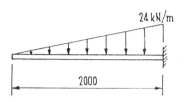

Problems 9.21 through 9.32: Determine the moment of inertia of the given sections about their centroidal horizontal *xx* axis and vertical *yy* axis. All dimensions are exact.

Problem 9.21

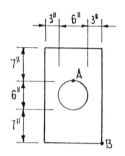

Problem 9.22

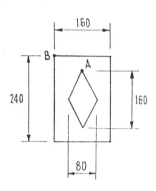

Problem 9.23

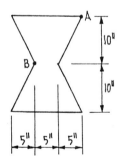

Problem 9.24

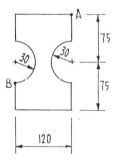

Problem 9.25

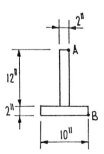

Problem 9.26

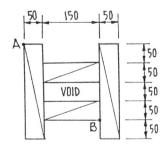

Problem 9.27   **Problem 9.28**

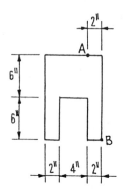   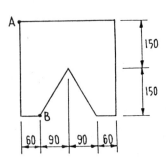

For the following problems, see Tables S-1 through S-9 for dimensions of the steel sections.

Problem 9.29   **Problem 9.30**

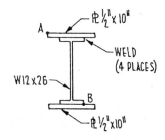   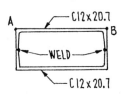

Problem 9.31   **Problem 9.32**

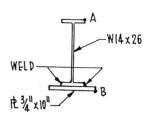   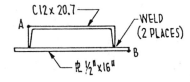

The following problems relate to the subject matter developed in Section 9.5.

Problem 9.33 For the section of Problem 9.21, find the stress at points *A* and *B* due to a moment of 15 kip•ft about the *xx* axis. Indicate whether the stress is tension or compression.

Problem 9.34 For the section of Problem 9.22, find the stress at points *A* and *B* due to a moment of 4kN•m about the *yy* axis. Indicate whether the stress is tension or compression.

Problem 9.35 For the section of Problem 9.23, find the stress at points *A* and *B* due to a moment of 30 kip•ft about the *yy* axis. Indicate whether the stress is tension or compression.

Problem 9.36 For the section of Problem 9.24, find the stress at points *A* and *B* due to a moment of 15kN•m about the *xx* axis. Indicate whether the stress is tension or compression.

Problem 9.37 For the section of Problem 9.29, find the stress at points *A* and *B* due to a moment of 80 kip•ft about the *xx* axis. Indicate whether the stress is tension or compression.

Problem 9.38 For the section of Problem 9.30, find the stress at points A and B due to:
   a) M_{xx} of 25 kip•ft about the xx axis
   b) M_{yy} of 50 kip•ft about the yy axis
   Indicate whether the stress is tension or compression.

Problem 9.39 For the section of Problem 9.31, find the stress at points A and B due to a moment of 50 kip•ft about the xx axis. Indicate whether the stress is tension or compression.

Problem 9.40 For the section of Problem 9.32, find the stress at points A and B due to a moment of 5 kip•ft about the xx axis. Indicate whether the stress is tension or compression.

The following problems relate to the subject matter developed in Section 9.6.

Problem 9.41 For the section of Problem 9.21, find the maximum shear flow and the maximum shear stress on the section for a shear force V of 4 kips.

Problem 9.42 For the section of Problem 9.22, find the maximum shear flow and the maximum shear stress on the section for a shear force V of 10 kN.

Problem 9.43 For the section of Problem 9.23, find the maximum shear flow and the maximum shear stress on the section for a shear force V of 8 kips.

Problem 9.44 For the section of Problem 9.24, find the maximum shear flow and the maximum shear stress on the section for a shear force V of 12 kN.

Problem 9.45 For the section of Problem 9.25, determine the shear stress where the members are joined if the horizontal member is glued to the vertical members. The shearing force is 1000 lb.

Problem 9.46 For the section of Problem 9.25, determine the load per nail where the members are nailed together with nails @ 3 in. o.c. The shearing force is 500 lb.

Problem 9.47 For the section of Problem 9.26, determine the shear stress if the two horizontal members are glued to the two vertical members. The shearing force is 8kN.

Problem 9.48 For the section of Problem 9.26, determine the load per nail where the members are nailed together with nails @ 50 mm o.c. The shearing force is 5kN.

Problem 9.49 For the combined section of Problem 9.29, find the shear flow in each weld due to a shear force of 16 kips.

Problem 9.50 For the combined section of Problem 9.30, find the shear flow in each weld due to a shear force of 15 kips.

Problem 9.51 For the combined section of Problem 9.31, find the shear flow in each weld due to a shear force of 12 kips.

Problem 9.52 For the combined section of Problem 9.32, find the shear flow in each weld due to a shear force of 7 kips.

CHAPTER 10
Strength of Materials

The following problems relate to the subject matter developed in Section 10.2.

Problems 10.1 to 10.24: Sketch the shear and moment diagrams for the given loads.

Problem 10.1

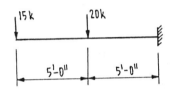

Problem 10.2

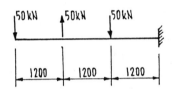

Problem 10.3

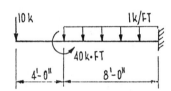

Problem 10.4

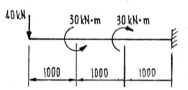

Problem 10.5

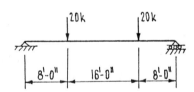

Problem 10.6

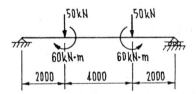

Problem 10.7

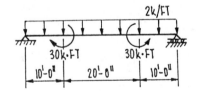

Problem 10.8

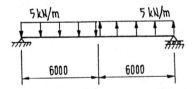

Problem 10.9

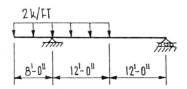

Problem 10.10

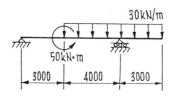

Problem 10.11

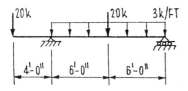

Problem 10.12

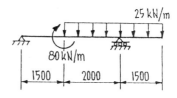

Problem 10.13

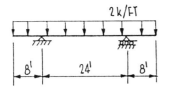

Problem 10.14

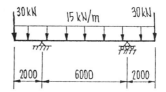

Problem 10.15

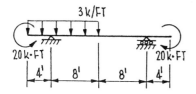

Problem 10.16

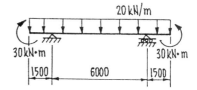

Problem 10.17

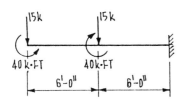

Problem 10.18

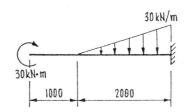

Problem 10.19

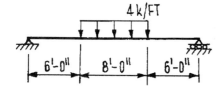

Problem 10.20

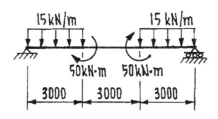

Problem 10.21

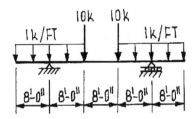

Problem 10.22

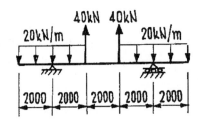

Problem 10.23

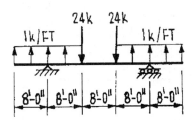

Problem 10.24

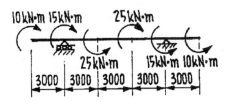

The following problems relate to the subject matter developed in Section 10.3.

Problems 10.25 through 10.32: The listed values for shear and moment are taken from shear and moment diagrams for the service load conditions. Select the most economical steel beam of the indicated type to carry the given moment and shear. In all cases, assume the compression flange is fully supported laterally over the entire span. ASTM A-36 steel.

Problem No.	Type of Section	Moment kip·ft	Shear kips
10.25	wide flange	40	50
10.26	wide flange	60	45
10.27	wide flange	80	40
10.28	wide flange	100	35
10.29	wide flange	120	50
10.30	wide flange	140	45
10.31	wide flange	160	40
10.32	wide flange	180	35
10.33	WT tee	15	10.0
10.34	WT tee	17.5	12.5
10.35	WT tee	20	15.0
10.36	WT tee	25	17.5
10.37	pipe	40	10.0
10.38	pipe	60	15.0
10.39	pipe	80	20.0
10.40	pipe	100	25.0

Problems 10.41 through 10.48: The listed values for shear and moment are taken from the shear and moment diagrams for the service load conditions for single members. Select the most economical wood beam of the indicated species and grade to carry the given moment and shear. In all cases, assume the compression side of the beam is fully supported laterally over the entire span or that suitable bracing can be provided.

Problem No.	Species	Grade	Moment kip·ft	Shear kips
10.41	Douglas fir-larch	No. 1	30	8
10.42	"	No. 2	16	7
10.43	"	select structural	22	8
10.44	"	construction	2	1
10.45	Southern pine	No. 1	28	9
10.46	"	No. 2	15	8
10.47	"	select structural	20	7
10.48	"	construction	2	1
10.49	Spruce-pine-fir	No. 1	20	5
10.50	"	No. 2	16	7
10.51	"	No. 2	14	6
10.52	"	select structural	22	7

The following problems relate to the subject matter developed in Section 10.4.

Problem 10.53 Find the critical buckling load on a 4 in. diameter standard steel pipe section 24'-0" long, simply supported both ends.

Problem 10.54 Find the critical buckling load on the pipe section of Problem 10.53 if the top of the column is fixed against rotation. Determine whether the column will yield first or buckle first under these conditions. Assume yield stress of 36 ksi.

Problem 10.55 Find the critical buckling load on a 6 in. × 6 in. wood column section 18'-0" long, simply supported at both ends. Southern pine, No. 2 grade.

Problem 10.56 Find the critical buckling load on a W8 × 31, 32-ft long, braced at midpoint on its weak axis only. Ends are hinged in both directions. Determine whether the column will yield first or buckle first. ASTM A-36 steel.

Problem 10.57 Find the critical buckling load on a square tube column, fixed at its base and unsupported at the top. Section is 8 × 8 × 3/16 in, 28'-0" long. Determine whether the column will yield first or buckle first. Assume yield stress of 46,000 psi.

Problem 10.58 Find the critical buckling load on a W14 × 48, 48'-0" long, hinged in both directions at its ends. Determine whether the column yields first or buckles first. ASTM A-36 steel.

Problem 10.59 For the column of Problem 10.58, find the critical buckling load if it is braced at midpoint on its weak axis only. Determine whether the column yields first or buckles first. ASTM A-36 steel.

Problem 10.60 For the column of Problem 10.58, find the critical buckling load if it is braced at its third point on its weak axis only. Determine whether the column yields first or buckles first. Determine also if failure for these conditions occurs on the weak axis or the strong axis. ASTM A-36 steel.

CHAPTER 11

Strength of Materials

The following problems relate to the subject matter developed in Section 11.3.

Problems 11.1 through 11.12: Using Mohr's stress circle, find the principal stress and maximum shear stress for the given states of stress at a point. Show your results on a sketch, to include the required angles of rotation to the maxima.

Problem 11.1

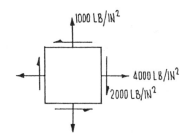

Problem 11.2

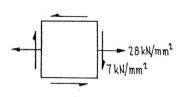

Problem 11.3

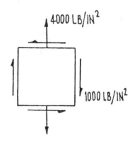

Problem 11.4

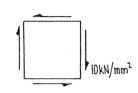

Problem 11.5

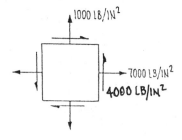

Problem 11.6

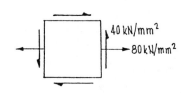

Problem 11.7

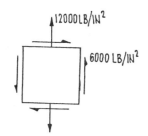

Problem 11.8

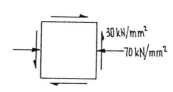

Problem 11.9

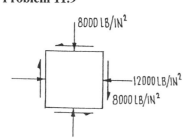

Problem 11.10

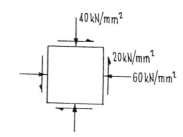

Problem 11.11

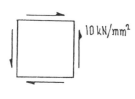

Problem 11.12

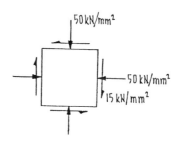

The following problems relate to the subject matter developed in Section 11.4. The following sketch applies to problems 11.13 through 11.16.

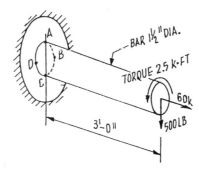

Problem 11.13 For the given cantilever, determine the principal stresses and maximum shear stress that will occur at Point *A*.

Problem 11.14 For the given cantilever, determine the principal stresses and maximum shear stress that will occur at Point *B*.

Problem 11.15 For the given cantilever, determine the principal stresses and maximum shear stress that will occur at Point *C*.

Problem 11.16 For the given cantilever, determine the principal stresses and maximum shear stress that will occur at Point *D*.

CHAPTER 12

Strength of Materials

The following problems relate to the subject matter developed in Section 12.3.

The following sketch applies to Problems 12.1 through 12.4.

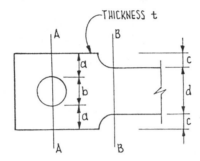

Problem 12.1 The strap in the sketch is subjected to a tensile force of 20 kips. Determine the average stress at Section *AA* and the stress levels at the edge of the hole. a = 12 in., b = 1 in., t = 0.25 in.

Problem 12.2 The strap in the sketch is subjected to a tensile force of 60 kN. Determine the average stress at Section *BB* and the stress levels at the edge of the fillet. c = 10 mm, d = 100 mm, t = 5 mm.

Problem 12.3 The strap in the sketch is subjected to a tensile force of 30 kips. Determine the average stress at Section *BB* and the stress levels at the edge of the fillet. c = 3/4 in., d = 4 1/2 in., t = 0.1875 in.

Problem 12.4 The strap in the sketch is subjected to a tensile force of 80 kN. Determine the average stress at Section *AA* and the stress levels at the edge of the hole. a = 50 mm, b = 20 mm, t = 5 mm.

The following problems relate to the subject matter developed in Section 12.6.

Problem 12.5 A plastic ball 12 in. diameter with a wall thickness of 1/32 in. is inflated to a pressure of 15 psi. Determine the tensile stress in the material.

Problem 12.6 A cylindrical pressure vessel of 12 gage uncoated steel is 24 in. diameter and has semicircular domed ends. Overall length is 6 ft. The vessel is inflated to 55 psi:
a) determine the longitudinal stress in the material in the cylindrical part of the tank;
b) determine the hoop stress in the material in the cylindrical part of the tank.

Problem 12.7 A cylindrical aluminum can containing pressurized gas is subject to an internal pressure of 500 N/mm^2. The can has a nominal length of 300 mm, a diameter of 100 mm and a wall thickness of 0.5 mm:
a) determine the hoop stress in the cylindrical part of the can;
b) determine the longitudinal stress in the cylindrical part of the can.

Problem 12.8 For the aluminum can of Problem 12.7, determine the principal stress and the maximum shear stress in the material, using a Mohr's circle solution.

The following problems relate to the subject matter developed in Section 12.7.

Problem 12.9 A 6 × 12 timber beam is subjected to a moment of 4.0 kip•ft about its xx axis and a moment of 3.0 kip•ft about its yy axis. Find the stress at the four corners and indicate whether the stress is tensile or compressive. Show your results in a sketch of the cross-section.

Problem 12.10 A steel beam W14 × 22 is subjected to a moment of 25 kip•ft about its xx axis and a moment of 2.5 kip•ft about its yy axis. Determine the maximum compressive stress that occurs anywhere on the section.

Problem 12.11 A steel tee section ST6 × 25 is subject to a moment M_{xx} = 5.0 kip•ft about the xx axis placing compression on the flange side and tension on the stem side. In addition, it is subject to a moment M_{yy} = 5.0 k°ft about its yy axis. Determine the largest tensile stress on the section and its location. Show your results on a sketch of the cross-section.

Problem 12.12 A steel channel section C12 × 20.7 is subjected to a positive moment M_{xx} = 15 kip•ft about its xx axis and a moment M_{yy} = 2 kip•ft about its yy axis; moment My places the free ends of the flanges in tension. Determine the maximum tensile and compression stresses.

Problem 12.13 A column load of 100 kips is to be placed slightly off-center on a 5 ft × 5 ft square concrete footing. How far off center can it be placed without lifting one edge off the supporting soil?

Problem 12.14 Determine the size and shape of the kern of a solid circular section.

Problem 12.15 Determine the size and shape of the kern of the section of a standard circular steel pipe, 8 in. diameter.

Problem 12.16 Determine the size and shape of the kern of the section of a structural steel tube 10 × 10 × 1/4.

CHAPTER 13

Structural Analysis

Problems 13.1 through 13.4: Sketch the shear and moment diagrams for the given uniform load to large and accurate scale. Then for each load case 13.1 through 13.4, superimpose their shear and moment diagrams on the diagrams for uniform load. Determine the percent difference in maximum values of shear and moment between the uniform load and each of the discrete load systems, noting that the total load in all cases remains the same.

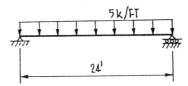

Problem 13.1

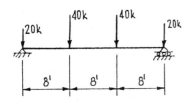

Problem 13.2

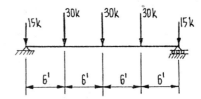

Problem 13.3

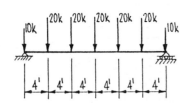

Problem 13.4

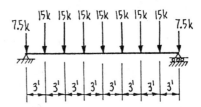

Problem 13.5 The upper floors in a concrete building consist of a floor slab 3" thick cast on tee beams placed 6'-0" on center. The stem of the tee beams is 8" wide; overall depth of tee beam and stem is 18 inches. The floor surface is terrazzo, 1 in. thick. The ceiling below is acoustic fiber tile in a furred channel system, suspended from the floor above. An allowance of 4 psf is estimated for electrical fixtures and wiring with an allowance of 3 psf is estimated for mechanical ductwork and equipment. Sketch the system and determine the dead load in psf.

Problem 13.6 The roof structure on a timber building is built with a slope of 9 vertical to 12 horizontal. The structure consists of 2 in. × 8 in. rafters spaced at 16 in. o.c., sheathed with 5/8 in. oriented strand board; roofing material is wood shingles over building paper. Insulation consists of 8 in. batt (loose) insulation placed between the joists. The lower face of the joists is faced with 1/2 in. gypsum plaster on 1/2 in. gypsum lathing. Sketch the system and determine the dead load per square foot on the horizontal projected area.

Problem 13.7 A bearing wall 12'-0" high is to be built of 8 in. lightweight concrete block, faced on the outside with 4 in. brick. Sketch the wall and determine the dead load on the foundation per foot of length.

Problem 13.8 A concrete bearing wall 10 in. thick and 12'-0" high is to be supported by a continuous footing 3'-0" wide and 15 in. thick. The wall supports a sustained vertical load of 4500 plf at the top of the wall. Sketch the system and determine the bearing pressure in pounds per square foot to be supported by the underlying soil.

Problem 13.9 The concrete roof structure of a commercial restaurant is built at a slope of 5 vertical to 12 horizontal. The roof structure consists of a 5 in. concrete slab supported by concrete tees 10 in. wide, spaced at 8'-0" o.c.; overall depth of slab and tees is 20 in. Roofing is terra cotta tile (curved) bedded in 1 1/2 in. of mortar. The under surface of the roof is sprayed with 3 in. of acoustic fiber insulation (1 psf) on all exposed concrete surfaces. Sketch the system and determine the dead load per foot of horizontal projected area.

Problem 13.10 The flat roof in a steel building is built of 2 1/2 in. lightweight concrete, cast on a 20 gage corrugated steel deck. The steel deck in turn is supported by lightweight steel joists weighing 9.1 lb/ft, spaced at 4'-0" o.c. Joists are 24'-0" long, supported at their ends by W21 × 50 beams. A part of the roof is set aside as a rooftop recreation area and patio. The surfacing on this area is 3/4 in. quarry tile bedded in 1 1/2 in. mortar. The ceiling below this area is acoustic tile on a suspended channel system. An allowance of 4 psf is to be made for the dead load of the mechanical equipment and an additional 2 1/2 psf for the electrical fixtures. Sketch the system at the patio area and determine the dead load per square foot of surface.

Problem 13.11 The flat roof structure of a concrete building consists of a 2 1/2 in. deck supported by concrete joists 7 in. wide spaced at 30 in. o.c., overall depth of the slab and joist system is 16 1/2 in. Roofing is hot mopped 3 ply felt and gravel. The entire underside of the system, including the vertical sides of the joists, is sprayed with acoustic spray weighing 1 1/2 lb/ft². Sketch the system and determine the dead load per square foot of area.

Problem 13.12 A steel roof structure slopes 9 vertical in 12 horizontal. The roofing is 20 gage steel deck, supported by channels C9 × 13.4 spaced at 4'-0" o.c., spanning 20'-0". Loose insulation batts 6 in. thick are placed between the channels. Sketch the roof structure and determine the dead load per foot of projected area.

The following problems relate to the subject matter developed in Section 13.4

Problems 13.13 through 13.28: Determine the live load per square foot to be used in the listed conditions.

Problem 13.13 Library stackrooms, upper floors

Problem 13.14 Theater auditorium, fixed seats

Problem 13.15 Hospital corridor, third floor

Problem 13.16 Gambling casino catwalks

Problem 13.17 School classrooms

Problem 13.18 Morgue floor

Problem 13.19 Public restrooms

Problem 13.20 Aircraft hangars

Problem 13.21 Guest room in a motel

Problem 13.22 Dining room in a hotel

Problem 13.23 Bleachers in a baseball stadium

Problem 13.24 Auditorium in a school

Problem 13.25 Floors in a light manufacturing plant

Problem 13.26 Corridors, general

Problem 13.27 Fire escapes in an apartment building

Problem 13.28 Floors in a disco

The following problems relate to the subject matter developed in Section 13.6.

Problem 13.29 Determine the basic wind stagnation pressure for the following locations:
    a) central New Mexico
    b) Chicago
    c) New Hampshire
    d) southern Arkansas
    e) northeastern California

Problem 13.30 For a wind velocity of 90 mph, plot a curve of wind stagnation pressures vs height up to a maximum height of 70 feet.

Problem 13.31 Determine the wind pressure and its distribution to be used in the design of the support posts of the given billboard, located in central Kansas.

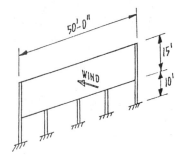

Problem 13.32 A typical truss panel in a rural timber bridge in central Pennsylvania is shown in the following sketch. Compute the "percent solid" for the typical panel and determine the wind pressure to be used in designing the bridge, assuming the wind "sees" an identical truss on the leeward side of the bridge. The average clear distance to the surface below is 40 feet.

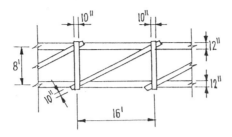

Problem 13.33 A silo near Lincoln, Nebraska is 16 ft. in diameter and 40 ft. tall. Determine the design wind pressure acting on 8 ft. increments of height over the full height of the silo, and compute the corresponding base shear and the overturning moments.

Problem 13.34 A church in the Missouri bootheel has the silhouette dimensions shown. Determine the design wind pressure acting on 10 ft. increments over the full height of the church. Determine also the base shear and overturning moments for wind acting in either direction and show your results on sketches of the silhouettes.

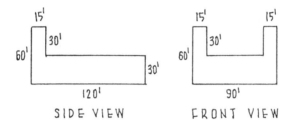

Problem 13.35 The bell towers of the church of Problem 13.34 have 6 ft. × 6 ft. wood shutters on all four sides of the tower, centered 5 ft. below the top of the tower. Determine the design wind pressures to be used to design the attachments for the shutters, both on the windward side and on the leeward side of the tower.

Problem 13.36 A hotel in LasVegas, Nevada, has a rooftop swimming pool at an elevation roughly 50 ft. above the surrounding terrain. The equipment for the pool is housed on the roof in a small flat-roofed shed, 12 ft. wide 20 ft. long and 8 ft. high with doors at both ends. Determine the design wind pressures to be used to design the shed and the design wind pressures to be used to design the attachments that hold the doors.

Problem 13.37 Determine the base shear and overturning moment acting on the given mill building for wind blowing against the long side. Include the effects of wind pressure and suction against the gable roof; surfaces and roof are sheathed with metal siding The building is located in central Ohio.

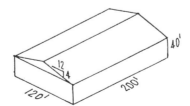

Problem 13.38   For the mill building of Problem 13.37, determine the base shear and overturning moment when the wind is blowing against the end of the building.

Problem 13.39   In the mill building of Problem 13.37, the middle 100 ft. of sheathing is to be removed from both sidewalls, leaving only 50 ft. of sheathing at the ends of the walls. Determine the base shear and overturning moment acting on this building for wind blowing against the long side.

Problem 13.40   In the mill building of Problem 13.37, the middle 60 ft. of sheathing is to be removed from both endwalls (but not from the gables) leaving only 30 ft. of sheathing at the ends of the walls. Determine the base shear and overturning moment acting on this building for wind blowing against the ends of the building.

The following problems relate to the subject matter developed in Section 13.7.

Problem 13.41   A five-story, flat-roofed reinforced concrete building located in the Missouri bootheel is 60 ft. wide, 200 ft. long and 58 ft. high and is generally symmetrical on two axes. The total dead weight of the above-grade mass is 3040 tons, with center of mass at 27.5 ft. above the ground floor slab. The building is a shearwall type structure where shearwalls carry only lateral load and the columns carry only vertical loads. Determine the base shear and overturning moment due to earthquake.

Problem 13.42   A single-story, gable-roofed rectangular steel mill building located near the coast in Oregon is 100 ft. wide, 300 ft. long and 40 ft. high to the eaves. Total dead weight of the above-grade mass is 241 tons, with center of mass at 41.6 ft. above grade. The building has cross-braced walls to carry the lateral load; columns at the cross-braced walls are loaded both by vertical load and by lateral loads. Determine the base shear and the overturning moment due to earthquake.

Problem 13.43   A steel silo in southeastern Oklahoma is 15 ft. in diameter and 40 ft. high, containing silage weighing about 25 lbs/ft^3. The dead weight of the steel silo is 12.6 kips, with its center of gravity 19.2 ft. above the base. Determine the base shear and overturning moment due to earthquake if the silo is three-quarters filled with silage. (Note: the value of R_w for the silage is taken to be the same as the tank).

Problem 13.44   A rectangular trussed steel water tower in northwestern Arizona is 12 ft. square and 36 ft. high. The tower supports a steel tank 12 ft. diameter and 12 ft. high at its top surface, completely filled with water. The dead weight of the tower is 8.3 tons with its center of gravity roughly at midheight. The dead weight of the tank is 2.2 tons with its center of gravity about 6 feet above the bottom of the tank. Determine the base shear and overturning moment due to earthquake.

Problem 13.45   A steel building having a moment-resistant frame is located near Los Angeles, California. The building is 60 ft. wide, 200 ft. long and 54 ft. high. The dead weight of the building is 1264 tons with its center of gravity at 29.4 ft. above ground level. Determine the base shear and overturning moment due to earthquake.

Problem 13.46   A concrete building near the California coast is built with shearwalls that carry only lateral loads; columns carry all the vertical load. A typical floor plan of the four-story building is shown in the sketch; mass is distributed reasonably uniformly around the floor plan. The total weight of the above-grade parts of the building is 1800 tons, with the center of gravity located at 21.6 feet above the ground floor. Determine the base shear due to earthquake acting on the building and its location within the building footprint. Determine also the overturning moment due to earthquake.

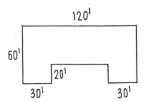

Problem 13.47 A concrete and masonry building located in central Colorado has masonry shear walls that also carry vertical loads. The building is rectangular, 80 ft. wide, 120 ft. long and 64 ft. high. The building weighs 1560 tons, with its center of gravity at 30.2 ft. above the ground floor level. Determine the base shear and overturning moment due to earthquake.

Problem 13.48 For the structure of Problem 13.47, determine the base shear and overturning moment due to earthquake if this structure were to be built along the northern California coast. Compare the increase in loads that accompanies this type of structural system in a high-risk earthquake area.

CHAPTER 14

Structural Analysis

The following problems relate to the subject matter developed in Section 14.3.

Problem 14.1 Determine the gravity loads acting on Column B1 in the building of Fig. 14-1. Show your results as indicated in Fig. 14-4, giving dead loads and live loads separately at each floor level.

Problem 14.2 Determine the gravity loads acting on Column C1 in the building of Fig. 14-1. Show your results as indicated in Fig. 14-4, giving dead loads and live loads separately at each floor level.

Problem 14.3 Determine the gravity loads acting on Column C2 in the building of Fig. 14-1. Show your results as indicated in Fig. 14-4, giving dead loads and live loads separately at each floor level.

Problem 14.4 Determine the gravity loads acting on Column B5 in the building of Fig. 14-1. Show your results as indicated in Fig. 14-4, giving dead loads and live loads separately at each floor level.

Problem 14.5 Determine the gravity loads acting on Column C5 in the building of Fig. 14-1. Show your results as indicated in Fig. 14-4, giving dead loads and live loads separately at each floor level.

Problem 14.6 In the building of Fig. 14-2, note that the gravity loads acting on columns B4 and C4 are the same. Determine the gravity loads acting on these columns. Show your results as indicated in Fig. 14-4, giving dead loads and live loads separately at each floor level.

Problem 14.7 In the building of Fig. 14-2, note that the gravity loads acting on columns A3, A4, D2 and D3 are the same. Determine the gravity loads acting on these columns. Show your results as indicated in Fig. 14-4, giving dead loads and live loads separately at each floor level.

Problem 14.8 In the building of Fig. 14-2, note that the gravity loads acting on columns A5 and D1 are the same. Determine the gravity loads acting on these columns. Show your results as indicated in Fig. 14-4, giving dead loads and live loads separately at each floor level.

The following problems relate to the subject matter developed in Section 14.4.

Problem 14.9 For the building of Fig. 14-2, determine the loads on the shear panels on lines 1 and 5 due to wind acting against the short side (the 80 ft. side) of the building. Show your results on a sketch of the panels, similar to the sketch of Example 14-17.

Problem 14.10 For the building of Fig. 14-2, determine the loads on the shear panels on lines 1 and 5 due to earthquake motions parallel to the long axis of the building. Show your results on a sketch of the panels, similar to the sketch of Example 14-18.

CHAPTER 15

Structural Analysis

The following problems relate to the subject matter developed in Section 15.1.

Problems 15.1 through 15.8: Determine the reactions at the supports for the given hinged beams. (An open circle denotes an interior hinge).

Problem 15.1

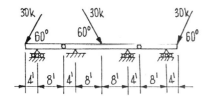

Problem 15.2

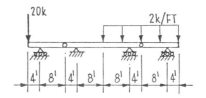

Problem 15.3

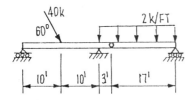

Problem 15.4

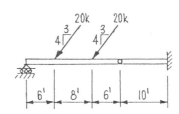

Problem 15.5

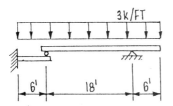

Problem 15.6

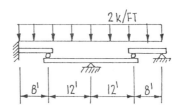

Problem 15.7

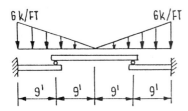

Problem 15.8

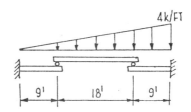

The following problems relate to the subject material developed in Sections 15.2 and 15.3.

Problems 15.9 through 15.16: Determine the midspan deflections of the indicated beams and loadings.

Problem 15.9
E = 29,000,000 psi
I = 385 in.^{4}

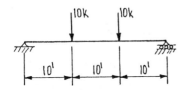

Problem 15.10
E = 29,000,000 psi
I = 2100 in.^{4}

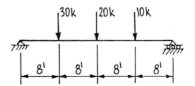

Problem 15.11
E = 10,000,000 psi
I = 1330 in.^{4}

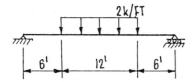

Problem 15.12
E = 1,500,000 psi
I = 678 in.^{4}

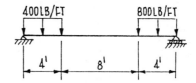

Problem 15.13
E = 1,600,000 psi
I = 415 in.^{4}

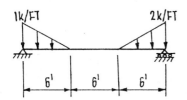

Problem 15.14
E = 29,000,000 psi
I = 890 in.^{4}

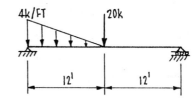

Problem 15.15
E = 29,000,000 psi
I = 1750 in.^{4}

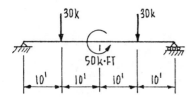

Problem 15.16
E = 14,000,000 psi
I = 170 in.^{4}

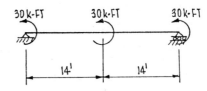

Problems 15.17 through 15.24: Determine the maximum deflections of the given beams, whether it occurs
  at midspan or at the quarter points.

Problem 15.17
E = 1,500,000 psi
I = 1706 in.⁴

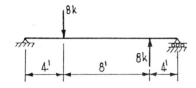

Problem 15.18
E = 1,400,000 psi
I = 1204 in.⁴

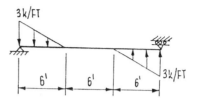

Problem 15.19
E = 10,000,000 psi
I = 612 in.⁴

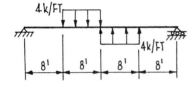

Problem 15.20
E = 29,000,000 psi
I = 170 in.⁴

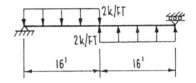

Problem 15.21
E = 29,000,000 psi
I = 291 in.⁴

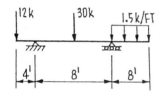

Problem 15.22
E = 1,400,000 psi
I = 2948 in.⁴

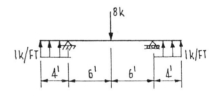

Problem 15.23
E = 29,000,000 psi
I = 1240 in.⁴

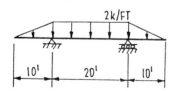

Problem 15.24
E = 1,300,000 psi
I = 2456 in.⁴

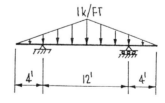

CHAPTER 17

Reinforced Concrete

The following problems relate to the subject matter developed in Section 17.10 and 17.11.

Problems 17.1 through 17.16: Determine the nominal ultimate design loads (M_n, V_n, P_n) and the service design loads (M_{sv}, V_{sv}, P_{sv}) for the given computed load cases. Moments are in kip•ft, shear in kips, axial loads in kips.

Prob. No.	M_{DL}	M_{LL}	M_W	M_E	V_{DL}	V_{LL}	V_W	V_E	P_{DL}	P_{LL}	P_W	P_E
17.1	154	108	-	-	75	50	-	-	-	-	-	-
17.2	166	115	-	-	82	48	-	-	-	-	-	-
17.3	175	123	80	-	94	59	46	-	50	41	31	-
17.4	191	143	86	-	115	51	59	-	57	49	35	-
17.5	120	72	-	-	66	45	-	-	-	-	-	-
17.6	128	102	-	-	79	48	-	-	-	-	-	-
17.7	135	108	61	-	92	42	40	-	41	36	39	-
17.8	149	134	70	-	104	50	51	-	45	42	46	-
17.9	104	63	-	56	59	40	-	32	-	-	-	-
17.10	112	78	-	64	68	47	-	39	-	-	-	-
17.11	118	95	-	71	77	54	-	43	36	35	-	49
17.12	124	112	-	74	90	59	-	50	39	32	-	54
17.13	77	49	41	34	45	31	29	26	26	19	24	22
17.14	85	60	44	39	55	36	34	29	29	22	26	27
17.15	96	77	48	54	65	41	41	46	33	21	24	49
17.16	103	93	54	59	75	48	43	49	31	18	25	55

The following problems relate to the subject matter developed in Section 17.12.

Problem 17.17 Using the ACI coefficients, draw the shear and moment envelopes for the end span of a multip-san continuous slab 6 in. thick where the slab is built integrally with all its supporting beams and all spans are 12'6" clear. Uniform live load is 120 lb/ft². Use working loads; that is, do not use load factors.

Problem 17.18 Draw the shear and moment envelopes for the first interior span of the slab of Problem 1.

Problem 17.19 Using the ACI coefficients, draw the shear and moment envelopes for the end span of a five-span continuous girder built integrally with the wall supports at the ends of the girder and being built integrally with the column supports elsewhere. Include the location of inflection points, both for the part due to negative moments and the part due to positive moments. Load is uniform, 42.50 lb/ft., including the dead load. End spans are 16'6" clear and all other spans are 18'6" clear. Use working loads; that is, do not use load factors.

Problem 17.20 Draw the shear and moment envelopes for the center span of the beam of Problem 17.19.

CHAPTER 18

Reinforced Concrete

The following problems relate to the subject matter developed in Sections 18.3, 18.4 and 18.5

Problem 18.1 A rectangular concrete beam, $f'_c = 3000$ psi, has a width of 14 in., a total height of 28 in., and an effective depth of 25.5 in; reinforcement consists of 4 no 7 bars, grade 40 steel. Sketch the cross section and compute the stress in the steel when the moment is 106 kip•ft.

Problem 18.2 A rectangular concrete beam, $f'_c = 5000$ psi, carries a moment of 84 kip•ft. The width of the beam is 12 in. and the effective depth is 16 in. Reinforcement on the tension side is 4 No. 8 bars and on the compression side is 2 No. 8 bars, all grade 60 steel. Cover over all reinforcement is 1 1/2 in. Sketch the cross section and determine the maximum stress in the concrete.

Problem 18.3 A rectangular concrete beam, $f'_c = 4000$ psi, is loaded in flexure in the laboratory. During the test, the elastic stress in the concrete is measured and found to be 1357 psi; at the same time, the stress in the steel is found to be 30,100 psi; if no compression steel was used, what was the tensile steel ratio used in building this beam?

Problem 18.4 A rectangular concrete beam having $f'_c = 4000$ psi has a tensile steel ratio $\rho = 0.012$ and a compression steel ratio of 0.0048. Sketch the stress diagram and determine the stress in both the tensile steel and the compressive steel when the maximum stress in the concrete is 1150 psi. $g = 0.125$.

For the following conditions, select a suitable size for a rectangular concrete beam and select the required reinforcement:

Problem No.	Concrete Strength f'_c	Allowable Stress in Steel, psi	Allowable Stress in Concrete, psi	Applied Moment in kip·ft
18.5	3000	20,000	900	60
18.6	3000	24,000	1200	70
18.7	4000	20,000	1200	80
18.8	4000	24,000	1600	90
18.9	5000	22,000	1600	100
18.10	5000	22,000	2000	110
18.11	3000	30,000	1100	60
18.12	3000	36,000	1350	70
18.13	4000	30,000	1500	80
18.14	4000	36,000	1800	90
18.15	5000	33,000	2100	100
18.16	5000	33,000	2400	110

Problem 18.17 A rectangular concrete beam, f'_c = 4000 psi, is limited to an overall depth of 16 in. The beam must sustain a moment of 102 kip•ft. Stress in the concrete is not to exceed 1500 psi and stress in the steel is not to exceed 33,000 psi. Clear cover over the reinforcement must not be less than 1.5 in. Select a suitable beam and its reinforcement.

Problem 18.18 For the conditions of Problem 18.17, select a suitable beam if, in addition, the width must not exceed 20 in.

Select a suitable rectangular concrete section and its reinforcement under service loads at balanced stress conditions for the given simply supported beams. Then determine the maximum deflection that will occur under the service load conditions.

Problem No.	Uniform Load kips/ft		Span ft	Concrete f'_c psi	Allowance f'_c psi	Steel Grade
	w_{DL}	w_{LL}				
18.19	1.9	1.8	26	3000	1200	40
18.20	2.2	2.8	18	5000	2200	60
18.21	1.8	2.1	24	4000	1800	50
18.22	2.1	2.6	18	4000	1600	40
18.23	2.0	2.3	22	5000	2000	60
18.24	1.9	2.4	20	3000	1350	50

The following problems relate to the subject matter developed in Sections 18.5, 18.6, 18.7, 18.8 and 18.9.

Problem 18.25 A rectangular concrete beam, f'_c = 3000 psi, has a width of 14 in., a total depth of 28 in., and an effective depth of 25.5 in. Tensile reinforcement consists of 4 No. 7 bars, grade 40 steel. Sketch the cross section and determine the ultimate moment M_n the section can sustain.

Problem 18.26 Compressive reinforcement consisting of 2 No. 7 bars is added to the beam of Problem 6.1. Sketch the cross section and determine the ultimate moment M_n the section can sustain.

Using the strength method for the following conditions, select a suitable rectangular concrete section and its reinforcement, keeping the steel ratio ρ at about half the maximum allowable value. Provide 1.5 in minimum cover over the reinforcement.

Problem No.	Concrete Strength f'_c psi	Steel Grade	Limits on Width b	Limits on Height h	Dead Load Moment kip·ft	Live Load Moment kip·ft	E'quake Moment kip·ft	Wind Moment kip·ft
18.27	3000	40	None	None	30	21	13	10
18.28	3000	50	None	None	59	72	40	44
18.29	3000	60	None	None	32	25	15	13
18.30	4000	40	None	None	56	65	37	38
18.31	4000	50	None	18 in.	35	28	15	21
18.32	4000	60	None	19 in.	52	58	27	30
18.33	5000	40	None	20 in.	38	32	17	18
18.34	5000	50	None	21 in.	49	54	33	32
18.35	5000	60	12 in.	20 in.	40	37	20	24
18.36	4000	40	12 in.	22 in.	47	49	33	30
18.37	4000	50	12 in.	24 in.	43	41	24	28
18.38	4000	60	12 in.	24 in.	45	45	25	28

Using strength method for the following conditions, select a suitable rectangular concrete section and its reinforcement, using tensile steel ratios above $0.75\rho_b$ where necessary. Provide 1.5 in minimum cover over the reinforcement.

Problem No.	Concrete Strength f_c' psi	Steel Grade	Limits on Width b	Limits on Height h	Dead Load Moment kip·ft	Live Load Moment kip·ft	E'quake Moment kip·ft	Wind Moment kip·ft
18.39	3000	60	18 in.	24 in.	168	192	0	0
18.40	3000	40	18 in.	27 in.	154	181	96	106
18.41	3000	60	24 in.	24 in.	194	206	0	0
18.42	4000	40	24 in.	26 in.	212	221	121	141
18.43	4000	50	16 in.	25 in.	216	204	0	0
18.44	4000	50	20 in.	22 in.	155	161	106	116
18.45	5000	60	16 in.	23 in.	201	216	0	0
18.46	5000	40	24 in.	22 in.	198	171	99	89

The following problems relate to the subject matter developed in Sections 18.5, 18.10, 18.11, 18.12 and 18.13.

For the following conditions, select a suitable rectangular concrete section and its reinforcement at the balanced stress condition. Provide 1.5-in. cover over the reinforcement. For the chosen section, indicate the stress in the concrete and steel under service conditions.

Problem No.	Concrete Strength f_c' psi	Steel Grade	Limits on Width b	Limits on Height h	Dead Load Moment kip·ft	Live Load Moment kip·ft	E'quake Moment kip·ft	Wind Moment kip·ft
18.47	3000	40	None	None	30	21	13	10
18.48	3000	50	None	None	59	72	40	44
18.49	3000	60	None	None	32	25	15	13
18.50	4000	40	None	None	56	65	37	38
18.51	4000	50	None	14 in.	35	28	15	21
18.52	4000	60	None	14 in.	52	58	27	30
18.53	5000	40	None	14 in.	38	32	17	18
18.54	5000	50	None	14 in.	49	54	33	32
18.55	5000	60	10 in.	16 in.	40	37	20	24
18.56	4000	40	14 in.	16 in.	47	49	33	30
18.57	4000	50	10 in.	16 in.	43	41	24	28
18.58	4000	60	12 in.	16 in.	45	45	25	28

Select a suitable rectangular concrete section under balanced stress conditions for the given simply supported beams, to include required reinforcement. Then check the maximum deflection (by computation) that will occur under service load conditions. Allowable deflection = L/240.

Problem No.	Uniform Load kips/ft w_{DL}	Uniform Load kips/ft w_{LL}	Span ft	Concrete f_c' psi	Steel Grade
18.59	1.9	1.8	26	3000	40
18.60	2.2	2.8	18	4000	50
18.61	1.8	2.1	24	5000	60
18.62	2.1	2.6	18	3000	60
18.63	2.0	2.3	22	4000	40
18.64	1.9	2.4	20	5000	50

CHAPTER 19

Reinforced Concrete

The following problems relate to the subject matter developed in Sections 19.4 and 19.5

An interior span in a continuous rectangular beam has the shear and moment diagrams indicated below. The size of the member and its flexural reinforcement have already been selected as given. Select the shear reinforcement for the given section using the strength method. Use grade 60 steel, $f'_c = 3000$ psi.

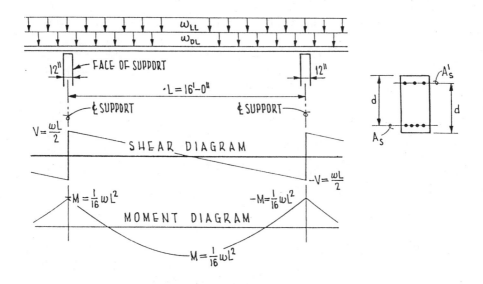

Problem No.	Uniform Load DL(kips/ft)	Load LL(kips/ft)	d (in.)	b_w (in.)	A_s (in.^{2})	A'_s (in.^{2})
19.1	3.5	15.75	36	18	3.14	1.50
19.2	5.25	15.75	34	20	4.00	2.09
19.3	7.00	14.00	32	22	4.68	2.53
19.4	8.75	14.00	30	24	5.57	2.88
19.5	10.50	12.25	28	26	6.33	3.12
19.6	12.25	12.25	28	28	7.33	3.76
19.7	14.00	10.50	26	28	7.57	3.88
19.8	15.75	10.50	24	30	8.78	4.10
19.9	17.50	8.75	24	30	9.37	5.07
19.10	19.50	8.75	24	32	10.30	5.57

CHAPTER 20

Reinforced Concrete

The following problems relate to the subject matter developed in Section 20.3.

Determine the required development length for straight deformed bars used as flexural tensile reinforcement in the given rectangular beams, used in an exterior exposure. Assume that the given clear cover is maintained both horizontally and vertically.

Problem No.	Concrete Strength f'_c psi	Steel Grade	Beam Width in.	Top or Other Bars	Bars	Required A_s in.^{2}	Clear Cover in.
20.1	3000	40	12"	Other	4 No. 5	1.19	1.5
20.2	3000	50	12"	Top	4 No. 6	1.66	1.5
20.3	3000	60	18'	Other	4 No. 9	3.91	2.5
20.4	3000	60	16"	Top	3 No. 6	0.64	2.5
20.5	4000	40	18"	Other	4 No. 8	3.01	2.5
20.6	4000	50	15"	Top	3 No. 10	3.06	2.5
20.7	4000	60	18"	Other	3 No. 6	0.61	3.0
20.8	4000	60	18"	Top	4 No. 8	2.09	3.0
20.9	5000	40	21"	Other	3 No. 11	4.60	3.0
20.10	5000	50	21"	Top	5 No. 9	4.88	2.5
20.11	5000	60	16"	Other	3 No. 8	1.09	2.5
20.12	5000	60	18"	Top	5 No. 10	6.09	2.5

The following problems relate to the subject matter developed in Section 20.4.

Determine the required development length for the given straight deformed bars used as compression reinforcement in beams in an exterior exposure. Assume the clear cover is maintained both horizontally and vertically.

Problem No.	Concrete Strength f_c psi	Steel Grade	Member Width in.	bars	Required A_s in.^{2}	Clear Cover in.
20.13	3000	40	10	4 No. 5	1.19	1.5
20.14	3000	50	10	4 No. 6	1.66	1.5
20.15	3000	60	14	4 No. 9	3.91	2.5
20.16	3000	60	9	3 No. 6	0.64	2.5
20.17	4000	40	12	4 No. 8	3.01	2.5
20.18	4000	50	12	3 No. 10	3.06	2.5
20.19	4000	60	10	3 No. 6	0.61	3.0
20.20	4000	60	14	4 No. 8	2.09	3.0
20.21	5000	40	13	3 No. 11	4.60	3.0
20.22	5000	50	16	5 No. 9	4.88	2.5
20.23	5000	60	12	3 No. 8	1.09	2.5
20.24	5000	60	18	5 No. 10	6.09	2.5

The following problems relate to the subject matter developed in Sections 20.5 and 20.6.

Problem 20.25   Flexural reinforcement in a beam consists of 12 No. 6 bars having at least $2d_b$ cover and at least $3d_b$ spacing. The bars are grade 60 top bars in tension, $f'_c = 3000$ psi. What would be the development length, cover requirements, and spacing requirements if the bars are collected into 6 bundles? 4 bundles? 3 bundles?

Problem 20.26   Compressive reinforcement placed uniformly around a circular pedestal consists of 15 No. 8 bars placed in a circle of 12 in. diameter. What would be the required cover if the bars were collected into 5 bundles uniformly spaced around the pedestal?

Problem 20.27-20.38   For Problems 20.1 through 20.12, respectively, determine the required development length of the bars if they were to be hooked. Assume that the given cover is maintained everywhere, both vertically and horizontally.

The following problems relate to the subject matter developed in Section 20.7 and 20.8.

For the following rectangular beams, determine the required areas of positive and negative reinforcement at the balanced stress condition. Find also the cutoff points (or anchorage requirements) for both positive and negative reinforcement. Use the abbreviated criteria for cutoffs and anchorages. Assume that the given dead load includes an allowance for the weight of the rectangular beam. It is not necessary to include a design for any required shear reinforcement, $f'_c = 4000$ psi, grade 60 steel.

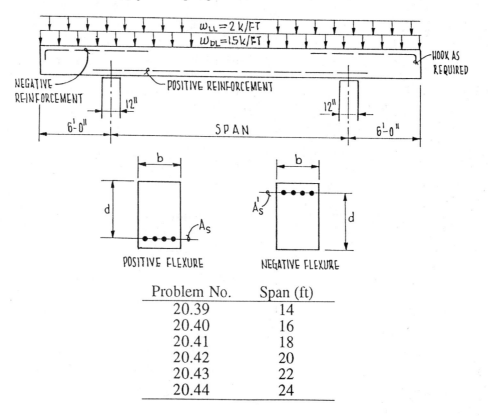

Problem No.	Span (ft)
20.39	14
20.40	16
20.41	18
20.42	20
20.43	22
20.44	24

The following problems relate to the subject matter developed in Sections 20.10, 20.211 and 20.12.

Problems 20.45 through 20.62. Determine the required splice length for beam reinforcement under the given conditions.

Problem No.	Concrete Strength f'_c psi	Steel Grade	Type of Splice	$\%A_s$ Spliced	Top or Other Location	Bars
20.45	3000	40	Tension	<25%	Top	#6
20.46	3000	50	Tension	50%	Other	3 No. 5 bundled
20.47	3000	60	Tension	100%	Top	#7
20.48	3000	40	Comp.			#9
20.49	3000	50	Comp.			4 No. 6 bundled
20.50	3000	60	Comp.			#8
20.51	4000	40	Tension	<25%	Other	#6
20.52	4000	50	Tension	50%	Top	3 No. 5 bundled
20.53	4000	60	Tension	100%	Other	#7
20.54	4000	40	Comp.			#9
20.55	4000	50	Comp.			4 No. 6 bundled
20.56	4000	60	Comp.			#8
20.57	5000	40	Tension	<25%	Top	#6
20.58	5000	50	Tension	50%	Other	3 No. 5 bundled
20.59	5000	60	Tension	100%	Top	#7
20.60	5000	40	Comp.			#9
20.61	5000	50	Comp.			4 No. 6 bundled
20.62	5000	60	Comp.			#6

CHAPTER 21

Reinforced Concrete

The following problems relate to the subject matter developed in Sections 21.7 and 21.8.

Design a one-way floor slab for the given loads and simply supported spans in an interior exposure.

Problem Number	Concrete Strength psi	Steel Grade	Live Load kips/ft^2	Clear Span Feet
21.1	3000	40	60	12
21.2	3000	50	100	12
21.3	3000	60	150	12
21.4	4000	40	75	14
21.5	4000	50	125	14
21.6	4000	60	200	14
21.7	5000	40	100	16
21.8	5000	50	150	16
21.9	5000	60	200	16
21.10	5000	40	75	16
21.11	4000	50	150	14
21.12	3000	60	300	12

Determine the amount of live load the given slabs can carry over the given simply supported span.

Problem No.	Concrete Strength f'_c	Steel Grade	Height h in.	Effective Depth d in.	Reinforce.	Span Feet
21.13	3000	40	6	$4^1/_2$	#5@12	10
21.14	4000	40	$6^1/_2$	5	#6@10	12
21.15	5000	50	7	6	#4@8	14
21.16	3000	50	$8^1/_2$	7	#5@10	15
21.17	4000	60	7	6	#4@6	14
21.18	5000	60	9	$7^1/_2$	#6@12	16

The following problems relate to the subject matter developed in Section 21.9.

Design a suitable continuous concrete slab for the given conditions under exterior exposures. Show a longitudinal sketch of your solution, to include bar locations and cutoff points. (It is usually desirable when designing for exterior exposures to add about 1 in. to the estimated slab thickness to account for the increased cover requirements.)

Prob. No.	No. of Spans	End Support Type	End Support Clear Span	Interior Supports Type	Interior Supports Clear Span	f'_c psi	Steel Grade	Live Load psf
21.19	2	Beam	12'6"	Beam		3000	40	160
21.20	3	Simple	13'6"	Beam	15'6"	3000	60	140
21.21	4	Beam	14'6"	Beam	16'6"	3000	40	120
21.22	5	Simple	16'6"	Beam	18'6"	3000	60	100
21.23	2	Simple	12'6"	Beam		4000	40	160
21.24	3	Beam	13'6"	Beam	15'6"	4000	60	140
21.25	4	Simple	14'6"	Beam	16'6"	4000	40	120
21.26	5	Beam	16'6"	Beam	18'6"	4000	60	100

The following problems relate to the subject matter developed in Section 21.11, 21.12, 21.13 and 21.14.

An interior span of a continuous tee beam has the shear and moment diagrams indicated below. The live load and the beam spacings are listed in the tabulation. The dead load of the stem must be estimated, then confirmed later when design is final. Using the strength method, select a suitable tee section for the given conditions, to include both positive and negative reinforcement and any shear reinforcement required. $f'_c =$ 3000 psi, grade 60 steel.

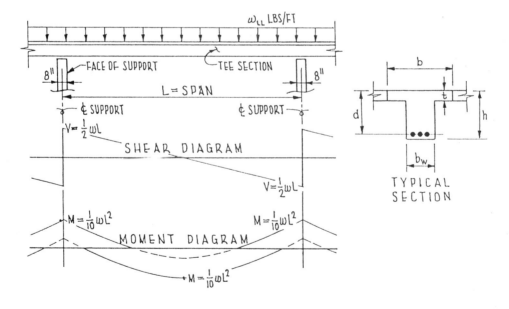

Problem No.	Span ft	LL psf	Slab Thickness in.	Tee Spacing ft
21.27	24	100	4	8
21.28	23	120	4	8
21.29	22	140	4	8
21.30	21	160	4	8
21.31	20	180	5	10
21.32	19	200	5	10
21.33	18	220	5	10
21.34	17	240	5	10
21.35	16	260	6	12
21.36	15	280	6	12
21.37	14	300	6	12
21.38	13	320	6	12

Design a tee-beam floor system for a simple span with the conditions and live loads shown below. The dead load of the system must also be included. Select the flexural reinforcement and any shear reinforcement required. f_c = 4000 psi, grade 60 steel.

Problem No.	Live Load psf	Slab Thickness in.	Tee Spacing	Clear Span
21.39	250	6	12'0"	18'0"
21.40	250	6	12'0"	16'0"
21.41	200	5	10'0"	22'0"
21.42	200	5	10'0"	18'0"
21.43	150	5	10'0"	24'0"
21.44	150	5	10'0"	20'0"
21.45	125	4	8'0"	28'0"
21.46	125	4	8'0"	24'0"
21.47	100	4	8'0"	30'0"
21.48	100	4	8'0"	26'0"
21.49	75	4	8'0"	32'0"
21.50	75	4	8'0"	28'0"

The following problems relate to the subject matter developed in Section 21.15.

Determine the required area of positive flexural reinforcement A_s for the following tee beams using the exact methods developed earlier. Then recalculate the area of reinforcement using the approximate equations (21.2). Compare the two sets of results.

Prob. No.	Concrete Strength f_c' psi	Steel Grade	Dead Load Moment kip·ft	Live Load Moment kip·ft	Stem Width b_w in.	Effective Depth d in.	Slab Thick. t in.	Clear Span ft	Tee Spacing ft
21.51	3000	40	58	39	14"	28"	5"	18'0"	6'0"
21.52	3000	50	82	73	16"	32"	6"	18'0"	6'0"
21.53	3000	60	99	106	18"	36"	7"	20'0"	6'0"
21.54	3000	60	118	141	20"	40"	8"	20'0"	8'0"
21.55	4000	40	58	39	12"	26"	5"	22'0"	6'0"
21.56	4000	50	82	73	14"	28"	5"	22'0"	6'0"
21.57	4000	60	99	106	14"	30"	6"	24'0"	8'0"
21.58	4000	60	118	141	16"	32"	8"	24'0"	12'0"
21.59	5000	40	58	39	12"	24"	5"	26'0"	8'0"
21.60	5000	50	82	73	14"	27"	5"	26'0"	8'0"
21.61	5000	60	99	106	16"	30"	6"	28'0"	8'0"
21.62	5000	60	118	141	18"	33"	8"	28'0"	12'0"

The following problems relate to the subject matter developed in Sections 21.16 through 21.22.

Design an isolated spread footing for the given columns under the given limitations.

Problem No.	Column Size in.	Axial P_{DL} kips	Load P_{LL} kips	Soil Pressure psf	f_c' psi	Steel Grade	Remarks
21.63	12 x 12	81	54	4000	3000	40	-
21.64	10 x 10	16	17	2000	3000	40	Property line 1.5 ft from centerline of columns
21.65	16 x 16	122	99	3000	4000	50	-
21.66	10 x 10	37	28	2000	4000	50	Obstruction 2 ft from centerline of column
21.67	12 x 12	40	27	2900	3000	60	Obstruction 1.5 ft from centerline of column
21.68	14 x 14	79	37	3000	4000	60	-
21.69	12 x 12	50	34	3000	4000	40	-
21.70	12 x 12	69	67	2000	3000	60	Property line 3.5 ft from centerline of column

Design a strip footing for the following wall loads.

Prob. No.	Wall Size in.	Axial P_{DL} k/ft	Load P_{LL} k/ft	Soil Pressure psf	f_c' psi	f_m' psi	Steel Grade	Remarks
21.71	8	14	12	4000	3000	1500	40	Masonry wall
21.72	12	7	9	2000	3000	-	50	Concrete wall
21.73	12	12	7	3000	4000	-	50	Concrete wall
21.74	16	15	12	3000	4000	1500	60	Masonry wall

Design a grade beam foundation for a repetitive column load as shown. The columns are square. The end column load may be assumed to be roughly half that of an interior column load.

Problem No.	Column Size in.	Bay Length ft	Axial P_{DL} kips	Load P_{LL} kips	Soil Pressure psf	f_c' psi	Steel Grade
21.75	12	20	32	36	3000	3000	40
21.76	12	16	24	32	5000	4000	50
21.77	10	25	14	20	4000	4000	40
21.78	16	30	46	46	5000	3000	60

CHAPTER 22

Reinforced Concrete

The following problems relate to the subject matter developed in Sections 22.1 through 22.8

Select a column section and its reinforcement for the conditions given below.

Prob. No.	Concrete Strength f_c' psi	Steel Grade	Axial Load in kips		Moment in kip·ft		L_u ft	Limits on h in.	End Conditions
			P_{DL}	P_{LL}	M_{LL}	M_{LL}			
22.1	3000	40	106	135	0	0	11'8"	None	
22.2·	3000	50	124	158	54	68	11'8"	16	
22.3	3000	60	148	180	86	74	11'8"	None	
22.4	4000	40	170	160	0	0	11'8"	16	
22.5	4000	50	231	193	102	120	11'8"	None	
22.6	4000	60	305	225	154	146	11'8"	16	
22.7	5000	40	190	206	0	0	11'8"	None	
22.8	5000	50	254	280	130	155	11'8"	16	
22.9	5000	60	405	351	202	195	11'8"	None	
22.10	3000	40	95	124	0	0	13'4"	None	
22.11	3000	50	110	143	58	72	13'4"	16	
22.12	3000	60	135	162	90	77	13'4"	None	
22.13	4000	40	161	144	0	0	13'4"	16	
22.14	4000	50	224	174	106	121	13'4"	None	
22.15	4000	60	291	203	159	140	13'4"	16	
22.16	5000	40	177	186	0	0	13'4"	None	
22.17	5000	50	248	172	133	159	13'4"	16	
22.18	5000	60	392	316	196	206	13'4"	None	
22.19	3000	40	91	119	0	0	10'4"	None	
22.20	3000	50	102	134	59	74	10'4"	15	
22.21	3000	60	130	159	92	81	10'4"	None	
22.22	3000	40	156	139	0	0	10'4"	15	
22.23	3000	50	219	168	110	123	10'4"	None	
22.24	3000	60	280	201	161	140	10'4"	15	
22.25	3000	40	172	187	0	0	10'4"	None	
22.26	3000	50	242	166	135	161	10'4"	15	
22.27	3000	60	384	304	201	209	10'4"	None	

The following problems relate to the subject matter developed in Section 22.9.

Determine the service moment M_{sv} for the given column sections. In all cases, the bars are concentrated at the flexure faces.

Problem No.	Concrete Strength f_c' psi	Steel Grade	Width b in.	Height h in.	Reinf. Bars (Total)	Ultimate P_n kips
22.28	3000	40	16	16	8 No. 8	460
22.29	3000	50	18	24	10 No. 9	644
22.30	3000	60	20	32	12 No. 10	916
22.31	4000	40	18	18	8 No. 9	770
22.32	4000	50	20	27	10 No. 10	1285
22.33	4000	60	22	36	12 No. 11	1800
22.34	5000	40	20	20	10 No. 10	1020
22.35	5000	50	22	30	12 No. 11	1814
22.36	5000	60	24	40	14 No. 11	2600

CHAPTER 23

Structural Masonry

The following problems relate to the subject matter developed throughout Chapter 23 for the empirical design method.

Problem 23.1 A value of f'_m = 1500 psi is specified for unreinforced fired clay masonry with Type M mortar (based on 10% voids). Determine the corresponding compressive strength to be used for walls designed using the empirical design method.

Problem 23.2 A compressive strength f'_m = 2000 psi is specified for unreinforced hollow CMU masonry with Type S mortar (based on net area). Determine the corresponding compressive strength to be used for 12 in. walls in the empirical design method (based on gross area).

Problem 23.3 A compressive strength of 4500 psi is specified for fired clay masonry units with 15% voids in the empirical design method. Determine the specified compressive strength f'_m to be used in the design tables for unreinforced masonry using Type M mortar.

Problem 23.4 A compressive strength of 1500 psi is specified for masonry of hollow CMU units in the empirical design method. Determine the specified compressive strength f'_m to be used in the design tables for 6 in walls of unreinforced masonry using Type N mortar.

Problem 23.5 Determine the allowable compressive stress on fired clay units having 10% voids to be used in the empirical design method for a value of f'_m = 2500 psi using Type S mortar.

Problem 23.6 An allowable compressive stress of 200 psi is to be used for the design masonry of fired clay units having 15% voids using Type M mortar in the empirical design method. Determine the value of f'_m to be specified.

Problem 23.7 An allowable compressive stress of 100 psi is to be used for the design of masonry of hollow ungrouted 12 in CMU units using Type M mortar in the empirical design method. Determine the value of f'_m to be specified.

Problem 23.8 Determine the allowable compressive stress to be used in the empirical design method for a hollow ungrouted 8 in. CMU masonry wall with f'_m = 2000 psi with Type M mortar.

Problem 23.9 Determine the linear modulus of elasticity and the shear modulus of elasticity to be used for CMU masonry in the empirical method of design, where the specified compressive strength f'_m = 2500 psi with Type S mortar.

Problem 23.10 Determine the shear modulus of elasticity to be used for fired clay brick masonry in the empirical method of design, where f'_m = 2000 psi with Type M mortar.

Problem 23.11 In the empirical method of design, determine the allowable load per foot of length along a hollow ungrouted 8 in. CMU wall with face shell bedding throughout. f'_m = 1500 psi with Type M mortar.

Problem 23.12 For the wall of Problem 23.11, determine the allowable load if full bed joints are used.

Problem 23.13 Determine the allowable in-plane shear stress in a masonry shear wall in a braced frame (no bearing load on the wall). $f'_m = 2000$ psi, running bond.

Problem 23.14 Determine the allowable in-plane shear stress in a combination shear and bearing masonry wall of hollow ungrouted 8 in. CMU masonry. $f'_m = 1500$ psi, running bond, vertical load 7 kips/ft.

Problem 23.15 Determine the allowable shear stress in the wall of Problem 23.13 for wind load transverse to the wall.

Problem 23.16 Determine the allowable transverse shear stress in the wall of Problem 23.14 if the wall is built with stack bond.

Problem 23.17 Determine the minimum wall thickness that may be used for an ungrouted combination shear and bearing wall spanning: a) 10'0" vertically from floor to roof; b) 18'0" vertically from floor to roof; and c) 24' vertically from floor to roof.

Problem 23.18 Determine the maximum pilaster spacing for a long, continuous fence built of ungrouted 8 in. CMU masonry.

Problem 23.19 Determine the minimum thickness of a fired clay masonry parapet wall 32 in. high.

Problem 23.20 Determine the minimum thickness of a CMU parapet wall 18 in. high.

Problem 23.21 A combination basement and foundation wall is to be built using ungrouted hollow CMU masonry, spanning 8'0" vertically from the floor slab at its base to the floor diaphragm above. Determine the minimum required wall thickness if the elevation of the outside fill is 6 ft. above the elevation of the base slab.

Problem 23.22 For the foundation wall of Problem 23.21, what would have to be done to increase the unbalanced fill depth to 7 ft.? To 8 ft.?

Problem 23.23 What is the minimum wall thickness of a masonry bearing wall in a three story apartment building?

Problem 23.24 What is the minimum wall thickness of a masonry bearing wall in a single story commercial building?

Problem 23.25 Determine the wire tie size and spacing for a 16 in. hollow unit CMU wall built of two 8 in. wythes.

Problem 23.26 A 3/4 in. diameter bolt is imbedded 6 in. into a bond beam in a vertical joint of a CMU wall. Determine the allowable vertical load (in shear) on the bolt.

CHAPTER 25

Structural Steel

The following problems relate to the subject matter developed in Section 25.1.

Problem 25.1 A single angle 6 × 4 × 3/8, short leg outstanding, is used as a diagonal in a truss. It is attached to a 1/2 in. gusset plate with eight A-325N bolts in two rows of four bolts of 7/8 in. diameter. Sketch the connection and determine the capacity of the member in tension. A36 steel.

Problem 25.2 A single angle truss member is to be attached to a 3/8 in. gusset plate with no more than three A325N bolts of 5/8 in. diameter. The angle is to carry a tensile load of 35 kips. Sketch the connection and select a suitable angle to carry the load. A36 steel.

Problem 25.3 A structural tee WT5 × 6 is used as a truss member with the stem outstanding. The tee is fastened to a 3/8 in gusset plate with four A325N bolts, 5/8 in. diameter, two bolts in a row at each side of the stem. Sketch the connection and determine the capacity of the member in tension. A36 steel.

Problem 25.4 A structural tee is used as a hanger, carrying a tensile load of 124 kips. The tee is attached by its flange to a 1/2 in. distribution plate by two rows of 7/8in. diameter A325N bolts, one row of 5 bolts at each side of the stem. Sketch the connection and select a suitable tee to carry the load. A36 steel.

Problem 25.5 Two steel plates 1/2 in. × 12 in. are spliced using two 5/16 in. × 12 in. plates each side. Bolts are 5/8 in. diameter, A325N, placed in a 3-bolt × 3-bolt pattern. Sketch the splice and determine the capacity of the connection. A36 steel.

Problem 25.6 Two steel plates 3/8 in. × 8 in. are to be lapped 6 in. and spliced with 3/4 in. diameter A325N bolts in single shear, using a pattern of 3 bolts wide (across the 8 in. direction) and 2 bolts long (along the 6 in. direction). Sketch the connection and determine the capacity of the connection. A36 steel.

Problem 25.7 A steel plate 5/16 in. × 8 in. is lapped 10 in. over a much larger base plate and welded along its two sides only. Welds are 1/4 in. fillet welds, each being 8 in. long. Sketch the connection and determine the capacity of the plate in tension. A36 steel.

Problem 25.8 A steel plate 1/4 in. × 6 in. is lapped 6 in. over a much larger base plate. It is welded to the base plate along its end and along the edge of the base plate, similar to that shown in Example 25-5. It is also welded along both sides of the plate. All welds are 3/16 fillet welds 5 in. long. Sketch the connection and determine the capacity of the plate in tension.

The following problems relate to the subject matter developed in Section 25.2.

Problem 25.9 Determine the required diameter of a threaded circular bar to sustain a load of 20 kips. (Use the minimum root area A_k rather than the AISC tensile stress area). A36 steel.

Problem 25.10 For the bar of Problem 25.9, determine the required diameter of the circular bar if the threaded ends are upset enough that the governing stress is in the shank of the bar rather than at the threads.

Problem 25.11 For the bar of Problem 25.9, determine the required threaded diameter of the upset end of the bar if the governing stress is to be in the shank of the bar.

Problem 25.12 For the bars of Problems 25.9 and 25.10, determine the required minimum lengths of the thread for the two cases (see Table S-20).

Problem 25.13 A splice has the trial layout shown. The applied load is 100 kips. ASTM A36 steel, sheared edges. The following problems relate to the subject matter developed in Section 25.2, 3, 4, and 5.A325N bolts, 3/4 in. diameter. Wrinkling of the plate is permitted. Determine the capacity of the splice in bearing.

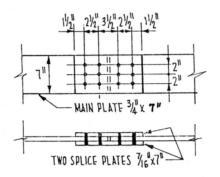

Problem 25.14 A splice has the trial layout shown. The applied load is to be 125 kips. ASTM A36 steel, sheared edges. A325N bolts, 7/8 in. diameter. Wrinkling of the plate is permitted. Determine the capacity of the splice in bearing.

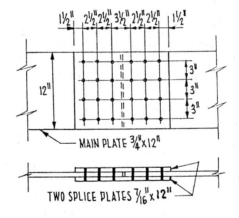

Problem 25.15 A single-plate splice on the tension side of a W18 × 35 is laid out as shown. Determine the capacity of the splice in bearing. ASTM 36 steel, sheared edges. A325N bolts, 7/8 in. diameter. No wrinkling of the plate is permitted.

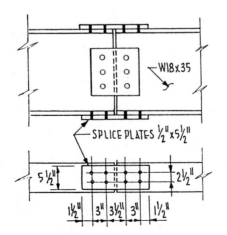

Problem 25.16 A single-plate splice on the tension side of a W12 × 65 beam is laid out as shown. Determine the capacity of the splice in bearing. ASTM A36 steel, sheared edges. A325N bolts, 3/4 in. diameter. No wrinkling of the plate is permitted.

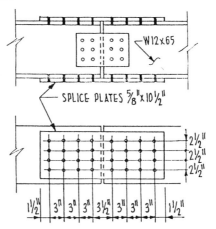

The following problems relate to the subject matter developed in Section 25.6.

Problem 25.17 Determine the capacity in block tearout in the splice plates and in the main plate of Problem 25.13. Show sketches of the two tearouts at failure. Assume a 1/4 in. gap between the main plates.

Problem 25.18 Determine the capacity in block tearout in the splice plates and in the main plate of Problem 25.14. Show sketches of the two tearouts at failure. Assume a 1/4 in. gap between the main plates.

Problem 25.19 Determine the capacity in block tearout in the splice plate and in the beam flange of Problem 25.15. Show sketches of the two tearouts at failure. Assume a 1/4 in. gap between the main plates.

Problem 25.20 Determine the capacity in block tearout in the splice plate and in the beam flange of Problem 25.16. Show sketches of the two tearouts at failure. Assume a 1/4 in. gap between the main plates.

Problem 25.21 Determine the capacity in block tearout of the connection of Example 25.14 if the bolt layout on the stem of the tee is changed to a single line of three 5/8 in. diameter bolts placed 2 in. above the bottom edge of the tee. Show a sketch of the tearout at failure.

Problem 25.22 Determine the capacity in block tearout of the connection of Example 25.14 if the bolt layout on the stem of the tee is changed to two 3/4 in. diameter bolts spaced at 3 in. on center, placed 3 in. above the bottom edge of the tee. Show a sketch of the tearout at failure.

The following problems relate to the subject matter developed in Section 25.7.

Problem 25.23 A double angle hanger with short legs outstanding is to be suspended from the downstanding stem of a WT10.5 × 25. Design load is 60 kips. A325N bolts, A36 steel. Angle is gas cut, tee stem is sawn. Select a suitable size for the hanger and design the bearing connection.

Problem 25.24 A WT with stem outstanding is to be used as vertical member in a truss. The tee is connected to a 5/8 in. gusset plate at a panel point. Load is 60 kips in tension. Layout is so restricted that the flange width of the tee may not exceed 7 in. A325N bolts, A36 steel. Gusset plate is gas cut. Select a suitable tee section and design the bearing connection.

Problem 25.25 A plate section not more than 12 in. wide is to be used as a hanger for a mezzanine. The plate is to be connected by two splice plates to the stem of a supporting tee section, WT12 × 52. The load on the hanger is 84 kips in tension. Tee stem is sawn, splice plates and hanger are gas cut. A325N bolts, A36 steel. Select a suitable size for the hanger that is compatible with the thickness of the stem of the tee and design the bearing connection.

Problem 25.26 A steel plate 3/8 in. thick with sheared edges is to be rolled into a cylindrical shape 8'0" long. The two edges are to be lapped. The length of the lap may be as much as 12 in. The lapped edges are to be bolted together using a staggered bolt pattern similar to that of Example 25.16. Using A325 bolts in a bearing connection, design the connection for maximum efficiency.

Problem 25.27 For the cylinder of Problem 25.27, design the connection if the two edges are butted together rather than being lapped. A gap of 1/8 in. is to be maintained between the edges. Splice plates may be as wide as 12 in.

The following problems relate to the subject matter developed in Section 25.8.

Problems 25.28 through 25.32: Design the connections of Problems 25.23 through 25.27, respectively, using slip-critical connections rather than bearing connections.

The following problems relate to the subject matter developed in Section 25.9.

Problems 25.33 through 25.37: Design the connections of Problems 25.23 through 25.27, respectively, using fillet welded connections rather than bearing connections.

CHAPTER 26

Structural Steel

The following problems relate to the subject matter developed in Section 26.1.

Problem 26.1 A square structural tube 5 in. × 5 in. with 3/16 in. walls is used as a column 16'0" long, hinged at both ends. Determine the critical load as buckling impends and the stress corresponding to that load. State whether the column will fail by yielding or by buckling. A500 steel.

Problem 26.2 A double angle strut consisting of 2 L6 × 4 × 3/8 spaced at 3/8 in. apart, short legs outstanding, is used as a truss member in compression. Ends are considered to be hinged, length between hinges is 18'7". Determine the critical load as buckling impends and the stress corresponding to that load. State whether the column will fail by yielding or by buckling and about which axis failure will occur. A36 steel.

Problem 26.3 A W8 × 31 is used as a column 40'0" long. The bottom of the column is hinged on both axes. The top of the column is hinged on the strong axis and fixed on the weak axis. Sketch the two possible buckling modes and determine the allowable load on the column for a factor of safety of 23/12. A36 steel.

Problems 26.4, 5, 6, 7 and 8: A W8 × 10 steel column is 34'0" long. For the following conditions, determine the critical load as buckling impends, the stress at buckling failure, whether the column fails in yield or in buckling and, if in buckling, determine the allowable load with a factor of safety of 23/12. Sketch the failure modes in buckling, both about the strong axis and about the weak axis.

Problem 26.4 Hinged ends in both axes at top and bottom, no interior bracing.

Problem 26.5 Hinged ends on both axes at top and bottom, laterally braced at midpoint on the weak axis only.

Problem 26.6 Hinged ends on both axes at top and bottom, laterally braced at third points on the weak axis only.

Problem 26.7 Hinged ends on both axes at top and bottom, laterally braced at quarter points on the weak axis only.

Problem 26.8 Hinged on the strong axis at top and bottom, fixed on the weak axis at top and bottom, laterally braced at midpoint on the weak axis only.

The following problems relate to the subject matter developed in Section 26.2 and 26.3.

Problems 26.9 through 26.16: Determine the allowable axial column load on the indicated sections for the given support conditions. Sketch the failure modes on both axes and state on which axis failure occurs.

Problem 26.9 Structural tube 6 in. × 6 in. with 1/4 in. walls, ASTM A500 steel, length 12'0", hinged top and bottom on both axes.

Problem 26.10 Wide flange W12 × 22, A36 steel, length 20'0", hinged top and bottom on both axis and laterally braced at quarter points on the weak axis only.

Problem 26.11 Standard 8 in. diameter pipe, A53 steel, length 16'0", fixed top and bottom, laterally braced at midpoint.

Problem 26.12 Double angle strut 2 L5 × 3 × 1/4, A36 steel, long legs back to back with 3/8 in. space, length 2'0", hinged both ends on both axes, no interior bracing.

Problem 26.13 American standard channel C4 × 5.4, A36 steel, length 8'0", hinged top and bottom on both axes, braced at midpoint on weak axis only.

Problem 26.14 American standard section S7 × 20, A36 steel, length 15'0", hinged top and bottom on strong axis, fixed at top and bottom on weak axis, braced at midpoint on weak axis only.

Problem 26.15 Two channels 5 × 6.7, welded together as shown, A36 steel, length 16'0", hinged top and bottom on both axes.

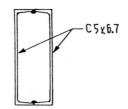

Problem 26.16 Two channels 5 × 6.7 welded together as shown, A36 steel, length 16'0", hinged top and bottom on both axes.

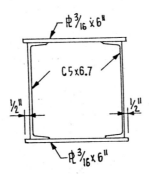

The following problems relate to the subject matter developed in Section 26.3

Problems 26.17 through 26.24: Select the lightest-weight column section of the shape specified to carry the indicated load under the given support conditions. State the allowable load on both axes of the selected section.

Problem 26.17 Wide flange section, A36 steel, unsupported length 16'0" on both axes, hinged top and bottom, concentric load 300 kips.

Problem 26.18 American standard section, A36 steel, unsupported length 16'0" on the strong axis, 8'0" on the weak axis, hinged top and bottom, concentric load 125 kips.

Problem 26.19 Standard pipe section, A53 steel, unsupported length 20'0", fixed top and bottom, concentric load 100 kips.

Problem 26.20 Extra strong pipe section A53 steel, unsupported length 20'0", fixed top and bottom, concentric load 100 kips.

Problem 26.21 Structural tee section, A36 steel, unsupported length 17', fixed at top on both axes, hinged at bottom on both axes, concentric load 110 kips.

Problem 26.22 Wide flange section, A36 steel, unsupported length 28', hinged top and bottom on strong axis, fixed top and bottom on weak axis, concentric load 175 kips.

Problem 26.23 Square structural tubing, A500 steel, unsupported length 17'0", hinged at bottom, fixed at top, concentric load 115 kips. State required wall thickness for the chosen section.

Problem 26.24 Wide flange section, A36 steel, unsupported length 28'0", fixed top and bottom on strong axis, laterally braced at midpoint on strong axis, hinged top and bottom on weak axis, laterally braced at quarter points on weak axis, concentric load 85 kips.

Problem 26.25 through 26.28: Select the lightest-weight column section of the shape specified to carry the indicated load on its strong axis. Then determine the required lateral bracing on the weak axis for the section to be able to carry the same load on the weak axis. Bracing can be placed only at midpoint, third points, and quarter points.

Problem 26.25 Wide flange section, A36 steel, unsupported length 24'0", hinged top and bottom, concentric load 100 kips.

Problem 26.26 American standard section, A36 steel, unsupported length 18'0", hinged top and bottom, concentric load 115 kips.

Problem 26.27 Structural tee section, A36 steel, unsupported length 24'0", hinged top and bottom, concentric load 190 kips.

Problem 26.28 Structural tee section, A36 steel, unsupported length 18'0", hinged top and bottom, concentric load 150 kips.

The following problems relate to the subject matter developed in Section 26.4.

Problem 26.29 Design a column splice for a column that is stepped from a W12 × 120 to a W12 × 65. A325 bolts, A36 steel. Column load is 300 kips. No computable moments or shears on the splice. Show a complete sketch of your design.

Problem 26.30 Design a column splice for a column that is stepped from a W12 × 152 to a W12 × 96. A325 bolts, A36 steel. Column load is 350 kips. No computable moments or shears on the splice. Show a complete sketch of your design.

Problem 26.31 Design a column splice for a column that is stepped from a W14 × 68 to a W12 × 45. A325 bolts, A36 steel. Column load is 220 kips. No computable moments or shears on the splice. Show a complete sketch of your design.

Problem 26.32 Design a column splice for a column that is stepped from a W12 × 65 to a W10 × 45. A325 bolts, A36 steel. Column load is 220 kips. No computable moments or shears on the splice. Show a complete sketch of your design.

The following problems relate to the subject matter developed in Section 26.5.

Problem 26.33 Design a column base plate for a W10 × 45 column carrying a concentric load of 200 kips, supported on a concrete spread footing 9'0" square. ASTM A36 steel, f'_c = 4000 psi.

Problem 26.34 Design a column base plate for a W12 × 45 column carrying a concentric load of 250 kips, supported on a concrete spread footing 9'6" square. ASTM A36 steel, f'_c = 3000 psi.

Problem 26.35 Design a column base plate for a W8 × 48 column carrying a concentric load of 225 kips, supported on a concrete spread footing 10'6" square. ASTM A36 steel, f'_c = 4000 psi.

Problem 26.36 Design a column base plate for a W14 × 43 column carrying a concentric load of 220 kips, supported on a concrete spread footing 9'0" square. ASTM A36 steel, f'_c = 3000 psi.

CHAPTER 27

Structural Steel

The following problems relate to the subject matter developed in Sections 27.1, 27.2 and 27.3.

Problems 27.1 through 27.8 The shear and moment diagrams shown below apply to Problems 27.1 through 27.8. In all cases, lateral support of the compression flange is provided at the lines of support. Other lines of support are provided as indicated in the individual problems.

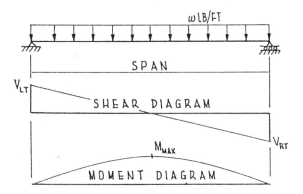

Problem 27.1 Maximum moment is 40 kip•ft. Span is 24 ft. Compression flange is supported continuously. Select the lowest-weight compact wide flange section in A36 steel.

Problem 27.2 Maximum moment is 80 kip•ft. Span is 28 ft. Compression flange is supported continuously. Select the lowest-weight wide flange section in A36 steel.

Problem 27.3 Maximum moment is 120 kip•ft. Span is 32 ft. Compression flange is supported continuously. Select the lowest-weight wide flange section in A36 steel.

Problem 27.4 Maximum moment is 160 kip•ft. Span is 24 ft. Compression flange is supported continuously. Select the lowest-weight compact wide flange section in A36 steel .

Problem 27.6 Maximum moment is 25 kip•ft. Span is 28 ft. Compression flange is laterally supported at ends and at quarter points. Select the lowest-weight wide flange tee section in A36 steel and compute the actual tensile and compressive stresses in the section.

Problem 27.7 Maximum moment is 120 kip•ft. Span is 32 ft. Compression flange is laterally supported at ends and at third points. Select the lowest-weight noncompact wide flange section in A36 steel with stress not less than $0.60F_Y$.

Problem 27.8 Maximum moment is 160 kip•ft. Span is 36 ft. Compression flange is laterally supported at ends and at third points. Select the lowest-weight noncompact wide flange section in A36 steel with stress not less than $0.60F_Y$.

Problems 27.9 through 27.16: The shear and moment diagrams shown below apply to Problems 27.9 through 27.16. In all cases, lateral support of the compression flange is provided at the free ends of the overhangs and at all lines of support. Other lines of support are provided as indicated in the individual problems.

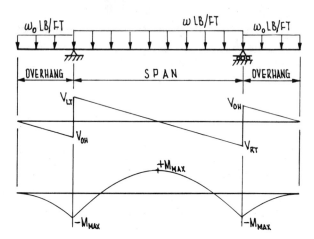

Problem 27.9 Maximum positive moment is 40 kip•ft. Maximum negative moment is 32 kip•ft. Span is 24 ft. Overhangs are 8 ft. Compression flange is laterally supported at ends, at supports and at midspan. Select the lowest-weight noncompact wide flange section in A36 steel with the stress not less than $0.60F_Y$.

Problem 27.10 Maximum positive moment is 80 kip•ft. Maximum negative moment is 92 kip•ft. Span is 28 ft. Overhangs are 10 ft. Compression flange is laterally supported at ends, at supports and at quarter points. Select the lowest-weight noncompact wide flange section in A36 steel with the stress not less than $0.60F_Y$.

Problem 27.11 Maximum positive moment is 120 kip•ft. Maximum negative moment is 100 kip•ft. Span is 32 ft. Overhangs are 10 ft. Compression flange is supported at ends, at supports and at midpoint. Select the lowest-weight American standard channel section in A36 steel and compute the actual flexural stress in the selected section.

Problem 27.12 Maximum positive moment is 42 kip•ft. Maximum negative moment is 36 kip•ft. Span is 18 ft. Overhangs are 4 ft. Compression flange is supported at ends at supports and at third points. Select the lowest-weight American standard section in A36 steel and compute the actual flexural stress in the selected section.

Problem 27.13 Maximum positive moment is 40 kip•ft. Maximum negative moment is 32 kip•ft. Span is 24 ft. Overhangs are 6 ft. Compression flange is supported at ends and at support lines. Select the lowest-weight wide flange section in A36 steel and compute the actual flexural stress in the selected section.

Problem 27.14 Maximum positive moment is 120 kip•ft. Maximum negative moment is 190 kip•ft. Span is 28 ft. Overhangs are 10 ft. Compression flange is supported at ends, at support lines and at midspan. Select the lowest-weight American standard section in A36 steel and compute the actual flexural stress in the selected section.

Problem 27.15 Maximum positive moment is 160 kip•ft. Maximum negative moment is 180 kip•ft. Span is 32 ft. Overhangs are 12 ft. Compression flange is supported at ends, at support lines and at midspan. Select the lowest-weight wide flange section in A36 steel and compute the actual flexural stress in the selected section.

Problem 27.16 Maximum positive moment is 80 kip•ft. Maximum negative moment is 80 kip•ft. Span is 36 ft. Overhangs are 12 ft. Compression flange is supported at ends, at support lines and at quarter points.

Select the lowest-weight wide flange section in A36 steel and compute the actual flexural stress in the selected section.

The following problems relate to the subject matter developed in Section 27.4.

Problem 27.17 through 27.24: For the listed sections, determine the exact maximum shear stress at the neutral axis using Equation 27-9. Then determine the average shear stress on the web as indicated in Fig. 27-5 and find the percent error by the following formula:

$$\% \text{ Error} = \frac{\text{Exact Stress - Average Stress}}{\text{Exact Stress}} \times 100$$

Problem Number	Section	Shear Kips
27.17	W 14 x 34	40
27.18	S8 x 18.4	20
27.19	C8 x 11.5	15
27.20	WT7 x 17	20
27.21	W16 x 40	50
27.22	S12 x 40.8	50
27.23	C12 X 20.7	30
27.24	WT8 x 20	20

The following problems relate to the subject matter developed in Section 27.5.

Problem 27.25 through 27.32: For the given simply supported sections and spans, determine whether the section is adequate for deflection limitations under two conditions:
1. When no brittle elements are to be supported.
2. When brittle elements are to be supported.

Problem	Section A36 Steel	Span ft.	Live Load lb/ft	Dead Load lb/ft	Position	Stress
27.25	W10 x 12	24	175	100	Floor	$0.66F_y$
27.26	W12 x 19	28	250	150	Floor	$0.66F_y$
27.27	S10 x 25.4	32	250	150	Floor	$0.60F_y$
27.28	S12 x 31.8	36	250	150	Floor	$0.60F_y$
27.29	C10 x 15.3	24	200	150	Roof	$0.60F_y$
27.30	C12 x 20.7	28	250	150	Roof	$0.60F_y$
27.31	S10 x 35	32	250	150	Roof	$0.60F_y$
27.32	W10 x 33	36	250	150	Roof	$0.66F_y$

The following problems relate to the subject matter developed in Section 27.6.

Problems 27.33 through 27.36: Determine the maximum tensile flexural stress and the maximum compressive flexural stress acting anywhere on the given sections.

Problem No.	Section	M_{xx} kip·ft	M_{yy} kip·ft
27.33	W18 x 50	60	12
27.34	S15 x 42.9	40	6
27.35	C12 x 20.7	15	2
27.36	WT9 x 25	5	6

Problems 27.37 through 27.40: Select the lightest-weight wide flange section to carry the indicated moments on two axes with the indicated spacing of lateral support on the compression flange.

Problem No.	M_{xx} kip·ft	M_{yy} kip·ft	Span ft	Lateral Support on Compression Flange
27.37	47	6.0	24	8'0"
27.38	56	8.0	28	7'0"
27.39	71	5.0	32	8'0"
27.40	86	11.0	36	Continuous

The following problems relate to the subject matter developed in Section 27.8.

Problems 27.41 through 27.44: The following sketches show the two types of connections to be analyzed in Problems 27.41 through 27.44.

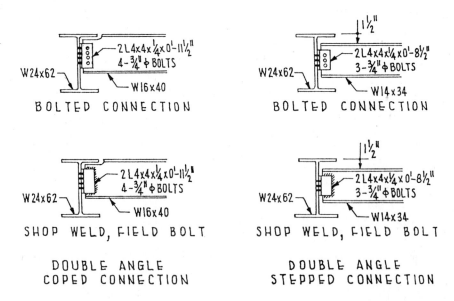

Problem 27.41 Establish the bolt layout and determine the allowable design load on the bolted coped connection shown in the foregoing sketch. A36 steel, A325N bolts. Bearing type connection. Assume the cope is 2 in. deep.

Problem 27.42 Determine the allowable design load on the shop weld, field bolt coped connection shown in the foregoing sketch. Weld size is limited to 3/16 in. A36 steel. A325N bolts. Bearing connections. $E60_{xx}$ electrodes. Assume the cope is 2 in. deep.

Problem 27.43 Establish the bolt layout and determine the allowable design load on the bolted stepped connection shown in the foregoing sketch. A36 steel, A325N bolts. Bearing type connections.

Problem 27.44 Determine the allowable design load on the shop weld, field bolt stepped connection shown in the foregoing sketch. Weld size is limited to 3/16 in. A36 steel. A325N bolts. Bearing connections. E60xx electrodes.

The following problems relate to the subject matter developed in Section 27.9.

Problems 27.45 through 27.52: The following sketch shows the two types of beam bearings to be designed in the following problems.

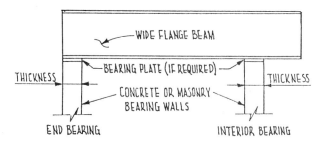

Determine whether a bearing plate is needed under the end bearing conditions shown in the sketch. If a bearing plate is needed, select one of suitable size. For concrete, use an allowable bearing stress of 900 psi. For masonry, use an allowable bearing stress of 500 psi. A36 steel.

Problem No.	Beam Section	End Bearing Wall	Interior Bearing Wall	Wall Thickness in.	Reaction kips
27.45	W16 x 40	Concrete	-	8	65
27.46	W18 x 50	Concrete	-	10	85
27.47	W18 x 55	Masonry	-	8	35
27.48	W21 x 62	Masonry	-	10	45
27.49	W16 x 40	-	Concrete	10	80
27.50	W18 x 50	-	Concrete	12	100
27.51	W18 x 55	-	Masonry	10	50
27.52	W21 x 62	-	Masonry	12	60

CHAPTER 28

Structural Steel

The following problems relate to the subject matter developed in Section 28.2.

Problem 28.1 For the given section, determine the following properties:
- a) L_c and L_u
- b) Moment capacity as a noncompact section
- c) Shear capacity where $F_v \leq 0.4F_y$ on the web of the beam
- d) Shear capacity as limited by the welds to the channel A36 steel, E60xx electrodes

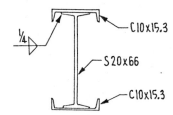

Problem 28.2 For the given section, determine the following properties:
- a) L_c and L_u, compression flange at top
- b) Moment capacity as a noncompact section
- c) Shear capacity where $F_v \leq 0.4F_y$ on the web of the vertical channel
- d) Shear capacity as limited by the welds to the horizontal channel A36 steel, E60xx electrodes

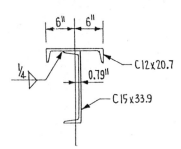

Problem 28.3 For the given section, determine the following properties:
- a) L_c and L_u, compression flange at top
- b) Moment capacity as a noncompact section
- c) Shear capacity where $F_v \leq 0.4F_y$ on the web of the channel
- d) Shear capacity as limited by the bolts to the angle A36 steel, E60xx electrodes

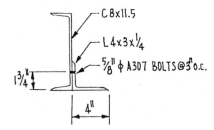

Problem 28.4 For the given section, determine the following properties:

    a) L_c and L_u

    b) Moment capacity as a noncompact section

    c) Shear capacity where $F_v \leq 0.4F_y$ on the web of the channel

    d) Shear capacity as limited by the welds to the angle A36 steel, E60xx electrodes

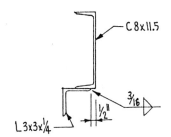

CHAPTER 30
Structural Timber

The following problems relate to the subject matter developed in Section 30.1

Problem 30.1 A 6 × 12 chord is spliced with two 4 × 10 splice plates as shown in the sketch. Bolts are 3/4 in. diameter in the pattern shown. No. 2 southern pine, load duration more than 10 years, moisture less than 19%. Determine the capacity of the chord member and of the two splice plates.

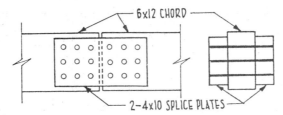

Problem 30.2 For the splice of Problem 30.1, determine the capacity of the splice if the moisture content can exceed 19% for extended periods of time.

Problem 30.3 A hanger for a mezzanine floor is made of 2-2 × 12 members, spaced 3 1/2 in. apart as shown in the sketch. The hanger supports one of the main girders of the mezzanine with 5/8 in. diameter bolts in the pattern shown. No. 2 Douglas fir-larch, duration of load more than 10 years, moisture content less than 19%. Determine the capacity of the hanger.

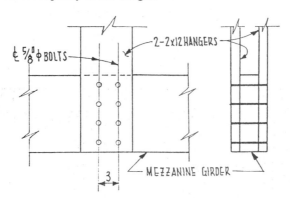

Problem 30.4 Determine the capacity of the hanger of Problem 30.3 if the hanger is changed to 2-4 × 6 members of spruce-pine-fir.

The following problems relate to the subject matter developed in Section 30.2.

Problems 30.5 through 30.8: The following connections are single lapped splices as indicated in the sketch. Determine the number and size of nails or spikes required to develop the given load. No. 2 spruce-pine-fir, duration of load less than two months. Subject to extended periods of high moisture.

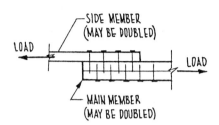

Problem No.	Side Member	Main Member	Tensile Load, kips
30.5	2 x 8	2 x 8	3
30.6	2 x 8	2 - 2 x 8	4
30.7	2 x 10	4 x 10	4
30.8	2 - 2 x 10	2 - 2 x 10	5

Problems 30.9 through 30.12: The following connections are butt spliced with two splice plates as shown in the sketch. Determine the number and size of nails or spikes required to develop the given load. No. 2 southern pine, duration of load more than 10 years, subject to high moisture for extended periods. (Dimensions of side members are exact. Dimensions of main member the nominal dimensions).

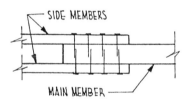

Problem No.	Side Members	Main Member	Tensile Load, kips
30.9	$^3/_4$ x $7^1/_4$	2 x 8	2
30.10	$^3/_4$ x $9^1/_4$	2 x 10	3
30.11	1 x $9^1/_4$	2 x 10	3
30.12	$1^1/_2$ x $9^1/_4$	4 x 10	5

The following problems relate to the subject matter developed in Section 30.3.

Problem 30.13 through 30.16: The following connections are single lapped splices as indicated in the sketch. Determine the capacity of the connection in tension. No. 2 Douglas fir-larch. A307 bolts. Duration of load less than two months. Dry conditions of service.

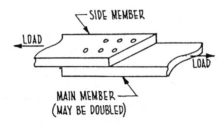

Problem No.	Side Member	Main Member	No. and Size of Bolts
30.13	2 x 10	2 x 10	8 - $^5/_8$ in. dia. in 2 rows
30.14	2 x 12	2-2 x 12	6 - $^5/_8$ in. dia. in 3 rows
30.15	4 x 8	4 x 12	6 - $^3/_4$ in. dia. in 2 rows
30.16	4 x 10	4 x 12	4 - $^7/_8$ in. dia. in 2 rows

Problem 30.17 A connection of the type shown in the following sketch is subject to a tensile load of 15 kips. Determine whether the connection is adequate for the load. No. 2 southern pine. A307 bolts. Duration of load more than 10 years. Dry conditions of service.

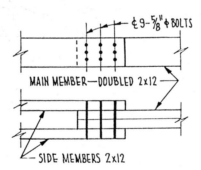

Problem 30.18 For the connection of Problem 30.17, determine the capacity if the main member is changed to a 4 × 12.

Problem 30.19 The end of a wood beam is supported by a double hanger as shown in the sketch. The beam reaction is 14 kips. Determine whether the connection is adequate for the given load. No. 2 spruce-pine-fir. Duration of load is less than two weeks. Dry conditions of service.

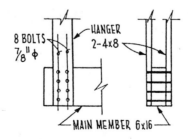

Problem 30.20 A truss diagonal is bolted to the lower chord as shown in the sketch. The diagonal is subject to a tensile load of 6 kips. Determine whether the connection is adequate for the given load. No. 2 Douglas-fir-larch. Duration of load more than 10 years. Subject to extended periods of high moisture.

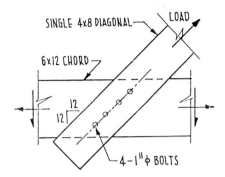

Problem 30.21 A 2 × 10 tension member is subject to a tensile load of 4 kips. The member is to be lap spliced to another 2 × 10. Design a bolted connection, to include bolt size and layout. No. 2 southern pine. Duration of load less than two weeks. Subject to extended periods of moist conditions.

Problem 30.22 A truss diagonal is subject to a tensile load of 12 kips as shown in the sketch. Design a bolted connection for a single diagonal member to carry the load. No. 2 spruce-pine-fir. Duration of load 1 year. Dry conditions of service.

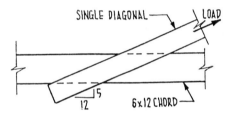

Problem 30.23 Design a bolted connection for the diagonal of Problem 30.22 if the diagonal is to be two members, one to either side of the chord.

Problem 30.24 Design a bolted connection for the diagonal of Problem 30.22 if the slope of the diagonal is 9:12 rather than 5:12.

Problem 30.25 Design a bolted connection for the diagonal of Problem 30.22 if the diagonal is to two members, one to either side of the chord, and the slope is to be 9:12 rather than 5:12.

Problem 30.26 A 4 × 12 truss chord carries a tensile load of 12 kips. Design a butt splice for the chord using steel side plates at the two sides of the chord. No. 2 southern pine. Normal duration of load, dry service conditions.

Problem 30.27 Design a wood ledger carrying 1 kip/ft, to be bolted to a grouted masonry bearing wall 12 in. thick. No. 2 spruce-pine-fir, normal conditions of service.

Problem 30.28 Design a wood ledger carrying 1 kip/ft, to be bolted to a concrete bearing wall 8 in. thick. No. 2 spruce-pine-fir, normal conditions of service.

CHAPTER **31**

Structural Timber

The following problems relate to the subject matter developed in Sections 31.1, 31.2, 31.3 and 31.4.

Problem 31.1 Determine the capacity of a 6 × 6 column section having hinged ends and an unsupported length of 16'0". No. 2 southern pine, permanent duration of load, dry conditions of use.

Problem 31.2 Determine the capacity of a 4 × 8 column section having hinged ends and an unsupported length of 16'0", braced at midheight on its weak axis only. No. 2 spruce-pine-fir, permanent duration of load, subject to wet service conditions over extended periods of time.

Problem 31.3 Determine the capacity of a 6 × 12 column section having hinged ends and an unsupported length of 20'0", braced at midheight on its weak axis only. No. 1 Douglas fir-larch, permanent duration of load, dry service conditions.

Problem 31.4 Determine the capacity of a 10 × 10 column section having hinged end and an unsupported length of 20'0". No. 2 southern pine, permanent duration of load, subject to wet conditions over extended periods of time.

Problems 31.5 through 31.16: Select a suitable P&B or J&P column section for the given loads and conditions. Permanent duration of load, dry service conditions. Rectangular sections may be braced at midheight on one or both axes if desired.

Problem No.	Load kips	Species and Grade	Length ft	Type of Section
31.5	24	No. 1 S-P-F	6	Square
31.6	30	No. 1 SoPine	8	Square
31.7	36	No. 1 DF-L	10	Rectangular
31.8	40	No. 2 S-P-F	12	Rectangular
31.9	36	No. 2 SoPine	10	Square
31.10	40	No. 2 DF-L	12	Square
31.11	44	Sel. Str.	14	Rectangular
31.12	48	Sel. Str.	16	Rectangular
31.13	44	Sel. Str.	14	Square
31.14	48	No. 1 S-P-F	16	Square
31.15	52	No. 1 SoPine	18	Rectangular
31.16	56	No. 1DF-L	20	Rectangular

CHAPTER 32

Structural Timber

The following problems relate to the subject matter developed in Sections 32.2 and 32.3.

Problems 32.1 through 32.8: The shear and moment diagrams shown below apply to Problems 32.1 through 32.8. In all cases, it may be assumed that lateral support of the compression edge meets minimum Code requirements given in Section 32.6.

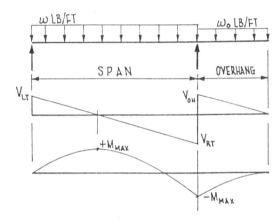

Select a suitable section to sustain the indicated moment and shear.

Problem No.	Species and Grade	M_{MAX} kip·ft	V_{MAX} kips
32.1	No. 2 DF-L	11	7
32.2	No. 2 SoPine	14	10
32.3	No. 2 S-P-F	10	6
32.4	No. 1 DF-L	28	11
32.5	No. 1 SoPine	31	14
32.6	No. 1 S-P-F	21	7
32.7	No. 2 DF-L	24	13
32.8	No. 2 SoPine	23	15
32.9	No. 2 S-P-F	17	9
32.10	No. 1 DF-L	45	15
32.11	No. 1 SoPine	45	20
32.12	No. 1 S-P-F	34	13

The following problems relate to the subject matter developed in Section 32.4.

Problem 32.13 A beam 6×12 is supported at its end by a 4×6 column, with the 6 in. dimensions matching. Determine the allowable end reaction as limited both by shear and by end bearing. No. 2 spruce-pine-fir. Normal conditions of service.

Problem 32.14 A joist 2×10 is supported by a steel joist hanger having a seat width of 1 in. Determine the allowable end reaction as limited both by shear and by end bearing. No. 2 southern pine. Duration of load less than 2 months, wet conditions of service.

Problem 32.15 A 6×16 timber beam is seated on the top flange of a steel girder $W12 \times 22$. Determine the allowable end reaction as limited both by shear and by end bearing. No. 2 Douglas fir-larch. Normal conditions of service.

Problem 32.16 A 4×12 wood beam of No. 2 Douglas fir-larch is seated on the top edge of a 6×16 timber girder of No. 2 spruce-pine-fir. Determine the allowable reaction as limited both by the beam and by the girder. Wet conditions of service, duration of load less than 2 months.

The following problems relate to the subject matter developed in Sections 32.2 through 32.6.

Problems 32.17 through 32.32 The following problems are based on the load systems shown in the following sketch:

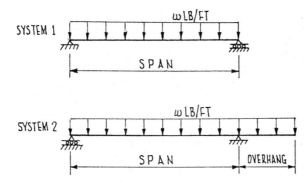

For the following problems, with the loads shown in the foregoing sketches:
1. Sketch the shear and moment diagrams.
2. Determine maximum moment and maximum shear.
3. Select a suitable section in the indicated species and grade to sustain the maximum moment and maximum shear.
4. Review the selected section for suitability in deflections; revise as necessary.
5. Review the selected section for suitability in lateral stability; and prescribe the measures to be taken in providing lateral stability.

Problem No.	System	w lbs/ft	Span ft.	Over-hang	Species and Grade	Appli-cation
32.17	1	1000	14	0	No. 2 DF-L	Floor
32.18	1	900	16	0	No. 2 SoPine	Roof
32.19	1	800	18	0	No. 2 S-P-F	Floor
32.20	1	700	20	0	No. 1 DF-L	Roof
32.21	1	1200	14	0	No. 1 SoPine	Floor
32.22	1	1000	16	0	No. 1 S-P-F	Roof
32.23	1	800	18	0	No. 2 DF-L	Floor
32.24	1	600	20	0	No. 2 SoPine	Roof
32.25	2	1000	14	6	No. 2 S-P-F	Floor
32.26	2	900	15	5	No. 1 DF-L	Roof
32.27	2	800	16	4	No. 1 SoPine	Floor
32.28	2	700	17	3	No. 1 S-P-F	Roof
32.29	2	1200	14	6	No. 2 DF-L	Floor
32.30	2	1000	15	5	No. 2 SoPine	Roof
32.31	2	800	16	4	No. 2 S-P-F	Floor
32.32	2	600	17	3	No. 2 S-P-F	Roof

CHAPTER 33
Structural Timber

The following problems relate to the subject matter developed in Section 33.1 and 33.2.

Problem 33.1 Select a suitable joist system and corresponding plywood sheathing to sustain a floor live load of 40 lbs/ft^2 on a span of 16'0". No. 2 Douglas fir-larch joists.

Problem 33.2 Select a suitable joist system and corresponding plywood sheathing to sustain a floor live load of 40 lb/ft^2 on a span of 18'0". No. 2 So. pine joists.

Problem 33.3 Select a suitable joist system and corresponding plywood sheathing to sustain a floor live load of 40 lb/ft^2 on a span of 20'0". No. 2 spruce-pine-fir joists.

Problem 33.4 Select a suitable rafter system in a low-slope roof to sustain a roof live load of 20 lb/ft^2 on a horizontal projected span of 16'0". Select also the corresponding plywood roof sheathing. No. 2 Douglas fir-larch rafters.

Problem 33.5 Select a suitable rafter system in a low-slope roof to sustain a roof live load of 20 lb/ft^2 on a horizontal projected span of 18'0". Select also the corresponding plywood roof sheathing. No. 2 southern pine rafters.

Problem 33.6 Select a suitable rafter system in a low-slope roof to sustain a roof live load of 20 lb/ft^2 on a horizontal projected span of 20'0". Select also the corresponding plywood roof sheathing. No. 2 spruce-pine-fir rafters.

Problem 33.7 Select a suitable rafter system in a high-slope roof to sustain a roof live load of 20 lb/ft^2 on a horizontal projected span of 16'0". Select also the corresponding plywood roof sheathing. No. 2 Douglas fir-larch rafters.

Problem 33.8 Select a suitable rafter system in a high-slope roof to sustain a roof live load of 20 lb/ft^2 on a horizontal projected span of 18'0". Select also the corresponding plywood roof sheathing. No. 2 southern pine rafters.

Problem 33.9 Select a suitable rafter system in a high-slope roof to sustain a roof live load of 20 lb/ft^2 on a horizontal projected span of 20'0". Select also the corresponding plywood roof sheathing. No. 2 spruce-pine-fir rafters.

PART 2
DESIGN TABLES

The design tables presented in the remainder of this volume have been selected (or prepared specifically) to suit the needs of this textbook. Though such a limitation may seem to limit the usefulness of the tables to a few homework problems, such is not the case. It is intended that the tables included here will, in fact, be useful for real problems in real design for the initial period after the student enters civil engineering practice. Eventually, of course, as the capacity of the designer increases, so must his references, and these design tables will very soon give way to more comprehensive ones. And, when it eventually comes time to purchase these more comprehensive references, the author recommends the manuals, codes and references from which the tables in this textbook are drawn.

The tables are presented in five parts:

Miscellaneous:	A-1 through A-3
Concrete:	C-1 through C-14
Masonry:	M-1 through M-14
Steel:	S-1 through S-32
Timber:	T-1 through T-23

Where appropriate, introductory or explanatory notes are included for certain tables. When the use of the table should be self evident, no additional explanations are provided.

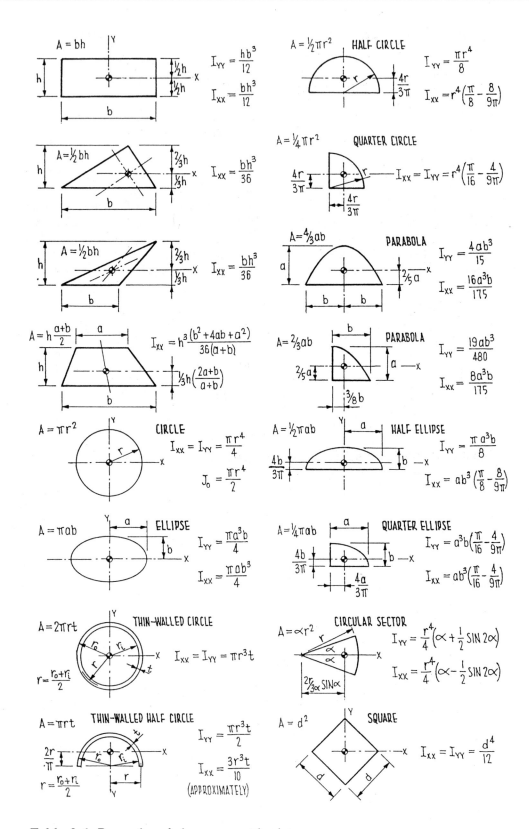

Table A-1 Properties of plane geometric shapes

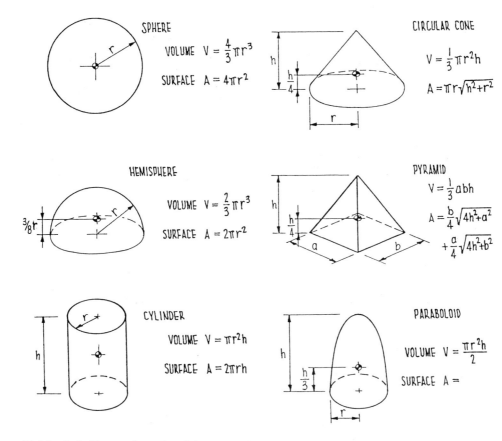

SPHERE

VOLUME $V = \frac{4}{3}\pi r^3$

SURFACE $A = 4\pi r^2$

CIRCULAR CONE

$V = \frac{1}{3}\pi r^2 h$

$A = \pi r \sqrt{h^2 + r^2}$

HEMISPHERE

VOLUME $V = \frac{2}{3}\pi r^3$

SURFACE $A = 2\pi r^2$

PYRAMID

$V = \frac{1}{3} abh$

$A = \frac{b}{4}\sqrt{4h^2 + a^2} + \frac{a}{4}\sqrt{4h^2 + b^2}$

CYLINDER

VOLUME $V = \pi r^2 h$

SURFACE $A = 2\pi rh$

PARABOLOID

VOLUME $V = \frac{\pi r^2 h}{2}$

SURFACE $A =$

Table A-2 Properties of solid geometric shapes

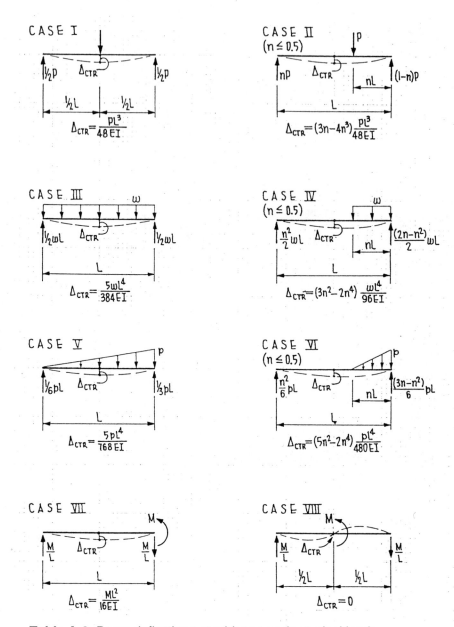

CASE I

$$\Delta_{CTR} = \frac{pL^3}{48EI}$$

CASE II (n ≤ 0.5)

$$\Delta_{CTR} = (3n - 4n^3)\frac{pL^3}{48EI}$$

CASE III

$$\Delta_{CTR} = \frac{5\omega L^4}{384EI}$$

CASE IV (n ≤ 0.5)

$$\Delta_{CTR} = (3n^2 - 2n^4)\frac{\omega L^4}{96EI}$$

CASE V

$$\Delta_{CTR} = \frac{5pL^4}{768EI}$$

CASE VI (n ≤ 0.5)

$$\Delta_{CTR} = (5n^2 - 2n^4)\frac{pL^4}{480EI}$$

CASE VII

$$\Delta_{CTR} = \frac{ML^2}{16EI}$$

CASE VIII

$$\Delta_{CTR} = 0$$

Table A-3 Beam deflections at midspan under typical loads

APPENDIX C

Concrete Design Tables

The data and tables in this section of the appendix are included in support of the design procedures for reinforced concrete given in this textbook. The initial entry is a listing of the standard symbols used in Chapters 16 through 22 for the development of the design equations and theory.

The following tables C-1 through C-14 provide design data for use in designing structural members in reinforced concrete. Concrete strengths of 3000, 4000 and 5000 psi are included in the tables along with steel grades 40, 50 and 60. The tables are adequate for the material covered in this text, to include the design of routine buildings and falsework less than about 65 feet high.

Tables C-1 through C-4 provide data on reinforcement, design criteria, bar spacing and minimum cover requirements. Tables C-5 through C-11 provide design constants for beams and columns. Tables C-12 through C-14 provide design data on anchorage, to include straight bars, standard hooks and imbedded bolts.

The design tables are taken from the author's more complete work on concrete, *Reinforced Concrete Technology*, Delmar Publishers, Inc., Albany, N.Y., 1994. The tables are reproduced here with the permission of the publisher.

STANDARD SYMBOLS USED IN REINFORCED CONCRETE

A = basic generic symbol for area

A_b = cross-sectional area of a reinforcing bar, square inches

A_s = area of tension reinforcement, square inches

A_g = gross area of column section, square inches

A'_s = area of compression reinforcement in beams, or, area of longitudinal reinforcement at the flexural faces of a column, square inches

A''_s = area of longitudinal reinforcement at the web faces of a column, square inches

A_{st} = total area of longitudinal column reinforcement, square inches

A_{tr} = area of transverse reinforcement for anchorage requirements

A_1 = loaded area

A_2 = maximum area of the portion of the supporting surface that is geometrically similar to and concentric with the loaded area

A_v = area of shear reinforcement within a distance s, square inches

b = width of compression face of member, inches

b' = effective width of a tee beam flange, inches

b_1 = required width of a rectangular beam to sustain only the flexural portion of loads when the beam is subject both to flexure and a small axial load

b_2 = required width of a rectangular beam to sustain only the axial load when the beam is subject both to flexure and a small axial load

b_3 = reduction or increase in width of rectangular beam to compensate for eccentricity of the axial load when the beam is subject both to flexure and a small axial load

b_m = least flexural width, $b_1 - b_3$, inches

b_o = length of perimeter subject to shear stress punch-out in footings, inches

b_w = width of the stem of a tee beam, inches

C = resultant of axial compressive forces acting on a section

c = perpendicular distance from the neutral axis of bending to the outermost compression fiber, inches

D = dead load effects

d = effective depth of a flexural section, measured from the outermost compression fiber to the center of tensile reinforcement, inches

d_b = diameter of a reinforcing bar, inches

E = earthquake load effects

E_c = elastic modulus of elasticity of concrete, psi

E_s = elastic modulus of elasticity of steel, psi

e = ratio of moment to axial force M/P, or, eccentricity of axial load

e_e = value of e in the elastic range of loads

e_n = value of e at ultimate levels of load

f = basic generic symbol for direct stress

f_c = elastic stress in concrete, psi

f'_c = specified ultimate compressive strength of concrete, psi

f_s = elastic stress in steel, psi

f_{sc} = elastic stress in steel that is located in a compression zone, psi

f_y = yield stress in steel, psi

g = factor defining concrete cover over longitudinal reinforcement

h = overall depth of a member, measured normal to the axis of bending, in.

I = basic generic symbol for moment of inertia in bending, inches^4

I_{cr} = moment of inertia of a cracked concrete section in bending, inches^4

I_e = effective moment of inertia of a section after corrections, inches^4

I_g = moment of inertia of the gross uncracked section of concrete neglecting reinforcement, inches^4

j = factor defining the distance from the center of compressive forces to the center of tensile forces in a flexural section

k = factor defining the distance from the outermost compression face to the neutral axis of bending, measured normal to the neutral axis of bending

k_b = value of k at the balanced strain condition

k_n = value of k at nominal ultimate levels of load

k_{sv} = value of k at service levels of stress

L = length of span, measured between centerlines of bearing

L = live load effects

L_n = length of clear span, measured face-to-face of supports

L_u = unsupported length of a compression member

ℓ_a = embedment length beyond center of support or point of inflection, inches

ℓ_d = development length of a reinforcing bar, inches

ℓ_{db} = basic development length of a reinforcing bar, before modifiers or multipliers are applied, inches

ℓ_{dh} = development length of standard hook in tension, inches

ℓ_{hb} = basic development length of a standard hook in tension, before modifiers or multipliers are applied, inches

M = basic generic symbol for moment acting on a section

M_a = moment acting on a section at the time deflections are being calculated

M_{DL} = that portion of the moment caused by dead loads

M_E = that portion of the moment caused by earthquake

M_e = moment at elastic levels of stress

M_{LL} = that portion of the moment caused by live loads

M_1 = value of smaller end moment on a compression member, positive if member is in single curvature, negative if in double curvature

M_2 = value of larger end moment on a compression member, always positive

M_n = nominal ultimate moment acting on a section, or, nominal ultimate resisting moment that the section must be able to develop

M_{sv} = moment acting on a section at service levels of load

M_u = factored moment acting on a section

M_w = that portion of the moment caused by wind

N = number of bars within a layer that are being spliced

P = basic generic symbol for axial load acting on a section

P_c = critical buckling load in the Euler formula

P_{DL} = that portion of the axial load caused by dead loads

P_E = that portion of the axial load caused by earthquake

P_e = axial load at elastic levels of stress

P_{LL} = that portion of the axial load caused by live loads

P_n = nominal ultimate axial load acting on a section

P_o = ultimate concentric axial load on a compression section at full plasticity

P_u = factored axial load acting on a section

P_w = that portion of the axial load caused by wind

p = basic generic symbol for pressure

p_a = allowable increase in soil pressure due to footing loads

p_{DL} = that portion of the soil pressure caused by dead loads

p_{LL} = that portion of the soil pressure caused by live loads

R_e = column stress ratio at elastic levels of stress

R_n = column stress ratio at ultimate load

S = basic generic symbol for section modulus, I/c, inches^3

S_c = section modulus taken to the outermost compression fiber, inches^3

s = stirrup spacing longitudinally, inches

T = resultant of axial tension loads acting on a section

T_{DL} = that portion of the axial tension caused by dead loads

T_{LL} = that portion of the axial tension caused by live loads

T_n = nominal ultimate axial tension load acting on a section

t = overall thickness of a concrete slab, inches

U_n = load components at nominal ultimate load

U_{sv} = load components at service levels of load

u = bond stress, either at elastic levels of stress or at ultimate levels of load

V = basic generic symbol for shear force acting on a section

V_c = shear force resisted by a concrete section unreinforced in shear

V_{DL} = that portion of the shear force caused by dead loads

V_E = that portion of the shear force caused by earthquake

V_{LL} = that portion of the shear force caused by live loads

V_s = that portion of the shear force that is carried by stirrups

V_{sv} = shear force acting on a section at service levels of load

V_n = nominal ultimate shear force acting on a section, or, nominal ultimate resistance to shearing force that the section must be able to develop

V_w = that portion of the shear force caused by wind

v = basic generic symbol for shear stress, psi

v_c = that portion of the shear stress carried only by concrete, psi

v_{sv} = average shear stress on a section, to include effects of reinforcement, under service levels of stress, psi

v_n = average shear stress on a section, to include effects of reinforcement, under nominal ultimate levels of load, psi

W = wind load effects

w = uniformly distributed load, force per unit length or per unit area

w_{DL} = that portion of the total uniform load caused by dead loads

w_{LL} = that portion of the total uniform load caused by live loads

y = factor defining the distance from the neutral axis to the center of compression loads on the concrete, measured normal to the neutral axis

y_n = factor y at nominal ultimate levels of load

y_{sv} = factor y at service levels of stress

Z = basic generic symbol for plastic section modulus, inches^3

Z_c = plastic section modulus at the outermost compression fiber, inches^3

β = ratio of the long side of a footing to the short side

β_1 = factor defining the height of the ACI equivalent stress block

γ = factor for correcting deflections in concrete for effects of time and for effects of compressive reinforcement

Δ = basic generic symbol for deflection

Δ_{DL} = deflections under dead load only

Δ_{LL} = deflections under live load only

Δ_{sv} = deflections under service levels of load

ε = basic generic symbol for strain, inches per inch

ε_c = strain in concrete, inches per inch

ε_s = strain in steel, inches per inch

ϕ = strength reduction factor

ρ = ratio of tensile steel area to effective cross-sectional area *bd* in a flexural section

ρ_b = value of ρ at balanced strain conditions

ρ' = ratio of compressive steel area to effective cross sectional area *bd* in a flexural section, or, ratio of longitudinal steel area at flexural faces of a column to the effective area *bd* of the column

ρ'' = ratio of longitudinal steel area at the web faces of a column to the effective area *bd* of the column

Σ = basic generic symbol for summation

ξ = empirical time factor for computing long-term deflections in concrete

Type of Stress Normal Weight Concrete: 145 pcf	ACI 318-89	Strength in psi		
		3000	4000	5000
Modulus of Elasticity, E_c (in psi)	$57000\sqrt{f_c'}$	3.1×10^6	3.6×10^6	4.0×10^6
Modular ratio $n = E_s/E_c$	$509/\sqrt{f_c'}$	9.29	8.04	7.20
Flexure Stress, f_c				
Extreme compression fiber	$0.45f_c'$	1350	1800	2250
Shear Stress, v				
Beams, walls, one-way slabs	$1.1\sqrt{f_c'}$	60	70	78
Joist floors, stems	$1.2\sqrt{f_c'}$	67	76	85
[1]Two-way slabs and footings	$(1 + 2/\beta_c)\sqrt{f_c'}$ but $\leq 2\sqrt{f_c'}$			
Bearing Stress, f_c				
On full area	$0.3f_c'$	900	1200	1500
[2]Less than full area (all sides)	$0.3f_c'\sqrt{A_2/A_1}$ but $\leq 0.6f_c'$			

[1]β_c is the ratio of long side to short side of the concentrated load or reaction area.

[2]A_1 is the loaded area.

A_2 is the maximum area of the portion of the supporting surface that is geometrically similar to and concentric with the loaded area.

Allowable Working Stresses in Reinforcement

Modulus of Elasticity of All Steel:	$E_s = 29,000,000$ psi
Tensile and Compressive Stresses:	
Grade 40 steel ($f_y = 40,000$ psi)	$f_s = 20,000$ psi
Grade 50 steel ($f_y = 50,000$ psi)	$f_s = 20,000$ psi
Grade 60 steel ($f_y = 60,000$ psi)	$f_s = 22,000$ psi

Table C-1 Allowable working stresses

Area of Steel (in.^{2}) in a Section 1 Ft. Wide

Spacing (in.)	Bar Sizes								
	#3	#4	#5	#6	#7	#8	#9	#10	#11
2	0.66	1.18	1.84	2.65	3.61	4.71			
2½	0.53	0.94	1.47	2.12	2.89	3.77	4.80		
3	0.44	0.79	1.23	1.77	2.41	3.14	4.00	5.07	6.25
3½	0.38	0.67	1.05	1.51	2.06	2.69	3.43	4.34	5.35
4	0.33	0.59	0.92	1.33	1.80	2.36	3.00	3.80	4.68
4½	0.29	0.52	0.82	1.18	1.60	2.09	2.66	3.38	4.16
5	0.27	0.47	0.74	1.06	1.44	1.88	2.40	3.04	3.75
5½	0.24	0.43	0.67	0.96	1.31	1.71	2.18	2.76	3.41
6	0.22	0.39	0.61	0.88	1.20	1.57	2.00	2.53	3.12
6½	0.20	0.36	0.57	0.82	1.11	1.45	1.84	2.34	2.88
7	0.19	0.34	0.53	0.76	1.03	1.35	1.71	2.17	2.68
7½	0.18	0.31	0.49	0.71	0.96	1.26	1.60	2.03	2.50
8	0.17	0.29	0.46	0.66	0.90	1.18	1.50	1.90	2.34
9	0.15	0.26	0.41	0.59	0.80	1.05	1.33	1.69	2.08
10	0.13	0.24	0.37	0.53	0.72	0.94	1.20	1.52	1.87
11	0.12	0.21	0.33	0.48	0.66	0.86	1.09	1.38	1.70
12	0.11	0.20	0.31	0.44	0.60	0.79	1.00	1.27	1.56
13	0.10	0.18	0.28	0.41	0.56	0.72	0.92	1.17	1.44
14	0.09	0.17	0.26	0.38	0.52	0.67	0.86	1.09	1.34
15	0.09	0.16	0.25	0.35	0.48	0.63	0.80	1.01	1.25
16	0.08	0.15	0.23	0.33	0.45	0.59	0.75	0.95	1.17
17	0.08	0.14	0.22	0.31	0.42	0.55	0.71	0.89	1.10
18	0.07	0.13	0.20	0.29	0.40	0.52	0.67	0.84	1.04

Properties of Reinforcing Bars

Bar Size	#3	#4	#5	#6	#7	#8	#9	#10	#11
Diameter, in.	0.375	0.500	0.625	0.750	0.875	1.000	1.125	1.270	1.410
Area, in.^{2}	0.11	0.20	0.31	0.44	0.60	0.79	1.00	1.27	1.56
Perimeter, in.	1.18	1.57	1.96	2.36	2.75	3.14	3.55	3.99	4.43
Weight, lb/ft	0.38	0.67	1.04	1.50	2.04	2.67	3.40	4.30	5.31

Table C-2 Steel areas per foot for spaced bars

No.	Area	Number of Bars				Number of Bars				Number of Bars			
		1	2	3	4	1	2	3	4	1	2	3	4
#3													
1	0.11												
2	0.22												
3	0.33												
4	0.44												
5	0.55												
6	0.66												
#4													
1	0.20												
2	0.39												
3	0.59												
4	0.79												
5	0.98												
6	1.18												
#5		**#5 plus #4**											
1	0.31	0.50	0.70	0.90	1.09								
2	0.61	0.81	1.01	1.20	1.40								
3	0.92	1.12	1.31	1.51	1.71								
4	1.23	1.42	1.62	1.82	2.01								
5	1.53	1.73	1.93	2.12	2.32								
6	1.84	2.04	2.23	2.43	2.63								
#6		**#6 plus #5**				**#6 plus #4**							
1	0.44	0.75	1.06	1.36	1.67	0.64	0.83	1.03	1.23				
2	0.88	1.19	1.50	1.80	2.11	1.08	1.28	1.47	1.67				
3	1.33	1.63	1.94	2.25	2.55	1.52	1.72	1.91	2.11				
4	1.77	2.07	2.38	2.69	2.99	1.96	2.16	2.36	2.55				
5	2.21	2.52	2.82	3.13	3.44	2.14	2.60	2.80	2.99				
6	2.65	2.96	3.26	3.57	3.88	2.85	3.04	3.24	3.44				
#7		**#7 plus #6**				**#7 plus #5**				**#7 plus #4**			
1	0.60	1.04	1.48	1.93	2.37	0.91	1.21	1.52	1.83	0.80	0.99	1.19	1.39
2	1.20	1.64	2.09	2.53	2.97	1.51	1.82	2.12	2.43	1.40	1.60	1.79	1.99
3	1.80	2.25	2.69	3.13	3.57	2.11	2.42	2.72	3.03	2.00	2.20	2.39	2.59
4	2.41	2.85	3.29	3.73	4.17	2.71	3.02	3.33	3.63	2.60	2.80	2.99	3.19
5	3.01	3.45	3.89	4.33	4.77	3.31	3.62	3.93	4.23	3.20	3.40	3.60	3.79
6	3.61	4.05	4.49	4.93	5.38	3.91	4.22	4.53	4.84	3.80	4.00	4.20	4.39
#8		**#8 plus #7**				**#8 plus #6**				**#8 plus #5**			
1	0.79	1.39	1.99	2.59	3.19	1.23	1.67	2.11	2.55	1.09	1.40	1.71	2.01
2	1.57	2.17	2.77	3.37	3.98	2.01	2.45	2.90	3.34	1.88	2.18	2.49	2.80
3	2.36	2.96	3.56	4.16	4.76	2.80	3.24	3.68	4.12	2.66	2.97	3.28	3.58
4	3.14	3.74	4.34	4.95	5.55	3.58	4.03	4.47	4.91	3.45	3.76	4.06	4.37
5	3.93	4.53	5.13	5.73	6.33	4.37	4.81	5.25	5.69	4.23	4.54	4.85	5.15
6	4.71	5.31	5.92	6.52	7.12	5.15	5.60	6.04	6.48	5.02	5.33	5.63	5.94
#9		**#9 plus #8**				**#9 plus #7**				**#9 plus #6**			
1	1.00	1.79	2.57	3.36	4.14	1.60	2.20	2.80	3.41	1.44	1.88	2.33	2.77
2	2.00	2.79	3.57	4.36	5.14	2.60	3.20	3.80	4.41	2.44	2.88	3.33	3.77
3	3.00	3.79	4.57	5.36	6.14	3.60	4.20	4.80	5.41	3.44	3.88	4.33	4.77
4	4.00	4.79	5.57	6.36	7.14	4.60	5.20	5.80	6.41	4.44	4.88	5.33	5.77
5	5.00	5.79	6.57	7.36	8.14	5.60	6.20	6.80	7.41	5.44	5.88	6.33	6.77
6	6.00	6.79	7.57	8.36	9.14	6.60	7.20	7.80	8.41	6.44	6.88	7.33	7.77
#10		**#10 plus #9**				**#10 plus #8**				**#10 plus #7**			
1	1.27	2.27	3.27	4.27	5.27	2.05	2.84	3.62	4.41	1.87	2.47	3.07	3.67
2	2.53	3.53	4.53	5.53	6.53	3.32	4.10	4.89	5.68	3.13	3.74	4.34	4.94
3	3.80	4.80	5.80	6.80	7.80	4.59	5.37	6.16	6.94	4.40	5.00	5.60	6.21
4	5.07	6.07	7.07	8.07	9.07	5.85	6.64	7.42	8.21	5.67	6.27	6.87	7.47
5	6.33	7.33	8.33	9.33	10.3	7.12	7.90	8.69	9.48	6.94	7.54	8.14	8.74
6	7.60	8.60	9.60	10.6	11.6	8.39	9.17	9.96	10.7	8.20	8.80	9.40	10.0
#11		**#11 plus #10**				**#11 plus #9**				**#11 plus #8**			
1	1.56	2.83	4.09	5.36	6.63	2.56	3.56	4.56	5.56	2.35	3.13	3.92	4.70
2	3.12	4.39	5.66	6.92	8.19	4.12	5.12	6.12	7.12	3.91	4.69	5.48	6.26
3	4.68	5.95	7.22	8.48	9.75	5.68	6.68	7.68	8.68	5.47	6.26	7.04	7.83
4	6.25	7.51	8.78	10.1	11.3	7.25	8.25	9.25	10.3	7.03	7.82	8.60	9.39
5	7.81	9.07	10.3	11.6	12.9	8.81	9.81	10.8	11.8	8.59	9.38	10.2	11.0
6	9.37	10.6	11.9	13.2	14.4	10.4	11.4	12.4	13.4	10.2	10.9	11.7	12.5

Table C-3 Steel areas for combinations of bar sizes

Design Guidelines. Minimum Width of Beams or Stems at
Interior Exposures[1] (Maximum aggregate size ¾ in.)

Bar Size	With #3 Stirrups[2] Number of Bars in a Single Layer							With No Stirrups No. of Bars	
	2	3	4	5	6	7	8	2	3
#4	6.1	7.6	9.1	10.6	12.1	13.6	15.1	5.00	6.50
#5	6.3	7.9	9.6	11.2	12.8	14.4	16.1	5.25	6.88
#6	6.5	8.3	10.0	11.8	13.5	15.3	17.0	5.50	7.25
#7	6.7	8.6	10.5	12.4	14.2	16.1	18.0	5.75	7.63
#8	6.9	8.9	10.9	12.9	14.9	16.9	18.9	6.00	8.00
#9	7.3	9.5	11.8	14.0	16.3	18.6	20.8	6.25	8.38
#10	7.7	10.2	12.8	15.3	17.8	20.4	22.9	6.50	8.75
#11	8.0	10.8	13.7	16.5	19.3	22.1	24.9	6.75	9.13

[1]For exterior exposures, bars #6 and larger, add ¾ inch.
[2]For #4 stirrups, add ¾ inch; for #5 stirrups add 1½ inch.

Clearances at Interior Exposures

In exterior exposures, for bar sizes #6 and larger, increase the minimum
clear concrete cover to 2 inches.

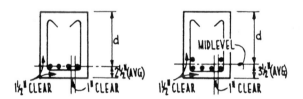

General Guidelines:

1. For rectangular beams, width of the beam should be between 0.5d
   and 0.6d if there are no constraints.
2. For tee beams, the depth d should be about 5 times the slab thickness and the width b should be about 0.5d.
3. Increments of overall dimensions for beams should not be less than
   ½ inch for sizes up to 18 inches and 1 inch for sizes above 18
   inches.
4. Increments of overall slab thickness should not be less than ½ inch
   with a minimum slab thickness of 3½ inches.
5. Minimum clear cover over reinforcement is 1½ inches, measured
   from outside of stirrups.
6. No fewer than 2 reinforcing bars should be used as longitudinal
   reinforcement in a beam.
7. Reinforcing bars should be arranged symmetrically about the vertical axis of a beam.
8. Bar sizes greater than #11 should not be used in beams and no more
   than two bar sizes should be used in a beam.
9. Bars should be placed in one layer if possible; if bars of two sizes
   are placed in multiple layers, the heavier sizes should be placed
   closest to the outer face.

Table C-4 Design guidelines for concrete sections

Coefficients for Section Constants. f'_c = 3000 psi (20.6 N/mm²); g = 0.125, β_1 = 0.85, n = 9.29

| | At Ultimate Loads | | | | | | | | | | | | At Service Loads | | | |
| | Grade 40 Steel f_y = 40000 psi | | | | Grade 50 Steel f_y = 50000 psi | | | | Grade 60 Steel f_y = 60000 psi | | | | Elastic Constants for All Steels | | | |
A_s/bd	Z_c	$\beta_1 k_n$	y_n	f_{sv}	Z_c	$\beta_1 k_n$	y_n	f_{sv}	Z_c	$\beta_1 k_n$	y_n	f_{sv}	S_c	k_{sv}	I_{cr}	f_s/f_c
					Compressive Steel Area 0% of Tensile Steel Area											
0.0005	0.008	.008	.559	263	0.010	.010	.558	329	0.012	.012	.557	394	.045	.092	.004	91.85
0.0010	0.016	.016	.555	383	0.019	.020	.553	478	0.023	.024	.551	572	.061	.127	.008	63.66
0.0015	0.023	.024	.551	479	0.029	.029	.548	597	0.035	.035	.545	714	.073	.154	.011	51.19
0.0020	0.031	.031	.547	562	0.038	.039	.543	700	0.046	.047	.539	836	.082	.175	.014	43.77
0.0025	0.038	.039	.543	637	0.048	.049	.538	792	0.057	.059	.533	946	.091	.194	.018	38.71
0.0030	0.046	.047	.539	706	0.057	.059	.533	878	0.068	.071	.527	1047	.098	.210	.020	34.98
0.0035	0.053	.055	.535	771	0.066	.069	.528	957	0.079	.082	.521	1140	.104	.225	.023	32.08
0.0040	0.061	.063	.531	832	0.075	.078	.523	1032	0.090	.094	.515	1228	.110	.238	.026	29.75
0.0050	0.075	.078	.523	946	0.093	.098	.513	1170	0.111	.118	.504	1390	.119	.262	.031	26.19
0.0060	0.090	.094	.515	1051	0.111	.118	.504	1297	0.131	.141	.492	1537	.128	.283	.036	23.56
0.0070	0.104	.110	.508	1148	0.128	.137	.494	1414	0.151	.165	.480	1672	.136	.301	.041	21.53
0.0080	0.118	.125	.500	1240	0.145	.157	.484	1524	0.171	.188	.468	1798	.142	.318	.045	19.89
0.0090	0.131	.141	.492	1327	0.161	.176	.474	1627	0.189	.212	.457	1915	.148	.334	.050	18.54
0.0100	0.145	.157	.484	1410	0.177	.196	.464	1725	0.208	.235	.445	2024	.154	.348	.054	17.40
0.0120	0.171	.188	.468	1564	0.208	.235	.445	1904	0.242	.282	.421	2224	.164	.374	.061	15.57
0.0140	0.195	.220	.453	1705	0.237	.275	.425	2066	0.275	.329	.398	2400	.172	.396	.068	14.15
0.0160	0.219	.251	.437	1836	0.265	.314	.406	2213	0.306	.376	.374	2556	.179	.416	.075	13.01
0.0180	0.242	.282	.421	1957	0.291	.353	.386	2346					.186	.435	.081	12.08
0.0200	0.265	.314	.406	2069	0.315	.392	.366	2466					.192	.451	.087	11.29
0.0220	0.286	.345	.390	2173	0.338	.431	.347	2575					.197	.467	.092	10.61
0.0240	0.306	.376	.374	2270									.202	.481	.097	10.02
0.0260	0.325	.408	.359	2359									.206	.494	.102	9.50
0.0280	0.343	.439	.343	2442									.211	.507	.107	9.05
	0.75 ρ_b = 0.0278				0.75 ρ_b = 0.0206				0.75 ρ_b = 0.0160							
					Compressive Steel Area 20% of Tensile Steel Area											
0.0035									0.079	.087	.519	1114	.107	.220	.024	32.94
0.0040					0.076	.085	.520	1005	0.090	.097	.514	1193	.113	.232	.026	30.68
0.0050	0.076	.086	.519	914	0.093	.100	.512	1126	0.111	.115	.505	1337	.124	.254	.032	27.23
0.0060	0.090	.098	.514	1003	0.111	.115	.505	1237	0.131	.133	.496	1467	.134	.273	.037	24.69
0.0070	0.104	.109	.508	1086	0.128	.129	.498	1339	0.152	.151	.487	1587	.143	.290	.042	22.73
0.0080	0.118	.119	.503	1164	0.145	.143	.491	1434	0.172	.168	.478	1698	.152	.305	.047	21.15
0.0090	0.131	.130	.498	1237	0.162	.157	.484	1523	0.191	.186	.470	1801	.159	.319	0.51	19.85
0.0100	0.145	.140	.492	1306	0.178	.170	.477	1607	0.211	.203	.461	1898	.166	.331	.056	18.75
0.0120	0.172	.161	.482	1434	0.211	.197	.464	1761	0.249	.237	.444	2075	.180	.353	.064	17.00
0.0140	0.198	.180	.472	1552	0.243	.224	.450	1901	0.285	.271	.427	2232	.192	.373	.072	15.64
0.0160	0.225	.200	.462	1660	0.274	.251	.437	2029	0.321	.305	.410	2373	.203	.390	.080	14.56
0.0180	0.250	.219	.450	1757	0.304	.277	.421	2140	0.355	.339	.393	2500	.213	.405	.087	13.67
0.0200	0.274	.238	.437	1845	0.333	.303	.406	2240	0.388	.373	.374	2608	.223	.418	.094	12.92
0.0220	0.298	.257	.424	1925	0.361	.330	.390	2330					.232	.431	.100	12.28
0.0240	0.322	.276	.412	1999	0.388	.356	.374	2411					.241	.442	.107	11.73
0.0260	0.344	.294	.399	2067	0.414	.382	.359	2484					.250	.452	.113	11.24
0.0280	0.367	.313	.387	2129									.258	.462	.119	10.82
0.0320	0.409	.350	.362	2237									.274	.479	.130	10.09
0.0360	0.449	.387	.337	2326									.289	.495	.141	9.50

Table C-5 Coefficient for section constants

| | At Ultimate Loads | | | | | | | | | | | | At Service Loads | | | |
| | Grade 40 Steel f_y = 40000 psi | | | | Grade 50 Steel f_y = 50000 psi | | | | Grade 60 Steel f_y = 60000 psi | | | | Elastic Constants for All Steels | | | |
A_s/bd	Z_c	$\beta_1 k_n$	y_n	f_{sv}	Z_c	$\beta_1 k_n$	y_n	f_{sv}	Z_c	$\beta_1 k_n$	y_n	f_{sv}	S_c	k_{sv}	I_{cr}	f_s/f_c
						Compressive Steel Area 40% of Tensile Steel Area										
0.0035									0.079	.091	.517	1089	.109	.216	.024	33.77
0.0040					0.076	.089	.518	979	0.090	.098	.513	1160	.116	.227	.027	31.58
0.0050	0.076	.090	.517	883	0.093	.101	.512	1087	0.111	.113	.506	1289	.129	.248	.033	28.24
0.0060	0.090	.100	.513	961	0.111	.113	.506	1184	0.131	.128	.499	1405	.140	.265	.038	25.78
0.0070	0.104	.108	.508	1032	0.128	.124	.501	1273	0.152	.141	.492	1510	.151	.280	.043	23.89
0.0080	0.118	.116	.504	1098	0.145	.134	.495	1355	0.172	.154	.485	1607	.161	.293	.048	22.37
0.0090	0.131	.124	.501	1160	0.162	.144	.490	1431	0.192	.167	.479	1697	.170	.305	.053	21.12
0.0100	0.145	.131	.497	1217	0.179	.154	.486	1502	0.212	.180	.473	1781	.179	.316	.058	20.07
0.0120	0.172	.145	.490	1323	0.213	.172	.476	1632	0.252	.204	.461	1933	.195	.336	.067	18.39
0.0140	0.199	.157	.484	1419	0.246	.190	.467	1749	0.291	.228	.449	2068	.211	.352	.076	17.10
0.0160	0.226	.170	.478	1506	0.279	.207	.459	1855	0.329	.251	.437	2190	.226	.366	.084	16.08
0.0180	0.253	.181	.472	1585	0.312	.224	.451	1951	0.367	.273	.426	2299	.240	.379	.093	15.23
0.0200	0.280	.192	.467	1659	0.344	.240	.442	2039	0.405	.296	.415	2399	.253	.390	.101	14.53
0.0220	0.306	.202	.459	1725	0.376	.256	.433	2119	0.442	.318	.404	2489	.266	.400	.109	13.93
0.0240	0.332	.213	.450	1784	0.407	.271	.421	2187	0.479	.340	.393	2572	.279	.409	.116	13.41
0.0260	0.358	.222	.440	1838	0.438	.287	.410	2249	0.514	.361	.379	2641	.292	.418	.124	12.96
0.0280	0.383	.232	.431	1887	0.467	.302	.398	2305					.304	.425	.131	12.56
0.0320	0.432	.250	.412	1973	0.515	.331	.374	2401					.328	.439	.145	11.88
0.0360	0.479	.268	.393	2045	0.581	.360	.351	2480					.351	.450	.159	11.34
0.0400	0.525	.285	.374	2105									.374	.461	.173	10.88
0.0440	0.570	.301	.355	2155									.397	.470	.186	10.49
0.0480	0.613	.317	.337	2197									.419	.477	.199	10.16
						Compressive Steel Area 60% of Tensile Steel Area										
0.0035									0.079	.093	.516	1066	.112	.212	.024	34.57
0.0040					0.076	.092	.517	955	0.090	.100	.513	1130	.119	.223	.027	32.45
0.0050	0.076	.093	.516	855	0.093	.102	.511	1051	0.111	.112	.506	1247	.133	.241	.033	29.21
0.0060	0.090	.101	.512	923	0.111	.112	.507	1137	0.131	.124	.501	1350	.146	.257	.039	26.84
0.0070	0.104	.108	.509	985	0.128	.120	.502	1215	0.152	.135	.495	1443	.158	.271	.044	25.02
0.0080	0.118	.114	.505	1042	0.145	.129	.498	1286	0.172	.145	.490	1527	.169	.283	.050	23.56
0.0090	0.131	.120	.502	1094	0.162	.136	.494	1351	0.193	.154	.485	1605	.180	.293	.055	22.36
0.0100	0.145	.125	.500	1142	0.179	.143	.491	1411	0.213	.164	.481	1677	.190	.303	.060	21.35
0.0120	0.172	.135	.495	1230	0.213	.156	.484	1520	0.253	.181	.472	1806	.210	.320	.070	19.75
0.0140	0.200	.144	.490	1307	0.247	.169	.478	1617	0.293	.197	.464	1920	.229	.334	.080	18.53
0.0160	0.227	.152	.486	1377	0.281	.180	.473	1703	0.333	.213	.456	2021	.247	.346	.089	17.55
0.0180	0.254	.159	.483	1439	0.135	.190	.467	1780	0.373	.227	.449	2112	.265	.357	.098	16.76
0.0200	0.281	.166	.479	1496	0.348	.200	.462	1851	0.413	.241	.442	2194	.282	.366	.107	16.09
0.0220	0.309	.172	.476	1548	0.382	.209	.458	1915	0.452	.255	.435	2269	.299	.374	.116	15.53
0.0240	0.336	.178	.473	1596	0.415	.218	.453	1974	0.492	.268	.428	2337	.316	.382	.125	15.05
0.0260	0.363	.184	.471	1641	0.449	.227	.449	2028	0.531	.281	.422	2400	.332	.388	.134	14.62
0.0280	0.390	.189	.468	1682	0.482	.235	.445	2079	0.570	.293	.416	2458	.348	.395	.142	14.25
0.0320	0.444	.198	.462	1755	0.549	.250	.437	2169	0.648	.317	.404	2562	.380	.405	.159	13.63
0.0360	0.497	.207	.450	1815	0.613	.265	.421	2240	0.726	.340	.392	2651	.411	.414	.176	13.12
0.0400	0.549	.214	.437	1866	0.676	.278	.406	2299	0.800	.362	.374	2718	.441	.422	.192	12.70
0.0440	0.600	.222	.424	1909	0.738	.291	.390	2348					.472	.429	.208	12.35
0.0480	0.651	.228	.412	1946	0.800	.303	.374	2391					.502	.435	.224	12.05

Tables values are coefficients for the section constants:

Z_c = coeff. $\times bd^2$; y_n = coeff. $\times d$; S_c = coeff. $\times bd^2$; I_{cr} = coeff $\times bd^3$; k_n and k_w are pure coefficients. Units for f_{sv} are in psi. Value of I_{cr} is for short-term live loads. For long-term loads, use 1/2 E_c with $I = S_c k_{sv} d$. Coefficients above the interior line apply only to tee shapes. Table is stopped when the sum of tensile and compressive steel exceeds 8% of gross area.

Table C-5 Coefficient for section constants *(continued)*

Coefficients for Section Constants. $f'_c = 4000$ psi (27.5 N/mm^2); $g = 0.125$, $\beta_1 = 0.85$, $n = 8.04$

| | At Ultimate Loads | | | | | | | | | | | | At Service Loads | | | |
| | Grade 40 Steel $f_y = 40000$ psi | | | | Grade 50 Steel $f_y = 50000$ psi | | | | Grade 60 Steel $f_y = 60000$ psi | | | | Elastic Constants for All Steels | | | |
A_s/bd	Z_c	$\beta_1 k_n$	y_n	f_{sv}	Z_c	$\beta_1 k_n$	y_n	f_{sv}	Z_c	$\beta_1 k_n$	y_n	f_{sv}	S_c	k_{sv}	I_{cr}	f_s/f_c
						Compressive Steel Area 0% of Tensile Steel Area										
0.0005	0.006	.006	.560	282	0.007	.007	.559	352	0.009	.009	.558	422	.042	.086	.004	85.76
0.0010	0.012	.012	.557	409	0.015	.015	.555	511	0.017	.018	.554	612	.057	.119	.007	59.53
0.0015	0.017	.018	.554	511	0.022	.022	.551	638	0.026	.026	.549	763	.068	.144	.010	47.92
0.0020	0.023	.024	.551	600	0.029	.029	.548	748	0.035	.035	.545	894	.078	.164	.013	41.00
0.0025	0.029	.029	.548	680	0.036	.037	.544	847	0.043	.044	.540	1012	.085	.181	.015	36.29
0.0030	0.035	.035	.545	754	0.043	.044	.540	938	0.052	.053	.536	1121	.092	.197	.018	32.81
0.0035	0.040	.041	.542	823	0.050	.051	.537	1023	0.060	.062	.532	1222	.098	.211	.021	30.12
0.0040	0.046	.047	.539	888	0.057	.059	.533	1104	0.068	.071	.532	1317	.103	.224	.023	27.94
0.0050	0.057	.059	.533	1010	0.071	.074	.526	1253	0.084	.088	.518	1493	.113	.246	.028	24.62
0.0060	0.068	.071	.527	1123	0.084	.088	.518	1391	0.100	.106	.510	1654	.121	.266	.032	22.18
0.0070	0.079	.082	.521	1229	0.098	.103	.511	1519	0.116	.124	.501	1803	.129	.284	.037	20.28
0.0080	0.090	.094	.515	1328	0.111	.118	.504	1640	0.131	.141	.492	1943	.135	.300	.041	18.76
0.0090	0.100	.106	.510	1423	0.124	.132	.496	1754	0.146	.159	.483	2075	.141	.315	.044	17.50
0.0100	0.111	.118	.504	1513	0.136	.147	.489	1862	0.161	.176	.474	2199	.146	.329	.048	16.43
0.0120	0.131	.141	.492	1684	0.161	.176	.474	2065	0.189	.212	.457	2430	.156	.353	.055	14.72
0.0140	0.151	.165	.480	1842	0.185	.206	.460	2251	0.217	.247	.439	2639	.164	.375	.062	13.40
0.0160	0.171	.188	.468	1990	0.208	.235	.445	2423	0.242	.282	.421	2830	.171	.395	.068	12.34
0.0180	0.189	.212	.457	2129	0.230	.265	.430	2582	0.267	.318	.404	3004	.178	.412	.073	11.46
0.0200	0.208	.235	.445	2260	0.251	.294	.415	2731	0.291	.353	.386	3164	.184	.429	.079	10.72
0.0240	0.242	.282	.421	2501	0.291	.353	.386	2998					.194	.458	.089	9.53
0.0280	0.275	.329	.398	2717	0.327	.412	.357	3229					.203	.483	.098	8.62
0.0320	0.306	.376	.374	2911									.210	.505	.106	7.89
0.0360	0.334	.424	.351	3085									.216	.525	.114	7.29
	0.75 $\rho_b = 0.0371$				0.75 $\rho_b = 0.0275$				0.75 $\rho_b = 0.0214$							
						Compressive Steel Area 20% of Tensile Steel Area										
0.0035									0.060	.071	.527	1207	.100	.207	.021	30.79
0.0040					0.058	.070	.528	1089	0.068	.078	.523	1291	.106	.219	.023	28.68
0.0050	0.058	.071	.527	991	0.071	.081	.522	1219	0.084	.092	.516	1447	.117	.240	.028	25.46
0.0060	0.069	.080	.522	1087	0.084	.093	.516	1339	0.100	.106	.510	1589	.126	.258	.033	23.09
0.0070	0.079	.089	.518	1176	0.098	.104	.511	1449	0.116	.120	.503	1720	.135	.275	.037	21.26
0.0080	0.090	.098	.514	1259	0.111	.115	.505	1552	0.131	.133	.496	1842	.143	.289	.042	19.78
0.0090	0.100	.106	.510	1338	0.124	.125	.500	1650	0.147	.146	.489	1956	.150	.302	.046	18.57
0.0100	0.111	.114	.505	1413	0.136	.136	.495	1742	0.162	.159	.483	2064	.157	.314	.050	17.54
0.0120	0.131	.130	.498	1553	0.162	.157	.484	1913	0.191	.186	.470	2262	.169	.336	.057	15.90
0.0140	0.152	.145	.490	1682	0.187	.177	.474	2069	0.220	.211	.457	2442	.180	.355	.065	14.63
0.0160	0.172	.161	.482	1802	0.211	.197	.464	2213	0.249	.237	.444	2606	.191	.371	.072	13.61
0.0180	0.192	.175	.475	1914	0.235	.218	.454	2346	0.276	.263	.431	2756	.200	.386	.078	12.77
0.200	0.212	.190	.467	2019	0.259	.238	.444	2470	0.303	.288	.418	2894	.210	.400	.084	12.07
0.0240	0.250	.219	.450	2205	0.304	.277	.421	2686	0.355	.339	.393	3138	.227	.424	.096	10.95
0.0280	0.286	.248	.431	2366	0.347	.317	.398	2868	0.404	.390	.365	3334	.242	.444	.107	10.09
0.0320	0.322	.276	.412	2507	0.388	.356	.374	3023					.257	.461	.118	9.40
0.0360	0.356	.304	.393	2629	0.426	.395	.351	3154					.270	.476	.128	8.84
0.0400	0.388	.331	.374	2735									.284	.490	.138	8.37
0.0480	0.449	.387	.337	2907									.309	.513	.155	7.63

Table C-6 Coefficients for section constants

| | At Ultimate Loads | | | | | | | | | | | | At Service Loads | | | |
| | Grade 40 Steel $f_y = 40000$ psi | | | | Grade 50 Steel $f_y = 50000$ psi | | | | Grade 60 Steel $f_y = 60000$ psi | | | | Elastic Constants for All Steels | | | |
A_s/bd	Z_c	$\beta_1 k_n$	y_n	f_{sv}	Z_c	$\beta_1 k_n$	y_n	f_{sv}	Z_c	$\beta_1 k_n$	y_n	f_{sv}	S_c	k_{sv}	I_{cr}	f_s/f_c
						Compressive Steel Area 40% of Tensile Steel Area										
0.0035									0.061	.076	.524	1188	.102	.204	.021	31.44
0.0040					0.058	.075	.525	1069	0.069	.082	.521	1265	.109	.215	.024	29.39
0.0050	0.058	.078	.524	966	0.071	.086	.520	1186	0.085	.095	.515	1405	.120	.234	.029	26.26
0.0060	0.069	.085	.520	1050	0.085	.095	.515	1291	0.100	.106	.509	1531	.131	.251	.034	23.97
0.0070	0.079	.093	.516	1127	0.098	.104	.510	1387	0.116	.117	.504	1646	.141	.266	.038	22.20
0.0080	0.090	.100	.513	1199	0.111	.113	.506	1477	0.131	.128	.499	1752	.150	.279	.043	20.78
0.0090	0.100	.106	.509	1266	0.124	.121	.502	1560	0.147	.138	.494	1851	.158	.291	.047	19.61
0.0100	0.111	.112	.506	1329	0.136	.129	.498	1638	0.162	.148	.489	1944	.167	.302	.051	18.62
0.0120	0.131	.124	.501	1444	0.162	.144	.490	1782	0.192	.167	.479	2114	.182	.321	.060	17.04
0.0140	0.152	.135	.495	1549	0.187	.159	.483	1912	0.222	.186	.470	2266	.196	.337	.068	15.83
0.0160	0.172	.145	.490	1645	0.213	.172	.476	2030	0.252	.204	.461	2404	.209	.351	.075	14.86
0.0180	0.193	.154	.485	1734	0.238	.186	.470	2138	0.281	.222	.452	2529	.222	.364	.083	14.07
0.0200	0.213	.164	.481	1816	0.262	.199	.463	2238	0.310	.239	.443	2645	.235	.375	.090	13.40
0.0240	0.253	.181	.472	1964	0.312	.224	.451	2417	0.367	.273	.426	2849	.258	.395	.104	12.34
0.0280	0.293	.197	.464	2095	0.361	.248	.439	2574	0.424	.307	.409	3024	.280	.411	.117	11.54
0.0320	0.332	.213	.450	2203	0.407	.271	.421	2701	0.479	.340	.393	3176	.302	.425	.130	10.90
0.0360	0.370	.227	.435	2296	0.452	.294	.404	2808	0.531	.372	.372	3294	.322	.437	.142	10.37
0.0400	0.407	.241	.421	2378	0.497	.316	.386	2899					.343	.447	.154	9.94
0.0480	0.479	.268	.393	2510	0.581	.360	.351	3043					.382	.465	.177	9.26
0.0560	0.548	.293	.365	2609									.420	.479	.200	8.74
						Compressive Steel Area 60% of Tensile Steel Area										
0.0035									0.061	.080	.523	1170	.104	.201	.021	32.06
0.0040					0.058	.079	.523	1049	0.069	.085	.520	1239	.111	.211	.024	30.07
0.0050	0.058	.082	.522	942	0.071	.089	.518	1154	0.085	.096	.514	1366	.124	.229	.029	27.04
0.0060	0.069	.089	.518	1016	0.085	.097	.514	1248	0.100	.106	.509	1479	.136	.245	.034	24.82
0.0070	0.079	.095	.515	1084	0.098	.105	.510	1333	0.116	.115	.505	1581	.147	.258	.039	23.11
0.0080	0.090	.101	.512	1145	0.111	.112	.507	1410	0.131	.124	.501	1674	.157	.270	.044	21.74
0.0090	0.100	.106	.509	1203	0.124	.118	.503	1482	0.147	.132	.497	1760	.167	.281	.049	20.62
0.0100	0.111	.111	.507	1256	0.136	.125	.500	1549	0.162	.140	.493	1840	.176	.290	.053	19.67
0.0120	0.131	.120	.502	1353	0.162	.136	.494	1671	0.193	.154	.485	1985	.194	.307	.062	18.16
0.0140	0.152	.128	.498	1439	0.188	.147	.489	1778	0.223	.168	.478	2113	.211	.321	.071	17.00
0.0160	0.172	.135	.495	1517	0.213	.156	.484	1875	0.253	.181	.472	2228	.227	.333	.079	16.08
0.0180	0.193	.142	.491	1587	0.239	.166	.480	1963	0.283	.193	.466	2331	.243	.344	.087	15.33
0.0200	0.213	.148	.488	1652	0.264	.174	.475	2043	0.313	.205	.460	2425	.258	.354	.095	14.70
0.0240	0.254	.159	.483	1766	0.315	.190	.467	2184	0.373	.227	.449	2591	.288	.370	.111	13.70
0.0280	0.295	.169	.478	1864	0.365	.205	.460	2305	0.433	.248	.438	2732	.317	.383	.126	12.94
0.0320	0.336	.178	.473	1950	0.415	.218	.453	2411	0.492	.268	.428	2855	.344	.394	.141	12.35
0.0360	0.377	.186	.469	2026	0.465	.231	.447	2504	0.551	.287	.419	2962	.372	.404	.155	11.86
0.0400	0.417	.194	.466	2093	0.516	.243	.441	2587	0.609	.305	.410	3056	.399	.412	.170	11.46
0.0480	0.497	.207	.450	2201	0.613	.265	.421	2716	0.726	.340	.392	3215	.451	.426	.198	10.84

Tables values are coefficients for the section constants:

Z_c = coeff. × bd^2; y_n = coeff. × d; S_c = coeff. × bd^2; I_{cr} = coeff × bd^3; k_n and k_w are pure coefficients. Units for f_{sv} are in psi. Value of I_{cr} is for short-term live loads. For long-term loads, use 1/2 E_c with $I=S_c k_{sv} d$. Coefficients above the interior line apply only to tee shapes. Table is stopped when the sum of tensile and compressive steel exceeds 8% of gross area.

Table C-6 Coefficients for section constants *(continued)*

Coefficients for Section Constants. f'_c = 5000 psi (34.4 N/mm²);
g = 0.125, β_1 = 0.80, n = 7.2

| | At Ultimate Loads | | | | | | | | | | | | At Service Loads | | | |
| | Grade 40 Steel f_y = 40000 psi | | | | Grade 50 Steel f_y = 50000 psi | | | | Grade 60 Steel f_y = 60000 psi | | | | Elastic Constants for All Steels | | | |
A_s/bd	Z_c	$\beta_1 k_n$	y_n	f_{sv}	Z_c	$\beta_1 k_n$	y_n	f_{sv}	Z_c	$\beta_1 k_n$	y_n	f_{sv}	S_c	k_{sv}	I_{cr}	f_s/f_c
	Compressive Steel Area 0% of Tensile Steel Area															
0.0005	0.005	.005	.560	297	0.006	.006	.560	371	0.007	.007	.559	445	.040	.081	.003	81.30
0.0010	0.009	.009	.558	431	0.012	.012	.557	538	0.014	.014	.555	645	.054	.113	.006	56.49
0.0015	0,014	.014	.555	538	0.017	.018	.554	671	0.021	.021	.552	804	.065	.137	.009	45.51
0.0020	0.019	.019	.553	631	0.023	.024	.551	787	0.028	.028	.548	942	.074	.156	.012	38.97
0.0025	0.023	.024	.551	715	0.029	.029	.548	891	0.035	.035	.545	1066	.081	.173	.014	34.51
0.0030	0.028	.028	.548	793	0.035	.035	.545	987	0.041	.042	.541	1180	.088	.187	.016	31.22
0.0035	0.032	.033	.546	865	0.040	.041	.542	1077	0.048	.049	.538	1287	.094	.201	.019	28.66
0.0040	0.037	.038	.544	934	0.046	.047	.539	1162	0.055	.056	.534	1387	.099	.213	.021	26.61
0.0050	0.046	.047	.539	1062	0.057	.059	.533	1320	0.068	.071	.527	1574	.108	.235	.025	23.47
0.0060	0.055	.056	.534	1181	0.068	.071	.527	1465	0.081	.085	.520	1746	.116	.254	.029	21.15
0.0070	0.064	.066	.530	1292	0.079	.082	.521	1602	0.094	.099	.513	1905	.123	.271	.033	19.36
0.0080	0.072	.075	.525	1398	0.090	.094	.515	1730	0.107	.113	.506	2056	.130	.287	.037	17.91
0.0090	0.081	.085	.520	1498	0.100	.106	.510	1852	0.119	.127	.499	2198	.135	.301	.041	16.72
0.0100	0.090	.094	.515	1594	0.111	.118	.504	1968	0.131	.141	.492	2333	.141	.314	.044	15.71
0.0120	0.107	.113	.506	1776	0.131	.141	.492	2187	0.155	.169	.478	2584	.150	.338	.051	14.09
0.0140	0.123	.132	.497	1946	0.151	.165	.480	2390	0.178	.198	.464	2816	.158	.359	.057	12.83
0.0160	0.139	.151	.487	2106	0.171	.188	.468	2579	0.200	.226	.450	3030	.165	.378	.063	11.82
0.0180	0.155	.169	.478	2257	0.189	.212	.457	2756	0.222	.254	.435	3229	.172	.396	.068	10.99
0.0200	0.171	.188	.468	2401	0.208	.235	.445	2923	0.242	.282	.421	3415	.178	.412	.073	10.29
0.0240	0.200	.226	.450	2669	0.242	.282	.421	3230	0.281	.339	.393	3749	.188	.440	.083	9.16
0.0280	0.229	.264	.431	2914	0.275	.329	.398	3504					.196	.465	.091	8.29
0.0320	0.256	.301	.412	3139	0.306	.376	.374	3750					.204	.486	.099	7.60
0.0360	0.281	.339	.393	3345									.210	.506	.106	7.03
0.0400	0.306	.376	.374	3535									.216	.524	.113	6.55
	0.75 ρ_b = 0.0371				0.75 ρ_b = 0.0275				0.75 ρ_b = 0.0214							
	Compressive Steel Area 20% of Tensile Steel Area															
0.0035									0.049	.059	.533	1282	.095	.198	.019	29.21
0.0040					0.047	.059	.533	1158	0.055	.065	.530	1372	.101	.209	.021	27.21
0.0050	0.047	.060	.532	1054	0.058	.068	.528	1296	0.068	.077	.524	1537	.111	.229	.026	24.16
0.0060	0.056	.068	.528	1155	0.068	.078	.524	1422	0.081	.088	.518	1688	.120	.247	.030	21.92
0.0070	0.064	.075	.525	1249	0.079	.087	.519	1540	0.094	.099	.513	1828	.129	.263	.034	20.18
0.0080	0.073	.082	.521	1338	0.090	.096	.515	1650	0.107	.110	.508	1958	.136	.277	.038	18.78
0.0090	0.081	.089	.518	1422	0.100	.104	.510	1753	0.119	.121	.502	2081	.143	.290	.042	17.63
0.0100	0.090	.096	.515	1502	0.111	.118	.506	1852	0.131	.131	.497	2197	.149	.302	.046	16.65
0.0120	0.107	.109	.508	1651	0.131	.130	.498	2036	0.156	.152	.486	2413	.161	.323	.053	15.09
0.0140	0.123	.122	.502	1789	0.152	.147	.489	2205	0.180	.173	.476	2610	.172	.341	.059	13.89
0.0160	0.140	.134	.496	1918	0.172	.163	.481	2361	0.203	.194	.465	2791	.182	.358	.066	12.92
0.0180	0.156	.146	.489	2038	0.192	.179	.473	2507	0.226	.215	.455	2958	.191	.372	.072	12.12
0.200	0.172	.158	.483	2152	0.211	.195	.465	2643	0.249	.235	.445	3113	.200	.386	.078	11.46
0.0240	0.204	.182	.472	2362	0.250	.227	.449	2892	0.293	.276	.424	3393	.216	.409	.089	10.39
0.0280	0.235	.205	.457	2546	0.286	.259	.431	3108	0.335	.317	.404	3637	.230	.429	.099	9.56
0.0320	0.265	.228	.442	2710	0.322	.291	.412	3294	0.375	.357	.382	3842	.244	.447	.109	8.91
0.0360	0.294	.250	.427	2856	0.356	.322	.393	3458					.257	.462	.118	8.37
0.0400	0.322	.273	.412	2986	0.388	.353	.374	3601					.269	.476	.127	7.92
0.0480	0.375	.317	.382	3207									.292	.499	.144	7.21
0.0560	0.425	.361	.352	3381									.314	.519	.160	6.67

Table C-7 Coefficients for section constants

A_s/bd	Grade 40 Steel $f_y = 40000$ psi				Grade 50 Steel $f_y = 50000$ psi				Grade 60 Steel $f_y = 60000$ psi				Elastic Constants for All Steels			
	Z_c	$\beta_1 k_n$	y_n	f_{sv}	Z_c	$\beta_1 k_n$	y_n	f_{sv}	Z_c	$\beta_1 k_n$	y_n	f_{sv}	S_c	k_{sv}	I_{cr}	f_s/f_c
Compressive Steel Area 40% of Tensile Steel Area																
0.0035									0.049	.065	.530	1270	.097	.195	.019	29.74
0.0040					0.047	.065	.530	1144	0.056	.070	.527	1351	.103	.206	.022	27.80
0.0050	0.047	.067	.529	1035	0.058	.074	.526	1268	0.069	.081	.522	1500	.114	.225	.026	24.84
0.0060	0.056	.074	.526	1124	0.069	.082	.522	1380	0.081	.090	.517	1635	.124	.241	.031	22.66
0.0070	0.064	.080	.522	1206	0.079	.089	.518	1483	0.094	.099	.513	1758	.134	.255	.035	20.98
0.0080	0.073	.086	.520	1283	0.090	.096	.514	1579	0.106	.108	.508	1873	.142	.268	.039	19.63
0.0090	0.081	.092	.517	1354	0.100	.103	.511	1668	0.119	.117	.504	1980	.150	.280	.043	18.51
0.0100	0.090	.097	.514	1422	0.111	.110	.507	1752	0.131	.125	.500	2080	.158	.290	.047	17.58
0.0120	0.106	.107	.509	1546	0.131	.123	.501	1907	0.156	.141	.492	2264	.172	.309	.055	16.07
0.0140	0.123	.116	.505	1659	0.152	.135	.495	2048	0.180	.156	.484	2430	.185	.325	.062	14.92
0.0160	0.140	.125	.500	1763	0.172	.147	.489	2176	0.204	.171	.477	2581	.198	.340	.069	14.00
0.0180	0.156	.133	.496	1859	0.193	.158	.484	2294	0.228	.186	.470	2720	.210	.352	.076	13.24
0.0200	0.172	.141	.492	1948	0.213	.169	.478	2404	0.252	.200	.462	2848	.221	.363	.082	12.60
0.0240	0.205	.156	.485	2110	0.253	.190	.468	2602	0.299	.228	.448	3077	.243	.383	.095	11.59
0.0280	0.237	.170	.478	2254	0.292	.210	.458	2777	0.345	.256	.435	3277	.263	.400	.107	10.81
0.0320	0.270	.183	.471	2383	0.332	.229	.448	2932	0.391	.282	.421	3452	.283	.414	.119	10.20
0.0360	0.301	.195	.461	2494	0.370	.248	.435	3064	0.436	.309	.408	3607	.302	.426	.130	9.70
0.0400	0.332	.207	.450	2592	0.407	.266	.421	3178	0.479	.335	.393	3738	.320	.437	.141	9.28
0.0480	0.392	.230	.427	2755	0.479	.302	.393	3363					.356	.455	.163	8.62
0.0560	0.451	.251	.404	2884	0.548	.337	.365	3505					.391	.470	.183	8.12
Compressive Steel Area 60% of Tensile Steel Area																
0.0035									0.049	.069	.528	1256	.099	.192	.019	30.25
0.0040					0.047	.069	.528	1128	0.056	.074	.526	1330	.105	.202	.022	28.36
0.0050	0.048	.072	.527	1015	0.058	.077	.524	1240	0.069	.083	.521	1466	.117	.220	.026	25.49
0.0060	0.056	.078	.524	1094	0.069	.084	.520	1341	0.081	.092	.517	1587	.128	.235	.031	23.38
0.0070	0.065	.083	.521	1166	0.079	.091	.517	1432	0.094	.099	.513	1696	.138	.249	.035	21.75
0.0080	0.073	.088	.518	12`3	0.090	.097	.514	1516	0.106	.107	.509	1797	.148	.260	.040	20.45
0.0090	0.081	.093	.516	1294	0.100	.103	.511	1593	0.119	.114	.506	1891	.157	.271	.044	19.37
0.0100	0.090	.097	.514	1352	0.111	.108	.508	1666	0.131	.120	.502	1978	.166	.280	.048	18.47
0.0120	0.106	.105	.510	1457	0.131	.118	.503	1797	0.156	.133	.496	2135	.183	.297	.056	17.03
0.0140	0.123	.113	.506	1551	0.152	.128	.499	1915	0.181	.145	.490	2276	.198	.311	.064	15.92
0.0160	0.140	.119	.503	1635	0.172	.136	.494	2021	0.205	.156	.485	2402	.213	.324	.072	15.04
0.0180	0.156	.125	.500	1712	0.193	.144	.490	2117	0.229	.166	.479	2517	.228	.334	.079	14.32
0.0200	0.173	.131	.497	1783	0.213	.152	.487	2206	0.254	.176	.474	2621	.242	.344	.087	13.72
0.0240	0.205	.141	.492	1909	0.254	.166	.480	2362	0.302	.195	.465	2807	.269	.361	.101	12.76
0.0280	0.238	.150	.488	2019	0.295	.178	.473	2498	0.350	.213	.456	2966	.295	.374	.115	12.03
0.0320	0.271	.158	.484	2115	0.335	.190	.467	2617	0.398	.230	.448	3105	.320	.386	.128	11.46
0.0360	0.303	.165	.480	2200	0.376	.201	.462	2722	0.445	.246	.440	3228	.345	.396	.141	10.99
0.0400	0.336	.172	.477	2277	0.416	.212	.457	2816	0.493	.261	.432	3336	.369	.404	.154	10.60
0.0480	0.401	.183	.471	2408	0.496	.231	.447	2977	0.587	.291	.417	3521	.417	.419	.180	9.99

Tables values are coefficients for the section constants:

Z_c = coeff. $\times bd^2$; y_n = coeff. $\times d$; S_c = coeff. $\times bd^2$; I_{cr} = coeff $\times bd^3$; k_n and k_w are pure coefficients. Units for f_{sv} are in psi. Value of I_{cr} is for short-term live loads. For long-term loads, use $1/2\ E_c$ with $I = S_c k_{sv} d$. Coefficients above the interior line apply only to tee shapes. Table is stopped when the sum of tensile and compressive steel exceeds 8% of gross area.

Table C-7 Coefficients for section constants *(continued)*

	Steel Yield and Concrete Stresses, psi			Tensile Steel Ratio	At Full Plastic Rotation to Ultimate Load			For Elastic Rotations Up to and Including the Elastic Limit				
	f'_c	f_{sv}	f_y	ρ	Z_c	$\beta_1 k_n$	y_n	S_c	k_{sv}	y_{sv}	I_{cr}	f_s/f_c
Compressive Steel Area 0% of Tensile Steel Area												
	3000	1553	40000	0.0119	0.169	0.186	0.470	0.163	0.372	0.439	0.0606	15.69
		1545	50000	0.0082	0.148	0.161	0.482	0.144	0.321	0.455	0.0461	19.61
		1539	60000	0.0060	0.132	0.142	0.492	0.128	0.283	0.468	0.0363	23.53
	4000	2078	40000	0.0173	0.182	0.203	0.461	0.176	0.406	0.427	0.0713	11.76
		2067	50000	0.0120	0.161	0.177	0.474	0.156	0.354	0.445	0.0551	14.71
		2058	60000	0.0089	0.144	0.157	0.484	0.140	0.313	0.458	0.0439	17.65
	5000	2605	40000	0.0230	0.193	0.217	0.454	0.185	0.433	0.418	0.0803	9.41
		2591	50000	0.0161	0.172	0.190	0.468	0.166	0.379	0.436	0.0629	11.76
		2579	60000	0.0120	0.155	0.169	0.478	0.150	0.338	0.450	0.0506	14.12
Compressive Steel Area 20% of Tensile Steel Area												
	3000	1548	40000	0.0139	0.197	0.180	0.473	0.191	0.372	0.439	0.0696	15.69
		1541	50000	0.0092	0.165	0.160	0.483	0.161	0.321	0.455	0.0510	19.61
		1536	60000	0.0066	0.143	0.143	0.491	0.139	0.283	0.468	0.0392	23.53
	4000	2068	40000	0.0210	0.221	0.198	0.464	0.214	0.406	0.427	0.0842	11.76
		2059	50000	0.0139	0.185	0.176	0.475	0.180	0.354	0.445	0.0623	14.71
		2052	60000	0.0099	0.160	0.158	0.484	0.156	0.313	0.458	0.0483	17.65
	5000	2583	40000	0.0289	0.241	0.217	0.454	0.233	0.433	0.418	0.0972	9.41
		2578	50000	0.0190	0.202	0.188	0.469	0.196	0.379	0.436	0.0725	11.76
		2570	60000	0.0136	0.175	0.169	0.478	0.170	0.338	0.450	0.0565	14.12
Compressive Steel Area 40% of Tensile Steel Area												
	3000	1541	40000	0.0169	0.238	0.175	0.475	0.232	0.372	0.439	0.0824	15.69
		1536	50000	0.0105	0.187	0.159	0.483	0.183	0.321	0.455	0.0573	19.61
		1532	60000	0.0072	0.156	0.144	0.490	0.153	0.283	0.468	0.0426	23.53
	4000	2056	40000	0.0268	0.281	0.192	0.466	0.273	0.406	0.427	0.1042	11.76
		2050	50000	0.0164	0.217	0.175	0.475	0.212	0.354	0.445	0.0722	14.71
		2046	60000	0.0112	0.180	0.159	0.483	0.176	0.313	0.458	0.0537	17.65
	5000	2561	40000	0.0387	0.322	0.218	0.453	0.314	0.433	0.418	0.1256	9.41
		2566	50000	0.0232	0.245	0.186	0.470	0.239	0.379	0.436	0.0863	11.76
		2560	60000	0.0157	0.201	0.169	0.478	0.196	0.338	0.450	0.0642	14.12
Compressive Steel Area 60% of Tensile Steel Area												
	3000	1534	40000	0.0214	0.301	0.171	0.477	0.294	0.372	0.439	0.1021	15.69
		1531	50000	0.0122	0.217	0.158	0.484	0.212	0.321	0.455	0.0656	19.60
		1529	60000	0.0080	0.173	0.145	0.490	0.170	0.283	0.468	0.0469	23.53
	4000	2042	40000	0.0369	0.386	0.188	0.469	0.378	0.406	0.427	0.1394	11.76
		2041	50000	0.0200	0.264	0.174	0.475	0.258	0.354	0.445	0.0863	14.71
		2039	60000	0.0128	0.205	0.160	0.482	0.201	0.313	0.458	0.0608	17.65
	5000		40000	Balanced stress condition is not feasible								
		2552	50000	0.0298	0.312	0.184	0.471	0.306	0.379	0.436	0.1078	11.76
		2551	60000	0.0186	0.237	0.169	0.478	0.232	0.338	0.450	0.0748	14.12
Compressive Steel Area 80% of Tensile Steel Area												
	3000	1525	40000	0.0293	0.409	0.167	0.479	0.402	0.372	0.439	0.1363	15.69
		1526	50000	0.0146	0.258	0.157	0.484	0.253	0.321	0.455	0.0772	19.61
		1526	60000	0.0090	0.193	0.146	0.490	0.190	0.283	0.468	0.0522	23.53
	4000		40000	Balanced stress condition is not feasible								
		2032	50000	0.0256	0.336	0.173	0.476	0.331	0.354	0.445	0.1084	14.71
		2033	60000	0.0151	0.240	0.161	0.482	0.236	0.313	0.458	0.0704	17.65
	5000		40000	Balanced stress condition is not feasible								
		2538	50000	0.0414	0.433	0.182	0.472	0.426	0.379	0.436	0.1463	11.76
		2541	60000	0.0229	0.290	0.170	0.478	0.285	0.338	0.450	0.0902	14.12

*Tabled values are coefficients for the section constants:

Z_c = coeff. $\times bd^2$; S_c = coeff. $\times bd^2$; I_{cr} = coeff. $\times bd^3$; y_n = coeff. $\times d$; y_{sv} = coeff. $\times d$; k_n and k_w are pure coefficients.

Balanced stress condition is considered not to be feasible if total area of tensile and compressive reinforcement would exceed 8% of gross area.

Table C-8 Coefficients for balanced stress conditions*

Stress Ratios for Columns. $f'_c = 3000$ psi; $n = 9.29$; $g = 0.125$

$$R_n = \frac{P_n/bh}{0.85f'_c}$$

$$R_e = \frac{P_e/bh}{f_c}$$

$$e_n = M_n P_n$$

$$e_e = M_e P_e$$

Steel at Flexure Faces Only

Steel Uniformly Distributed

Steel ratio A_s/bh

Grade 40 Steel — Allowable R_n at ultimate load

e_n/h	0.01	0.02	0.03	0.04	0.05	0.06	0.07	0.08	0.01	0.02	0.03	0.04	0.05	0.06	0.07	0.08
0.10	0.91	1.03	1.15	1.26	1.38	1.50	1.61	1.72	0.91	1.02	1.14	1.25	1.36	1.47	1.59	1.70
0.20	0.72	0.83	0.93	1.03	1.13	1.23	1.33	1.42	0.70	0.80	0.90	0.99	1.08	1.18	1.27	1.36
0.30	0.58	0.68	0.77	0.86	0.95	1.03	1.12	1.20	0.54	0.64	0.72	0.80	0.88	0.96	1.04	1.11
0.40	0.46	0.57	0.65	0.73	0.81	0.88	0.96	1.03	0.41	0.51	0.59	0.67	0.74	0.80	0.87	0.94
0.50	0.35	0.48	0.56	0.64	0.71	0.77	0.84	0.90	0.31	0.41	0.50	0.57	0.63	0.69	0.75	0.81
0.60	0.26	0.40	0.50	0.56	0.62	0.69	0.75	0.81	0.23	0.34	0.42	0.48	0.55	0.60	0.66	0.71
0.70	0.20	0.33	0.43	0.50	0.56	0.62	0.67	0.72	0.19	0.28	0.35	0.42	0.47	0.53	0.58	0.63
0.80	0.17	0.28	0.37	0.45	0.51	0.56	0.61	0.66		0.24	0.30	0.36	0.42	0.47	0.52	0.56
0.90		0.24	0.32	0.40	0.46	0.51	0.56	0.60		0.20	0.27	0.32	0.37	0.42	0.46	0.51
1.00		0.20	0.28	0.35	0.42	0.47	0.52	0.56		0.18	0.24	0.29	0.33	0.38	0.42	0.46
2.00					0.19	0.23	0.26	0.29						0.18	0.21	0.23
3.00								0.19								
4.00																

Grade 50 Steel — Allowable R_n at ultimate load

e_n/h	0.01	0.02	0.03	0.04	0.05	0.06	0.07	0.08	0.01	0.02	0.03	0.04	0.05	0.06	0.07	0.08
0.10	0.95	1.09	1.24	1.39	1.53	1.68	1.83	1.97	0.94	1.08	1.22	1.36	1.50	1.64	1.78	1.92
0.20	0.75	0.89	1.01	1.14	1.26	1.38	1.51	1.63	0.72	0.84	0.96	1.08	1.19	1.30	1.42	1.53
0.30	0.60	0.72	0.84	0.95	1.06	1.16	1.27	1.37	0.56	0.67	0.77	0.87	0.96	1.06	1.15	1.25
0.40	0.49	0.61	0.71	0.81	0.90	1.00	1.09	1.18	0.43	0.54	0.63	0.72	0.80	0.88	0.97	1.05
0.50	0.38	0.52	0.61	0.70	0.79	0.87	0.95	1.04	0.33	0.45	0.54	0.61	0.69	0.76	0.83	0.90
0.60	0.30	0.45	0.54	0.62	0.70	0.77	0.85	0.92	0.26	0.37	0.46	0.53	0.60	0.66	0.73	0.79
0.70	0.24	0.38	0.48	0.56	0.63	0.70	0.76	0.83	0.21	0.31	0.39	0.47	0.53	0.59	0.64	0.70
0.80	0.20	0.32	0.43	0.50	0.57	0.63	0.69	0.76	0.18	0.27	0.34	0.41	0.47	0.53	0.58	0.63
0.90	0.18	0.28	0.38	0.46	0.52	0.58	0.64	0.69		0.23	0.30	0.36	0.42	0.48	0.53	0.57
1.00		0.24	0.34	0.42	0.48	0.53	0.59	0.64		0.21	0.27	0.33	0.38	0.43	0.48	0.52
2.00				0.20	0.23	0.28	0.32	0.36					0.19	0.21	0.24	0.27
3.00						0.18	0.20	0.23								0.18
4.00								0.17								

Grade 60 Steel — Allowable R_n at ultimate load

e_n/h	0.01	0.02	0.03	0.04	0.05	0.06	0.07	0.08	0.01	0.02	0.03	0.04	0.05	0.06	0.07	0.08
0.10	0.98	1.15	1.33	1.51	1.69	1.86	2.04	2.22	0.96	1.12	1.29	1.46	1.62	1.79	1.96	2.13
0.20	0.78	0.94	1.09	1.24	1.39	1.54	1.69	1.84	0.74	0.88	1.02	1.15	1.29	1.42	1.55	1.69
0.30	0.62	0.77	0.90	1.04	1.17	1.30	1.42	1.55	0.57	0.70	0.81	0.93	1.04	1.15	1.26	1.37
0.40	0.51	0.64	0.77	0.88	1.00	1.11	1.22	1.33	0.45	0.57	0.67	0.77	0.86	0.96	1.05	1.15
0.50	0.42	0.55	0.66	0.77	0.87	0.97	1.07	1.17	0.36	0.47	0.57	0.65	0.74	0.82	0.90	0.98
0.60	0.33	0.48	0.58	0.68	0.77	0.86	0.95	1.04	0.29	0.40	0.49	0.57	0.64	0.71	0.79	0.86
0.70	0.28	0.43	0.52	0.61	0.69	0.77	0.86	0.94	0.24	0.34	0.43	0.50	0.57	0.63	0.70	0.76
0.80	0.24	0.37	0.47	0.55	0.63	0.70	0.78	0.85	0.20	0.30	0.38	0.45	0.51	0.57	0.63	0.68
0.90	0.21	0.32	0.43	0.50	0.57	0.64	0.71	0.78	0.17	0.26	0.33	0.40	0.46	0.51	0.57	0.62
1.00	0.19	0.29	0.39	0.46	0.53	0.59	0.66	0.72		0.23	0.30	0.36	0.42	0.47	0.52	0.57
2.00			0.19	0.23	0.28	0.32	0.37	0.41				0.18	0.21	0.24	0.27	0.30
3.00					0.19	0.22	0.24	0.27							0.18	0.20
4.00							0.19	0.21								

For All Steel — Elastic stress ratio R_e for any elastic stress

e_e/h	0.01	0.02	0.03	0.04	0.05	0.06	0.07	0.08	0.01	0.02	0.03	0.04	0.05	0.06	0.07	0.08
0.10	0.76	0.90	1.03	1.16	1.29	1.42	1.55	1.69	0.74	0.85	0.97	1.09	1.20	1.31	1.43	1.54
0.20	0.56	0.67	0.78	0.88	0.98	1.09	1.19	1.30	0.54	0.63	0.71	0.79	0.88	0.97	1.05	1.14
0.30	0.42	0.51	0.61	0.70	0.79	0.87	0.97	1.05	0.38	0.45	0.52	0.59	0.66	0.73	0.80	0.86
0.40	0.31	0.40	0.48	0.56	0.63	0.71	0.78	0.86	0.27	0.34	0.40	0.46	0.51	0.57	0.62	0.68
0.50	0.24	0.32	0.39	0.46	0.52	0.59	0.65	0.72	0.21	0.27	0.32	0.37	0.42	0.46	0.51	0.56
0.60	0.19	0.27	0.33	0.39	0.45	0.51	0.56	0.62	0.17	0.22	0.26	0.31	0.35	0.39	0.43	0.47
0.70	0.16	0.23	0.28	0.34	0.39	0.44	0.49	0.54	0.14	0.19	0.23	0.27	0.30	0.34	0.37	0.41
0.80	0.14	0.20	0.25	0.30	0.35	0.39	0.44	0.48	0.12	0.16	0.20	0.23	0.27	0.30	0.33	0.36
0.90	0.13	0.18	0.23	0.27	0.31	0.35	0.40	0.44	0.11	0.14	0.18	0.21	0.24	0.27	0.30	0.32
1.00	0.11	0.16	0.20	0.24	0.28	0.32	0.36	0.40	0.09	0.13	0.16	0.19	0.21	0.24	0.27	0.29
2.00	0.06	0.08	0.11	0.13	0.15	0.17	0.19	0.21	0.05	0.06	0.08	0.10	0.11	0.12	0.14	0.15
3.00	0.04	0.06	0.07	0.09	0.10	0.12	0.13	0.14	0.03	0.04	0.05	0.06	0.07	0.08	0.09	0.10
4.00	0.03	0.04	0.05	0.07	0.08	0.09	0.10	0.11	0.02	0.03	0.04	0.05	0.06	0.06	0.07	0.08

There is no entry in the table for values of R_n less than 0.168. Such members may be designed either as a beam carrying a small axial load or as a column carrying a dominant flexural load. The design procedures for a beam carrying a small axial load are simpler and are much preferred

Table C-9 Stress ratios for columns

Stress Ratios for Columns. $f'_c = 4000$ psi; $n = 8.04$; $g = 0.125$

$$R_n = \frac{P_n/bh}{0.85f'_c}$$

$$R_e = \frac{P_e/bh}{f_c}$$

$$e_n = M_n P_n$$

$$e_e = M_e P_e$$

Steel at Flexure Faces Only

Steel Uniformly Distributed

Steel ratio A_s/bh

		0.01	0.02	0.03	0.04	0.05	0.06	0.07	0.08	0.01	0.02	0.03	0.04	0.05	0.06	0.07	0.08
	e_n/h				Allowable R_n at ultimate load												
Grade 40 Steel	0.10	0.88	0.97	1.05	1.14	1.22	1.31	1.39	1.48	0.88	0.96	1.04	1.13	1.21	1.29	1.37	1.46
	0.20	0.69	0.77	0.85	0.92	1.00	1.07	1.14	1.21	0.67	0.75	0.82	0.89	0.96	1.03	1.09	1.16
	0.30	0.54	0.62	0.70	.76	0.83	0.89	0.95	1.02	0.51	0.58	0.65	0.71	0.77	0.83	0.89	0.95
	0.40	0.41	0.52	0.58	0.65	0.70	0.76	0.82	0.87	0.37	0.46	0.53	0.59	0.64	0.69	0.74	0.79
	0.50	0.30	0.42	0.50	0.56	0.61	0.66	0.71	0.76	0.27	0.36	0.43	0.49	0.55	0.59	0.64	0.68
	0.60	0.22	0.34	0.43	0.49	0.54	0.59	0.63	0.68	0.20	0.29	0.36	0.41	0.46	0.51	0.56	0.60
	0.70	0.17	0.27	0.36	0.43	0.48	0.52	0.57	0.61		0.24	0.30	0.35	0.40	0.44	0.49	0.53
	0.80		0.23	0.30	0.37	0.43	0.48	0.51	0.55		0.20	0.26	0.30	0.35	0.39	0.43	0.47
	0.90		0.19	0.26	0.32	0.38	0.43	0.47	0.50		0.17	0.22	0.27	0.31	0.35	0.38	0.42
	1.00			0.23	0.29	0.34	0.39	0.43	0.47			0.20	0.24	0.27	0.31	0.34	0.37
	2.00						0.18	0.20	0.23								0.18
	3.00																
	4.00																
	e_n/h				Allowable R_n at ultimate load												
Grade 50 Steel	0.10	0.90	1.01	1.12	1.23	1.34	1.45	1.55	1.66	0.90	1.00	1.10	1.21	1.31	1.42	1.52	1.63
	0.20	0.71	0.81	0.91	1.00	1.10	1.19	1.28	1.37	0.69	0.78	0.87	0.95	1.04	1.12	1.21	1.29
	0.30	0.56	0.66	0.75	0.83	0.91	0.99	1.07	1.15	0.52	0.61	0.69	0.76	0.83	0.90	0.98	1.05
	0.40	0.45	0.54	0.63	0.70	0.77	0.85	0.92	0.99	0.40	0.49	0.56	0.63	0.69	0.75	0.81	0.87
	0.50	0.34	0.46	0.54	0.61	0.67	0.74	0.80	0.86	0.30	0.40	0.47	0.53	0.59	0.64	0.70	0.75
	0.60	0.25	0.38	0.47	0.53	0.59	0.65	0.71	0.76	0.23	0.32	0.40	0.46	0.51	0.56	0.61	0.65
	0.70	0.21	0.32	0.41	0.48	0.53	0.58	0.64	0.69	0.18	0.27	0.34	0.39	0.45	0.49	0.54	0.58
	0.80	0.17	0.26	0.35	0.43	0.48	0.53	0.58	0.62		0.23	0.29	0.34	0.39	0.44	0.48	0.52
	0.90		0.23	0.31	0.38	0.44	0.48	0.53	0.57		0.19	0.25	0.30	0.35	0.39	0.43	0.47
	1.00		0.20	0.27	0.34	0.40	0.44	0.49	0.53		0.17	0.22	0.27	0.31	0.35	0.39	0.43
	2.00					0.18	0.22	0.25	0.28						0.17	0.19	0.21
	3.00								0.18								
	4.00																
	e_n/h				Allowable R_n at ultimate load												
Grade 60 Steel	0.10	0.93	1.06	1.19	1.32	1.45	1.59	1.72	1.85	0.91	1.04	1.16	1.28	1.40	1.53	1.65	1.78
	0.20	0.73	0.85	0.97	1.08	1.19	1.30	1.41	1.52	0.70	0.81	0.91	1.01	1.11	1.21	1.31	1.41
	0.30	0.58	0.69	0.80	0.90	0.99	1.09	1.19	1.28	0.54	0.63	0.72	0.81	0.89	0.97	1.06	1.14
	0.40	0.46	0.57	0.67	0.76	0.84	0.93	1.02	1.10	0.41	0.51	0.59	0.66	0.74	0.81	0.88	0.95
	0.50	0.37	0.49	0.57	0.65	0.73	0.81	0.88	0.96	0.32	0.42	0.49	0.56	0.62	0.69	0.75	0.81
	0.60	0.29	0.42	0.50	0.58	0.65	0.72	0.78	0.85	0.25	0.35	0.42	0.48	0.54	0.60	0.65	0.71
	0.70	0.24	0.36	0.45	0.51	0.58	0.64	0.70	0.76	0.20	0.29	0.37	0.43	0.48	0.53	0.58	0.62
	0.80	0.20	0.30	0.40	0.46	0.52	0.58	0.64	0.69	0.17	0.25	0.32	0.38	0.43	0.47	0.52	0.56
	0.90	0.18	0.27	0.35	0.42	0.48	0.53	0.58	0.64		0.22	0.28	0.33	0.38	0.43	0.47	0.51
	1.00		0.24	0.31	0.39	0.44	0.49	0.54	0.59		0.19	0.25	0.30	0.35	0.39	0.43	0.47
	2.00				0.19	0.22	0.26	0.29	0.32					0.17	0.19	0.22	0.24
	3.00						0.17	0.19	0.22								
	4.00																
	e_e/h				Elastic stress ratio R_e for any elastic stress												
For All Steel	0.10	0.74	0.86	0.97	1.09	1.20	1.31	1.42	1.54	0.72	0.82	0.92	1.02	1.12	1.22	1.31	1.41
	0.20	0.55	0.64	0.73	0.82	0.91	1.00	1.09	1.18	0.53	0.60	0.67	0.75	0.82	0.89	0.97	1.04
	0.30	0.40	0.49	0.57	0.65	0.72	0.80	0.88	0.96	0.37	0.43	0.49	0.55	0.61	0.67	0.73	0.79
	0.40	0.29	0.37	0.44	0.51	0.58	0.64	0.71	0.77	0.26	0.32	0.37	0.42	0.47	0.52	0.57	0.62
	0.50	0.22	0.30	0.36	0.42	0.48	0.53	0.59	0.65	0.20	0.25	0.30	0.34	0.38	0.42	0.46	0.50
	0.60	0.18	0.25	0.30	0.36	0.41	0.46	0.51	0.55	0.16	0.21	0.25	0.28	0.32	0.36	0.39	0.43
	0.70	0.15	0.21	0.26	0.31	0.35	0.40	0.44	0.49	0.13	0.17	0.21	0.24	0.28	0.31	0.34	0.37
	0.80	0.13	0.19	0.23	0.27	0.31	0.35	0.39	0.43	0.11	0.15	0.18	0.21	0.24	0.27	0.30	0.33
	0.90	0.12	0.16	0.21	0.25	0.28	0.32	0.36	0.39	0.10	0.13	0.16	0.19	0.22	0.24	0.27	0.29
	1.00	0.10	0.15	0.19	0.22	0.26	0.29	0.32	0.36	0.09	0.12	0.15	0.17	0.20	0.22	0.24	0.26
	2.00	0.05	0.08	0.10	0.12	0.13	0.15	0.17	0.19	0.04	0.06	0.07	0.09	0.10	0.11	0.12	0.14
	3.00	0.03	0.05	0.07	0.08	0.09	0.10	0.12	0.13	0.03	0.04	0.05	0.06	0.07	0.08	0.08	0.09
	4.00	0.03	0.04	0.05	0.06	0.07	0.08	0.09	0.10	0.02	0.03	0.04	0.04	0.05	0.06	0.06	0.07

There is no entry in the table for values of R_n less than 0.168. Such members may be designed either as a beam carrying a small axial load or as a column carrying a dominant flexural load. The design procedures for a beam carrying a small axial load are simpler and are much preferred

Table C-10 Stress ratios for columns

Stress Ratios for Columns. $f'_c = 5000$ psi; $n = 7.2$; $g = 0.125$

$$R_n = \frac{P_n/bh}{0.85f'_c}$$

$$R_e = \frac{P_e/bh}{f_c}$$

$$e_n = M_nP_n$$

$$e_e = M_eP_e$$

Steel at Flexure Faces Only | Steel Uniformly Distributed

		Steel ratio A_s/bh								Steel ratio A_s/bh							
		0.01	0.02	0.03	0.04	0.05	0.06	0.07	0.08	0.01	0.02	0.03	0.04	0.05	0.06	0.07	0.08
	e_n/h	Allowable R_n at ultimate load															
Grade 40 Steel	0.10	0.86	0.93	0.99	1.06	1.13	1.19	1.26	1.33	0.85	0.92	0.98	1.05	1.11	1.18	1.25	1.31
	0.20	0.67	0.74	0.80	0.86	0.91	0.97	1.03	1.08	0.65	0.71	0.77	0.82	0.88	0.93	0.99	1.04
	0.30	0.52	0.59	0.65	0.70	0.75	0.81	0.86	0.91	0.49	0.55	0.61	0.66	0.70	0.75	0.80	0.84
	0.40	0.38	0.48	0.54	0.59	0.64	0.68	0.73	0.77	0.35	0.43	0.49	0.54	0.58	0.62	0.66	0.70
	0.50	0.27	0.38	0.46	0.51	0.55	0.59	0.63	0.67	0.25	0.33	0.39	0.44	0.49	0.53	0.57	0.60
	0.60	0.19	0.30	0.38	0.44	0.48	0.52	0.56	0.60	0.18	0.26	0.32	0.37	0.41	0.45	0.49	0.53
	0.70		0.24	0.31	0.37	0.43	0.47	0.50	0.53		0.21	0.26	0.31	0.35	0.39	0.42	0.46
	0.80		0.19	0.26	0.32	0.37	0.42	0.45	0.48		0.17	0.22	0.27	0.30	0.34	0.37	0.40
	0.90			0.22	0.28	0.33	0.37	0.41	0.44			0.19	0.23	0.27	0.30	0.33	0.36
	1.00			0.19	0.24	0.29	0.33	0.37	0.41			0.17	0.21	0.24	0.27	0.30	0.32
	2.00							0.17	0.19								
	3.00																
	4.00																
	e_n/h	Allowable R_n at ultimate load															
Grade 50 Steel	0.10	0.88	0.96	1.05	1.14	1.22	1.31	1.39	1.48	0.87	0.95	1.03	1.12	1.20	1.28	1.36	1.44
	0.20	0.69	0.77	0.84	0.92	0.99	1.07	1.14	1.21	0.67	0.74	0.81	0.88	0.94	1.01	1.08	1.14
	0.30	0.53	0.61	0.69	0.76	0.82	0.89	0.95	1.01	0.50	0.57	0.64	0.70	0.75	0.81	0.87	0.92
	0.40	0.41	0.50	0.57	0.63	0.69	0.75	0.81	0.86	0.37	0.46	0.52	0.57	0.62	0.67	0.72	0.77
	0.50	0.30	0.42	0.49	0.55	0.60	0.65	0.70	0.75	0.27	0.36	0.43	0.48	0.53	0.57	0.61	0.66
	0.60	0.22	0.34	0.42	0.48	0.53	0.57	0.62	0.67	0.20	0.29	0.35	0.41	0.45	0.49	0.53	0.57
	0.70	0.18	0.27	0.36	0.43	0.47	0.51	0.56	0.60		0.24	0.30	0.35	0.39	0.43	0.47	0.51
	0.80		0.23	0.30	0.37	0.42	0.46	0.50	0.54		0.20	0.25	0.30	0.34	0.38	0.42	0.45
	0.90		0.20	0.26	0.32	0.38	0.42	0.46	0.49		0.17	0.22	0.26	0.30	0.34	0.37	0.41
	1.00		0.18	0.23	0.29	0.34	0.39	0.42	0.45			0.19	0.23	0.27	0.30	0.34	0.37
	2.00						0.18	0.20	0.23								0.18
	3.00																
	4.00																
	e_n/h	Allowable R_n at ultimate load															
Grade 60 Steel	0.10	0.89	1.00	1.11	1.21	1.31	1.42	1.52	1.62	0.88	0.98	1.08	1.18	1.27	1.37	1.47	1.57
	0.20	0.70	0.80	0.89	0.98	1.07	1.16	1.25	1.34	0.68	0.76	0.84	0.92	1.00	1.08	1.16	1.24
	0.30	0.55	0.64	0.73	0.81	0.89	0.96	1.04	1.12	0.51	0.59	0.66	0.73	0.80	0.87	0.93	1.00
	0.40	0.43	0.53	0.61	0.68	0.75	0.82	0.89	0.95	0.39	0.47	0.54	0.60	0.66	0.71	0.77	0.83
	0.50	0.33	0.44	0.52	0.58	0.65	0.71	0.77	0.83	0.29	0.38	0.45	0.50	0.55	0.60	0.65	0.70
	0.60	0.26	0.37	0.45	0.51	0.57	0.63	0.68	0.74	0.22	0.31	0.38	0.43	0.48	0.52	0.57	0.61
	0.70	0.21	0.31	0.40	0.46	0.51	0.56	0.61	0.66	0.18	0.26	0.32	0.38	0.42	0 46	0.50	0.54
	0.80	0.18	0.27	0.34	0.41	0.46	0.51	0.55	0.60		0.22	0.28	0.33	0.37	0.41	0.45	0.49
	0.90		0.23	0.30	0.37	0.42	0.46	0.50	0.55		0.19	0.24	0.29	0.33	0.37	0.41	0.44
	1.00		0.21	0.27	0.33	0.38	0.42	0.46	0.50		0.17	0.22	0.26	0.30	0.34	0.37	0.40
	2.00					0.19	0.21	0.24	0.27							0.18	0.20
	3.00								0.18								
	4.00																
	e_e/h	Elastic stress ratio R_e for any elastic stress															
For All Steel	0.10	0.73	0.83	0.93	1.03	1.14	1.24	1.34	1.43	0.71	0.80	0.89	0.97	1.06	1.15	1.24	1.32
	0.20	0.54	0.62	0.70	0.78	0.86	0.94	1.02	1.10	0.52	0.58	0.65	0.72	0.78	0.84	0.91	0.97
	0.30	0.39	0.47	0.54	0.61	0.68	0.75	0.82	0.88	0.36	0.42	0.47	0.53	0.58	0.63	0.68	0.73
	0.40	0.28	0.35	0.42	0.48	0.54	0.60	0.66	0.71	0.25	0.31	0.36	0.40	0.45	0.49	0.53	0.57
	0.50	0.22	0.28	0.34	0.39	0.45	0.50	0.55	0.60	0.19	0.24	0.28	0.32	0.36	0.40	0.43	0.47
	0.60	0.17	0.23	0.28	0.33	0.38	0.42	0.47	0.51	0.15	0.20	0.23	0.27	0.30	0.33	0.36	0.40
	0.70	0.15	0.20	0.25	0.29	0.33	0.37	0.41	0.45	0.13	0.17	0.20	0.23	0.26	0.29	0.32	0.34
	0.80	0.13	0.17	0.22	0.25	0.29	0.33	0.36	0.40	0.11	0.14	0.17	0.20	0.23	0.25	0.28	0.30
	0.90	0.11	0.16	0.19	0.23	0.26	0.30	0.33	0.36	0.09	0.13	0.15	0.18	0.20	0.23	0.25	0.27
	1.00	0.10	0.14	0.17	0.21	0.24	0.27	0.30	0.33	0.08	0.11	0.14	0.16	0.18	0.20	0.22	0.25
	2.00	0.05	0.07	0.09	0.11	0.12	0.14	0.16	0.17	0.04	0.06	0.07	0.08	0.09	0.10	0.12	0.13
	3.00	0.03	0.05	0.06	0.07	0.08	0.10	0.11	0.12	0.03	0.04	0.05	0.05	0.06	0.07	0.08	0.08
	4.00	0.02	0.04	0.05	0.05	0.06	0.07	0.08	0.09	0.02	0.03	0.03	0.04	0.05	0.05	0.06	0.06

There is no entry in the table for values of R_n less than 0.168. Such members may be designed either as a beam carrying a small axial load or as a column carrying a dominant flexural load. The design procedures for a beam carrying a small axial load are simpler and are much preferred

Table C-11 Stress ratios for columns

ALLOWABLE LOADS ON BOLTS EMBEDDED IN CONCRETE[a,b,c,d]

Diameter in.	Minimum[e] Embedment in.	Concrete Strength, psi			
		2000	3000	4000	2000 to 4000
		Shear Load, lbs			Tension Load, lbs
$1/4$	$2 1/2$	500	500	500	200
$3/8$	3	1100	1100	1100	500
$1/2$	4	1250	1250	1250	950
$5/8$	$4 1/2$	2750	2750	2750	1500
$3/4$	5	2940	3560	3560	2250
$7/8$	6	3350	4050	4050	2550
1^f	7	3750	4500	4500	2850
$1 1/8$	8	4750	4750	4750	3400
$1 1/4$	9	5800	5800	5800	4000

[a]Source: Uniform Building Code, 1994

[b]Bolt strength at least equal to A307. Bolts shall have a standard head or equal deformity in the imbedded portion.

[c]Bolt spacing at least 12 diameters, edge distance at least 6 diameters. Spacing and edge distance may be reduced 50 percent with an equal reduction in bolt values; intermediate values may be interpolated linearly.

[d]Values may be increased by 33% when loading includes wind or earthquake.

[e]An additional 2 inches of embedment is required for bolts located in the top of building columns located in Seismic Zones 2,3 and 4.

[f]Values are sometimes conservative for 1 in. diameter bolts.

Table C-12 Allowable service load on embedded bolts

Required Imbedment Length in Inches to Develop Full Strength

Ultimate Concrete Strength	Bar Size No.	Grade of Steel	Development Length			Tension Splice Length				Compression Splice Length
			Tension		Compression	Class A		Class B		
			Top	Other	All Bars	Top	Other	Top	Other	All Bars
		40	12	12	8	12	12	14	12	12
	3	50	13	12	8	13	12	17	13	12
		60	16	12	8	16	12	21	16	12
		40	14	12	8	14	12	19	14	12
	4	50	18	14	9	18	14	23	18	13
		60	21	16	11	21	16	28	21	15
		40	18	14	9	18	14	23	18	13
	5	50	22	17	11	22	17	29	22	16
		60	27	21	14	27	21	35	27	19
		40	21	16	11	21	16	28	21	15
	6	50	27	21	14	27	21	35	27	19
		60	32	25	16	32	25	42	32	23
		40	25	19	13	25	19	32	25	18
3000 psi	7	50	31	24	16	31	24	40	31	22
		60	37	29	19	37	29	49	37	26
		40	32	23	15	30	23	39	30	20
	8	50	37	29	18	37	29	48	37	25
		60	45	34	22	45	34	58	45	30
		40	38	29	16	38	29	49	38	23
	9	50	47	36	21	47	36	62	47	28
		60	57	44	25	57	44	74	57	34
		40	48	37	19	48	37	63	48	25
	10	50	60	46	23	60	46	78	60	32
		60	72	56	28	72	56	94	72	38
		40	59	46	21	59	46	77	59	28
	11	50	74	57	26	74	57	96	74	35
		60	89	68	31	89	68	116	89	42

Table C-13 Development length of straight bars

Required Imbedment Length in Inches to Develop Full Strength

Ultimate Concrete Strength	Bar Size No.	Grade of Steel	Development Length			Tension Splice Length				Compression Splice Length All Bars
			Tension		Compression All Bars	Class A		Class B		
			Top	Other		Top	Other	Top	Other	
	3	40	12	12	8	12	12	12	12	12
		50	12	12	8	12	12	15	12	12
		60	14	12	8	14	12	18	14	12
	4	40	12	12	8	12	12	16	12	12
		50	15	12	8	15	12	20	15	13
		60	18	14	9	18	14	24	18	15
	5	40	15	12	8	15	12	20	15	13
		50	19	15	10	19	15	25	19	16
		60	23	18	12	23	18	30	23	19
	6	40	18	14	9	18	14	24	18	15
		50	23	18	12	23	18	30	23	19
		60	28	21	14	28	21	36	28	23
4000 psi	7	40	22	17	11	22	17	28	22	18
		50	27	21	14	27	21	35	27	22
		60	32	25	17	32	25	42	32	26
	8	40	26	20	13	26	20	34	26	20
		50	32	25	16	32	25	42	32	25
		60	39	30	19	39	30	50	39	30
	9	40	33	25	14	33	25	43	33	23
		50	41	32	18	41	32	53	41	28
		60	49	38	21	49	38	64	49	34
	10	40	42	32	16	42	32	54	42	25
		50	52	40	20	52	40	68	52	32
		60	62	48	24	62	48	81	62	38
	11	40	51	40	18	51	40	67	51	28
		50	64	49	22	64	49	83	64	35
		60	77	59	27	77	59	100	77	42

Table C-13 Development length of straight bars *(continued)*

Required Imbedment Length in Inches to Develop Full Strength

Ultimate Concrete Strength	Bar Size No.	Grade of Steel	Development Length			Tension Splice Length				Compression Splice Length
			Tension		Compression	Class A		Class B		
			Top	Other	All Bars	Top	Other	Top	Other	All Bars
		40	12	12	8	12	12	12	12	12
	3	50	12	12	8	12	12	13	12	12
		60	12	12	8	12	12	16	12	12
		40	12	12	8	12	12	14	12	12
	4	50	14	12	8	14	12	18	14	13
		60	17	13	9	17	13	22	17	15
		40	14	12	8	14	12	18	14	13
	5	50	17	13	9	17	13	22	17	16
		60	21	16	11	21	16	27	21	19
		40	17	13	9	17	13	22	17	15
	6	50	21	16	11	21	16	27	21	19
		60	25	19	14	25	19	32	25	23
5000 psi		40	19	15	11	19	15	25	19	18
	7	50	24	19	13	24	19	31	24	22
		60	29	22	16	29	22	38	29	26
		40	23	18	12	23	18	30	23	20
	8	50	29	22	15	29	22	38	29	25
		60	35	27	18	35	27	45	35	30
		40	29	23	14	29	23	38	29	23
	9	50	37	28	17	37	28	48	37	28
		60	44	34	20	44	34	57	44	34
		40	37	29	15	37	29	48	37	25
	10	50	47	36	19	47	36	61	47	32
		60	56	43	23	56	43	73	56	38
		40	46	35	17	46	35	60	46	28
	11	50	57	44	21	57	44	75	57	35
		60	69	53	25	69	53	90	69	42

Table C-13 Development length of straight bars *(continued)*

Concrete Strength	Steel Grade	Development Length of Back of Hooks in Inches								
		#3	#4	#5	#6	#7	#8	#9	#10	#11
3000 psi	40	6	7	9	11	13	15	16	19	21
	50	7	9	11	14	16	18	21	23	26
	60	8	11	14	16	19	22	25	28	31
4000 psi	40	6	6	8	9	11	13	14	16	18
	50	6	8	10	12	14	16	18	20	22
	60	7	9	12	14	17	19	21	24	27
5000 psi	40	6	6	7	8	10	11	13	14	16
	50	6	7	9	11	12	14	16	18	20
	60	6	8	11	13	15	17	19	22	24

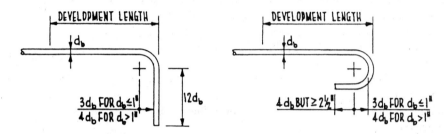

Table C-14 Development length of hooked bars

Masonry
Design Tables

The data and tables in this section are included in support of the design procedures for structural masonry given in this textbook. The initial entry is a listing of the standard symbols used in Chapter 23 for the development of the design equations and theory.

The following tables M-1 through M-14 provide design data for use in designing structural members in masonry using the empirical design method. The tables are adequate for the material covered in this text, to include the design of routine buildings and falsework less than 35 feet high.

The design data are extracted from three primary sources:

1. *Building Code Requirements for Masonry Structures*, Masonry Standards Joint Committee, 1992, to include the commentary on the Code provisions
2. *Specifications for Masonry Structures*, Masonry Standards Joint Committee, 1992, to include the commentary on the Specifications
3. *Uniform Building Code*, published by the International Conference of Building Officials, Whittier, California

The two sets of criteria adopted by the Masonry Standards Joint Committee are published jointly by the American Concrete Institute (ACI 530-92 and 530.1-92 respectively), The American Society of Civil Engineers (ASCE 5-92 and 6-92 respectively), and The Masonry Society (TMS 402-92 and 603-93 respectively). The data is reproduced here with the permission of the publishers.

STANDARD SYMBOLS USED IN STRUCTURAL MASONRY

A_b = cross sectional area of an anchor bolt, in.^{2}
a_n = net cross sectional area of masonry, in.^{2}
A_1 = bearing area, in.^{2}
A_2 = effective bearing area, in.^{2}
B_v = allowable shear force on an anchor bolt, lb
D = dead load, or related internal moments and forces
E = load effects of earthquake, or related internal moments and forces
E_m = modulus of elasticity of masonry in compression, psi
E_s = modulus of elasticity of steel, psi
E_v = modulus of elasticity in shear of masonry, psi
F_a = allowable compressive stress due to axial load only, psi
F_s = allowable tensile or compressive stress in reinforcement, psi
F_v = allowable shear stress in masonry, psi
I = moment of inertia, in.^{4}
L = live load, or related internal moments and forces
N_v = force acting normal to shear surface, lb
P = design axial load, lb
Q = first moment about the neutral axis of that portion of the cross section lying between the neutral axis and extreme fiber, in.^{3}
T = forces and moments caused by restraint of temperature, shrinkage and creep strains or differential movements

V	=	design shear force, lb
W	=	wind load, or related internal moments and forces
b	=	width of section, in.
b_v	=	total applied design shear force on an anchor bolt, lb
b_w	=	width of wall beam, in.
d_b	=	nominal diameter of reinforcement, in.
d_v	=	actual depth of masonry in direction of shear considered, in.
e	=	eccentricity of axial load, in.
f_a	=	calculated compressive stress in masonry due to axial load only, psi
f_g	=	compressive strength of grout, psi
f_m'	=	specified compressive strength of masonry, psi
f_s	=	calculated tensile or compressive stress in reinforcement, psi
f_v	=	calculated shear stress in masonry, psi
f_y	=	specified yield stress of steel for reinforcement and anchors, psi
h	=	effective height of column, wall or pilaster, in.
k_t	=	coefficient of thermal expansion of masonry per degree Fahrenheit
l	=	clear span between supports, in.
l_b	=	effective embedment length of headed or bent anchor bolt, in.
l_d	=	embedment length of straight reinforcement, in.
le	=	equivalent embedment length provided by a standard hook, in.
s	=	spacing of reinforcement, in.
t	=	nominal thickness of wall, given in Table M-7, in.
v	=	shear stress, given in Table M-6, psi

Mortar Type	Minimum Compressive Strength at 28 days psi	Portland Cement	Hydrated Lime or Lime Putty		Masonry Cement	Loose Aggregate
			Min.	Max.		
M	2500	1	–	$1/4$	–	Not less than $2^1/4$ and not more than 3 times the sum of the volumes of the cement and lime used
		1	–	–	1	
S	1800	1	$1/4$	$1/2$	–	
		$1/2$	–	–	1	
N	750	1	$1/2$	$1^1/4$	–	
		–	–	–	1	
O	350	1	$1^1/4$	$2^1/2$	–	

[a] Source: ASTM C270 Property Specification Requirements

Table M-1 Mortar proportions by volume[a]

Type	Parts by volume of Portland cement or blended cement	Parts by volume of hydrated lime or lime putty	Aggregate measured in a damp, loose condition	
			Fine	Coarse
Fine grout	1	0 to $1/10$	$2^1/4$ to 3 times the sum of the volumes of the cementitious materials	
Coarse grout	1	0 to $1/10$	$2^1/4$ to 3 times the sum of the volumes of the cementitious materials	1 to 2 times the sum of the volumes of the cementitious materials

[a] Source: Uniform Building Code
[b] Grout shall attain a minimum compressive strength of 2000 psi at 28 days

Table M-2 Group proportions by volume [a,b]

SPECIFIED COMPRESSIVE STRENGTH f_m' FOR
UNREINFORCED MASONRY OF FIRED CLAY UNITS[a]

Type M or S Mortar		Type N Mortar	
Minimum Net Area Compressive Strength of Masonry Units, psi	f_m' psi	Minimum Net Area Compressive Strength of Masonry Units, psi	f_m' psi
2400	1000	3000	1000
4400	1500	5500	1500
6400	2000	8000	2000
8400	2500	10500	2500
10400	3000	13000	3000
12400	3500		
14400	4000		

SPECIFIED COMPRESSIVE STRENGTH f_m' FOR
UNREINFORCED MASONRY OF CONCRETE MASONRY UNITS

Type M or S Mortar		Type N Mortar	
Minimum Net Area Compressive Strength of Masonry Units, psi	f_m' psi	Minimum Net Area Compressive Strength of Masonry Units, psi	f_m' psi
1250	1000	1300	1000
1900	1500	2150	1500
2800	2000	3050	2000
3750	2500	4050	2500
4800	3000	5250	3000

[a] Source: Specifications for Masonry Structures

Table M-3 Specified compressive strength of masonry

ALLOWABLE COMPRESSIVE STRESS BASED ON GROSS AREA OF UNITS[a]
FOR EMPIRICAL DESIGN OF MASONRY

Type of Masonry	Compressive strength of Masonry Units, Based on Gross Area psi	Allowable compressive Stresses on Gross Cross Sectional Area[b] psi	
		Type M or S mortar	Type N mortar
Solid masonry of fired clay brick or concrete brick	8000 or greater 4500 2500 1500	350 225 160 115	300 200 140 100
Grouted masonry of fired clay or concrete	4500 or greater 2500 1500	225 160 115	200 140 100
Solid masonry of solid concrete masonry units	3000 or greater 2000 1200	225 160 115	200 140 100
Masonry of hollow load bearing units	2000 or greater 1500 1000 700	140 115 75 60	120 100 70 55
Hollow wall masonry of solid units	2500 or greater 1500	160 115	140 100
Hollow wall masonry of hollow units		75	70

[a] Source: Building Code Requirements for Masonry Structures
[b] Linear interpolation is permitted for masonry units having strengths between the listed values.

Table M-4 Allowable compressive stress in the empirical method

LINEAR MODULUS OF ELASTICITY E_m
AND SHEAR MODULUS OF ELASTICITY E_v
FOR UNREINFORCED MASONRY CONSTRUCTION[a]

a) Masonry of Fired Clay Units

Net area compressive strength of units psi	Modulus of Elasticity E_m, psi[b]		
	Type M Mortar	Type S mortar	Type N Mortar
12000 or more	3,000,000	3,000,000	2,800,000
10000	3,000,000	2,900,000	2,400,000
8000	2,800,000	2,400,000	2,000,000
6000	2,200,000	1,900,000	1,600,000
4000	1,600,000	1,400,000	1,200,000
2000	1,000,000	900,000	800,000

b) Masonry of Concrete Masonry Units

Net area compressive strength of units psi	Modulus of Elasticity E_m, psi[b]		
	Type M Mortar	Type S mortar	Type N Mortar
6000 or more	3,500,000	3,500,000	——
5000	3,200,000	3,200,000	2,800,000
4000	2,900,000	2,900,000	2,600,000
3000	2,500,000	2,500,000	2,300,000
2500	2,400,000	2,400,000	2,200,000
2000	2,200,000	2,200,000	1,800,000
1500	1,600,000	1,600,000	1,500,000

c) For all Masonry, $E_v = 0.4E_m$

[a] Source: Building Code Requirements for Masonry Structures
[b] Linear interpolation permitted

Table M-5 Linear modules and shear modulus of elasticity

MAXIMUM ALLOWABLE SHEAR STRESS ON UNREINFORCED MASONRY[a]

· In-plane Shear Stress (shear only)

Allowable $F_v \leq 1.5 \sqrt{f_m'}$ but ≤ 120 psi

· In-plane Shear Stress (with bearing)

Allowable $F_v \leq v + 0.45(N_v/A_n)$ but ≤ 120 psi

where: $v = 37$ psi for masonry in running bond, not solidly grouted

$v = 37$ psi for masonry in stack bond with the open end units grouted solid

$v = 60$ psi for masonry in running bond, grouted solid

N_v = Dead load force acting normal to the surface being sheared, Lb.

A_v = Net cross sectional area of the surface being sheared, in.^{2}

· In-plane Shear Stress (Any loading)

Allowable $F_v \leq 15$ psi for masonry in any bond other than running bond with other than end units grouted solid

· Transverse shear stresses for any shear wall may not be greater than the in-plane shear stresses for that wall.

[a] Source: Building Code Requirements for Masonry Structures

Table M-6 Allowable shear stress on masonry

LATERAL SUPPORT REQUIREMENTS FOR MASONRY WALLS[a]
FOR HORIZONTAL LENGTH/THICKNESS RATIO (l/t)
OR FOR VERTICAL HEIGHT/THICKNESS RATIO (h/t)

Description of Wall	Maximum l/t or h/t[b]
· Bearing walls and shear walls	
Solid or solid grouted	20
All other	18
· Nonbearing walls	
Exterior	18
Interior	36
· Cantilever walls except parapets	
Solid or solid grouted	6
All other	4
· Parapet walls, minimum t = 8 in.	3

[a] Source: Building Code Requirements for Masonry Structures
[b] Only one of these support requirements must be satisfied.

Table M-7 Lateral support requirements for masonry walls

Diaphragm Construction	Maximum ratio Shearwall spacing[b] / Shearwall length[c]
Cast-in-place concrete	5 : 1
Precast concrete	4 : 1
Metal deck with concrete fill	3 : 1
Metal deck with no fill	2 : 1
Wood panels	2 : 1

[a] Source: Building Code Requirements for Masonry Structures
[b] Shearwall spacing is the perpendicular distance between a pair of shearwalls that lie parallel to the load.
[c] Shearwall length is the in-plane length of any loaded segment of a shearwall. Segments having a length less than the story height are usually ignored.

Table M-8 Maximum spacing of segments[a] of a shear walls

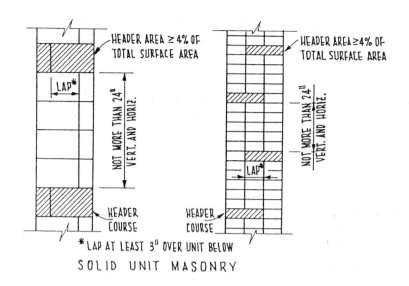

SOLID UNIT MASONRY

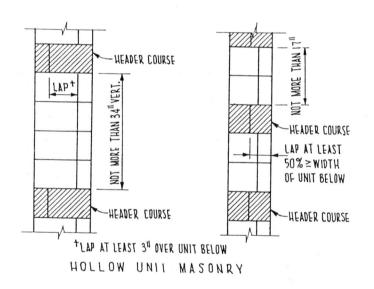

HOLLOW UNIT MASONRY

a Source: Building Code Requirements for Masonry Structures (commentary)

Table M-9 Requirements for lateral bonding using masonry headers[a]

Description of Wall	Nominal Minimum Thickness at Floor Level in.	Maximum Depth of Unbalanced Fill ft.
· Bearing walls		
Single story structures	6	NA
Two or more stories	8	NA
· Rubble stone walls		
Rough or random courses	16	NA
· Foundation walls		
Hollow units, ungrouted	8	4
	10	5
	12	6
Hollow units, grouted	8	5
	10	6
	12	7
Solid units	8	7
	10	8
	12	8
Hollow units, reinforced (Grouted #4 bars vertical @ 24"o.c., located not less than $4^1/_2$ in. from the pressure face)	8	7

[a] Source: Building Code Requirements for Masonry Structures

Mortar for foundation walls must be M or S.

Minimum thicknesses of foundation walls listed in Table M-10 apply only to walls having unbalanced fill extending not more than 8 feet in height above the inside grade, having lateral supports spaced vertically not more than 8 feet apart (cantilever walls are not permitted), and where the equivalent fluid weight of the soil is not more than 30 lb/ft^3.

The minimum thickness required at the lower levels of a floor must be maintained throughout the height of the floor; reductions may be made only at floor lines.

Where thickness is reduced by a step in the thickness, the top course of the lower wall must be solid masonry in order to receive and transmit the loads wythe-to-wythe from the upper wall to the lower wall.

Table M-10 Minimum wall thickness for masonry walls[a]

· Solid masonry unit construction
    Minimum wire size for ties: W2.8 embedded in mortar joints[b]
    Maximum wall surface area per tie: $4^1/_2$ ft^2
    Maximum vertical spacing of ties: 24 in.
    Maximum horizontal spacing of ties: 36 in.

· Hollow masonry unit construction
    Minimum wire size for ties: W2.8 embedded in mortar joints[b]
    Units laid with cells aligned vertically
    Ties rectangular, tying outside wythe to outside wythe
    Maximum wall surface area per tie: $4^1/_2$ ft^2
    Maximum vertical spacing of ties: 24 in.
    Maximum horizontal spacing of ties: 36 in.

· Additional ties required at all openings in masonry construction
    Minimum wire size for ties: W2.8 embedded in mortar joints[b]
    Maximum vertical spacing of ties: 36 in.
    Maximum horizontal spacing of ties: 36 in.
    Maximum distance of ties from opening: 12 in.

[a] Source: Building Code Requirements for Masonry Structures

[b] Physical properties of steel reinforcing wire:

Designation	Diameter in.	Area in^2	Perimeter in.
W 1.1(11 ga)	0.121	0.011	0.380
W 1.7(9 ga)	0.148	0.017	0.465
W 2.1(8 ga)	0.162	0.020	0.509
W 2.8($^3/_{16}$"wire)	0.187	0.027	0.587
W 4.9($^1/_4$" wire)	0.250	0.049	0.785

Table M-11 Requirements for lateral bonding using steel wire wall ties[a]

- Lateral bonding for all masonry construction
  Longitudinal wires embedded in mortar joints
  Minimum wire size for cross ties: W 1.7[b]
  Maximum wall surface area per cross tie: $2^2/_3$ ft^2
  Maximum vertical spacing of reinforcement: 16 in.

- Longitudinal bonding for all masonry construction
  Wythes in running bond
  No requirements

  Wythes in all other bonds
  Longitudinal wires embedded in mortar joints
  Minimum steel area: 0.0003 x cross sectional wall area
  Maximum vertical spacing of reinforcement: 48 in.

[a] Source: Building Code Requirements for Masonry Structures
[b] See Table M-11 for wire sizes

Table M-12 Requirements for lateral and longitudinal bonding using prefabricated joint reinforcement[a]

ALLOWABLE SHEAR ON BOLTS FOR
EMPIRICALLY DESIGNED MASONRY
EXCEPT UNBURNED CLAY UNITS[a]

Bolt diameter in.	Embedment[b] in.	Allowable load in shear	
		Solid masonry lbs.	Grouted masonry lbs.
$1/_2$	4	350	550
$5/_8$	4	500	750
$3/_4$	5	750	1100
$7/_8$	6	1000	1500
1	7	1250	1850[c]
$1^1/_8$	8	1500	2250[c]

[a] Source: Uniform Building Code, 1994

[b] An additional 2 in. of imbedment shall be provided for anchor bolts located in the top of columns for buildings located in Seismic Zones 2, 3 and 4.

[c] Permitted only with units having strengths greater than 2500 lbs/in^2

Table M-13 Allowable shear on bolts imbedded in masonry

· Masonry veneer over structural masonry
  Masonry units used as veneer may not be used to resist loads other than their own weight or their own earthquake shears.

· Anchorage of structural masonry walls to adjoining elements
  Positive anchorage to all adjoining elements is required.
  Anchorage must be capable of transmitting all computed earthquake loads.
  Minimum anchorage capacity in shear may not be less than 200 lbs/ft.
  Spacing of anchorages may not exceed 4 feet.
  Anchors must be imbedded in bond beams or in grouted and reinforced vertically aligned cells.

· Anchorage of floor joists to masonry walls
  Wood joists are to be anchored by metal strap anchors.
  Steel joists are to be anchored by $3/8$ in. diameter joist anchors.
  Continuous blocking or bridging is required at all lines of anchorage.

· Anchorage of roof structures to masonry walls
  Provide bond beams at the anchorage lines.
  Provide bond beam reinforcement not less than 2 - #4 bars.
  Place bond beam reinforcement not less than 6 in. from top of wall.
  Provide anchorage of roof to wall by one of the following alternatives:
    a) Bolts $1/2$ in. diameter imbedded at least 15 in. into the masonry
    b) Bolts $1/2$ in. diameter imbedded in the bond beam with a standard hook
    c) Bolts $1/2$ in. diameter welded to the bond beam reinforcement

· Requirements for reinforcement of masonry walls with running bond
  Provide vertical reinforcement not less than 0.2 in^2:
    a) At each corner
    b) At each side of each opening
    c) At ends of walls
  Provide horizontal reinforcement not less than 0.2 in^2:
    a) At bottom and top of wall openings, extending past the opening at least 24 in. but not less than 40 bar diameters
    b) Continuously at connected floors and roof and at tops of walls
    c) At bottom of wall, or, at top of foundations that are dowelled to the wall
    d) At a maximum spacing of 10 feet vertically

· Requirements for reinforcement of walls with bonds other than running bond
  Provide minimum horizontal reinforcement not less than 0.0007 times the gross cross sectional area of the masonry.
  Horizontal reinforcement may be uniformly distributed joint reinforcement or bond beams spaced not more than 4 feet apart vertically
  Provide minimum vertical reinforcement as prescribed for running bond.

[a] Source: Building Code Requirements for Masonry Structures

Table M-14 Requirements for earthquake loadings on masonry structural elements in earthquake risk zones 2[a] and 2b[a]

APPENDIX S

Steel Design Tables

The data and tables in this section of the appendix are included in support of the design procedures for structural steel given in this textbook. The initial entry is a listing of the standard symbols used in Chapters 24 through 28 for the development of the design equations and theory.

The following tables S-1 through S-9 provide physical dimensions and properties of wide flange shapes, American Standard shapes, tee shapes, round pipe and square tube. These tables are taken from the AISC *Manual of Steel Construction*, Eighth Edition, American Institute of Steel Construction, Chicago, IL, 1989. They are reprinted here with the permission of AISC. The tables are only a fraction of the sections included in the AISC manual; for the complete listing, the reader is referred to the manual.

Tables S-10 through S-18 are design tables; they have been originated for this textbook. In general, they provide the allowable axial load or allowable moment on sections for which the compression flange is intermittently supported. Except for pipes and tubes, the capacities listed in these tables are for ASTM A36 steel. For pipes and tubes, the ASTM designations are given with the appropriate table.

Tables S-19 through S-33 are generally physical design data and limitations extracted from various sources in the AISC manual, ninth edition. They are included here in support of the design of connections, both bolted and welded.

STANDARD SYMBOLS USED IN STRUCTURAL STEEL

A = Gross area of an axially loaded compression member, in.^{2}

A_b = Nominal body area of a fastener, in.^{2}; Area of an upset rod based upon the major diameter of its threads, in.^{2}

A_e = Effective net area of an axially loaded tension member, in.^{2}

A_f = Area of compression flange, in.^{2}

A_g = Gross area of member, in.^{2}

A_n = Net area of an axially loaded tension member, in.^{2}

A_t = Net tension area, in.^{2}

A_v = Net shear area, in.^{2}

A_w = Area of girder web, in.^{2}

A_1 = Area of steel concentrically bearing on a concrete support, in.^{2}

A_2 = Maximum area of the portion of the supporting surface that is geometrically similar to and concentric with the loaded area, in.^{2}

C_c = Column slenderness ratio that separates elastic and inelastic buckling

C_c' = Slenderness ratio of compression elements

D = Outside diameter of tubular member, in.

E = Modulus of elasticity of steel: 29,000 ksi

E_c = Modulus of elasticity of concrete, ksi

F_a = Axial compressive stress permitted in a prismatic member in the absence of bending moment, ksi

F_b = Bending stress permitted in a prismatic member in the absence of axial force, ksi

F_p = Allowable bearing stress, ksi

F_t = Allowable axial tensile stress, ksi

F_u = Specified minimum tensile strength of the type of steel or fastener being used, ksi

F_v = Allowable shear stress, ksi

F_y = Specified minimum yield stress of the type of steel being used, ksi

L = Unbraced length of tensile members, in.; actual unbraced length of a column, in.; unbraced length of a member measured between the centers of gravity of the bracing members, in.

L_c = Maximum unbraced length of the compression flange at which the allowable bending stress may be taken at $0.66F_y$, ft.

L_e = Distance from free edge to center of the bolt, in.

M = Moment, kip-ft.

N = Length of bearing of applied load, in.

P = Force transmitted by a fastener, kips; normal force, kips; axial load, kips

P_{cr} = Euler buckling load in an axially loaded compression member, kips

Q_s = Axial stress reduction factor where width-thickness ratio of unstiffened elements exceeds noncompact section limits

R = Reaction or concentrated load applied to a beam or girder, kips; radius, in.

U = Reduction coefficient used in calculating effective net area

V = Shear force, kips

a = Dimension parallel to the direction of stress, in.

b = Actual width of projecting compression element, in.; dimension normal to the direction of stress, in.

b_f = Flange width of rolled beam, in.

d = Depth of beam or girder, in.; nominal diameter of a fastener, in.

d_c = Web depth clear of fillets, in.

f = Axial compressive stress on member based on effective area, ksi

f_a = Computed axial stress, ksi

f_b = Computed bending stress, ksi

f_c' = Specified compressive strength of concrete, ksi

f_t = Computed tensile stress, ksi

f_v = Computed shear stress, ksi

g = Transverse spacing between fastener gage lines, in.

h = Clear distance between flanges of a beam or girder

k = Distance from outer face of flange to web toe of fillet of rolled shape

l = For beams, distance between cross sections braced against twist or lateral displacement of the compression flange, in.; for columns, actual unbraced length of member, in.; weld length, in.; largest laterally unbraced length along either flange at the point of load, in.

l_b = Actual unbraced length in plane of bending, in.

r = Governing radius of gyration, in.

r_T = Radius of gyration of a section comprising the compression flange plus 1/3 of the compression web area, taken about an axis in the plane of the web, in.

s = Longitudinal center-to-center spacing (pitch) of any two consecutive holes, in.

t = Thickness of a connected part, in.

t_f = Flange thickness, in.

t_w = Web thickness, in.

w = Plate width (distance between welds), in.

x = Subscript relating symbol to strong axis bending

y = Subscript relating symbol to weak axis bending

Δ = Displacement of the neutral axis of a loaded member from its position when the member is not loaded, in.

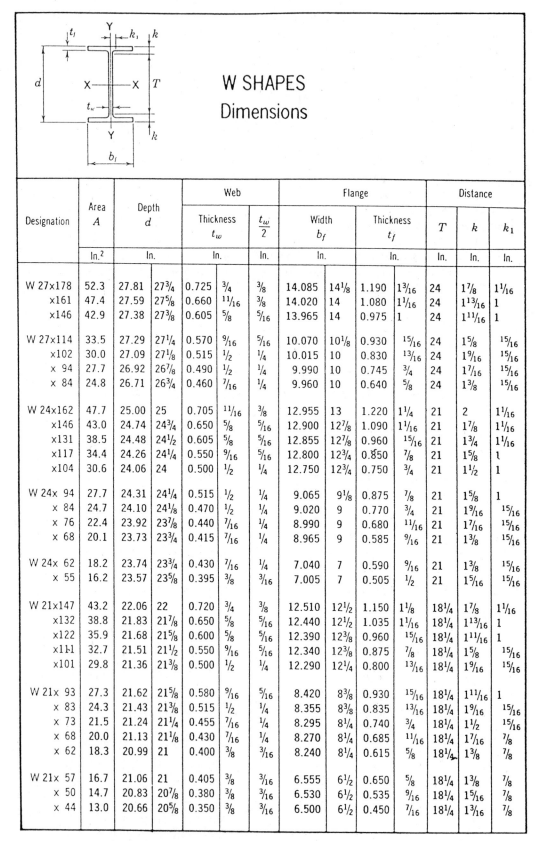

W SHAPES
Dimensions

Designation	Area A	Depth d	Web Thickness t_w	Web $\frac{t_w}{2}$	Flange Width b_f	Flange Thickness t_f	T	k	k_1
	In.²	In.	In.	In.	In.	In.	In.	In.	In.
W 27x178	52.3	27.81　27¾	0.725　¾	⅜	14.085　14⅛	1.190　1³/₁₆	24	1⅞	1¹/₁₆
x161	47.4	27.59　27⅝	0.660　¹¹/₁₆	⅜	14.020　14	1.080　1¹/₁₆	24	1¹³/₁₆	1
x146	42.9	27.38　27⅜	0.605　⅝	⁵/₁₆	13.965　14	0.975　1	24	1¹¹/₁₆	1
W 27x114	33.5	27.29　27¼	0.570　⁹/₁₆	⁵/₁₆	10.070　10⅛	0.930　¹⁵/₁₆	24	1⅝	¹⁵/₁₆
x102	30.0	27.09　27⅛	0.515　½	¼	10.015　10	0.830　¹³/₁₆	24	1⁹/₁₆	¹⁵/₁₆
x 94	27.7	26.92　26⅞	0.490　½	¼	9.990　10	0.745　¾	24	1⁷/₁₆	¹⁵/₁₆
x 84	24.8	26.71　26¾	0.460　⁷/₁₆	¼	9.960　10	0.640　⅝	24	1⅜	¹⁵/₁₆
W 24x162	47.7	25.00　25	0.705　¹¹/₁₆	⅜	12.955　13	1.220　1¼	21	2	1¹/₁₆
x146	43.0	24.74　24¾	0.650　⅝	⁵/₁₆	12.900　12⅞	1.090　1¹/₁₆	21	1⅞	1¹/₁₆
x131	38.5	24.48　24½	0.605　⅝	⁵/₁₆	12.855　12⅞	0.960　¹⁵/₁₆	21	1¾	1¹/₁₆
x117	34.4	24.26　24¼	0.550　⁹/₁₆	⁵/₁₆	12.800　12¾	0.850　⅞	21	1⅝	1
x104	30.6	24.06　24	0.500　½	¼	12.750　12¾	0.750　¾	21	1½	1
W 24x 94	27.7	24.31　24¼	0.515　½	¼	9.065　9⅛	0.875　⅞	21	1⅝	1
x 84	24.7	24.10　24⅛	0.470　½	¼	9.020　9	0.770　¾	21	1⁹/₁₆	¹⁵/₁₆
x 76	22.4	23.92　23⅞	0.440　⁷/₁₆	¼	8.990　9	0.680　¹¹/₁₆	21	1⁷/₁₆	¹⁵/₁₆
x 68	20.1	23.73　23¾	0.415　⁷/₁₆	¼	8.965　9	0.585　⁹/₁₆	21	1⅜	¹⁵/₁₆
W 24x 62	18.2	23.74　23¾	0.430　⁷/₁₆	¼	7.040　7	0.590　⁹/₁₆	21	1⅜	¹⁵/₁₆
x 55	16.2	23.57　23⅝	0.395　⅜	³/₁₆	7.005　7	0.505　½	21	1⁵/₁₆	¹⁵/₁₆
W 21x147	43.2	22.06　22	0.720　¾	⅜	12.510　12½	1.150　1⅛	18¼	1⅞	1¹/₁₆
x132	38.8	21.83　21⅞	0.650　⅝	⁵/₁₆	12.440　12½	1.035　1¹/₁₆	18¼	1¹³/₁₆	1
x122	35.9	21.68　21⅝	0.600　⅝	⁵/₁₆	12.390　12⅜	0.960　¹⁵/₁₆	18¼	1¹¹/₁₆	1
x111	32.7	21.51　21½	0.550　⁹/₁₆	⁵/₁₆	12.340　12⅜	0.875　⅞	18¼	1⅝	¹⁵/₁₆
x101	29.8	21.36　21⅜	0.500　½	¼	12.290　12¼	0.800　¹³/₁₆	18¼	1⁹/₁₆	¹⁵/₁₆
W 21x 93	27.3	21.62　21⅝	0.580　⁹/₁₆	⁵/₁₆	8.420　8⅜	0.930　¹⁵/₁₆	18¼	1¹¹/₁₆	1
x 83	24.3	21.43　21⅜	0.515　½	¼	8.355　8⅜	0.835　¹³/₁₆	18¼	1⁹/₁₆	¹⁵/₁₆
x 73	21.5	21.24　21¼	0.455　⁷/₁₆	¼	8.295　8¼	0.740　¾	18¼	1½	¹⁵/₁₆
x 68	20.0	21.13　21⅛	0.430　⁷/₁₆	¼	8.270　8¼	0.685　¹¹/₁₆	18¼	1⁷/₁₆	⅞
x 62	18.3	20.99　21	0.400　⅜	³/₁₆	8.240　8¼	0.615　⅝	18¼	1⅜	⅞
W 21x 57	16.7	21.06　21	0.405　⅜	³/₁₆	6.555　6½	0.650　⅝	18¼	1⅜	⅞
x 50	14.7	20.83　20⅞	0.380　⅜	³/₁₆	6.530　6½	0.535　⁹/₁₆	18¼	1⁵/₁₆	⅞
x 44	13.0	20.66　20⅝	0.350　⅜	³/₁₆	6.500　6½	0.450　⁷/₁₆	18¼	1³/₁₆	⅞

AMERICAN INSTITUTE OF STEEL CONSTRUCTION

Table S-1a Dimensions and physical properties of W shapes

W SHAPES
Properties

Nominal Wt. per Ft.	Compact Section Criteria				r_T	$\dfrac{d}{A_f}$	Elastic Properties						Torsional constant J	Plastic Modulus	
							Axis X-X			Axis Y-Y					
	$\dfrac{b_f}{2t_f}$	F_y'	$\dfrac{d}{t_w}$	F_y'''			I	S	r	I	S	r		Z_x	Z_y
Lb.		Ksi		Ksi	In.		In.4	In.3	In.	In.4	In.3	In.	In.4	In.3	In.3
178	5.9	—	38.4	44.9	3.72	1.66	6990	502	11.6	555	78.8	3.26	19.5	567	122
161	6.5	—	41.8	37.8	3.70	1.82	6280	455	11.5	497	70.9	3.24	14.7	512	109
146	7.2	—	45.3	32.2	3.68	2.01	5630	411	11.4	443	63.5	3.21	10.9	461	97.5
114	5.4	—	47.9	28.8	2.58	2.91	4090	299	11.0	159	31.5	2.18	7.33	343	49.3
102	6.0	—	52.6	23.9	2.56	3.26	3620	267	11.0	139	27.8	2.15	5.29	305	43.4
94	6.7	—	54.9	21.9	2.53	3.62	3270	243	10.9	124	24.8	2.12	4.03	278	38.8
84	7.8	—	58.1	19.6	2.49	4.19	2850	213	10.7	106	21.2	2.07	2.81	244	33.2
162	5.3	—	35.5	52.5	3.45	1.58	5170	414	10.4	443	68.4	3.05	18.5	468	105.
146	5.9	—	38.1	45.6	3.43	1.76	4580	371	10.3	391	60.5	3.01	13.4	418	93.2
131	6.7	—	40.5	40.3	3.40	1.98	4020	329	10.2	340	53.0	2.97	9.50	370	81.5
117	7.5	—	44.1	33.9	3.37	2.23	3540	291	10.1	297	46.5	2.94	6.72	327	71.4
104	8.5	58.5	48.1	28.5	3.35	2.52	3100	258	10.1	259	40.7	2.91	4.72	289	62.4
94	5.2	—	47.2	29.6	2.33	3.06	2700	222	9.87	109	24.0	1.98	5.26	254	37.5
84	5.9	—	51.3	25.1	2.31	3.47	2370	196	9.79	94.4	20.9	1.95	3.70	224	32.6
76	6.6	—	54.4	22.3	2.29	3.91	2100	176	9.69	82.5	18.4	1.92	2.68	200	28.6
68	7.7	—	57.2	20.2	2.26	4.52	1830	154	9.55	70.4	15.7	1.87	1.87	177	24.5
62	6.0	—	55.2	21.7	1.71	5.72	1550	131	9.23	34.5	9.80	1.38	1.71	153	15.7
55	6.9	—	59.7	18.5	1.68	6.66	1350	114	9.11	29.1	8.30	1.34	1.18	134	13.3
147	5.4	—	30.6	—	3.34	1.53	3630	329	9.17	376	60.1	2.95	15.4	373	92.6
132	6.0	—	33.6	58.6	3.31	1.70	3220	295	9.12	333	53.5	2.93	11.3	333	82.3
122	6.5	—	36.1	50.6	3.30	1.82	2960	273	9.09	305	49.2	2.92	8.98	307	75.6
111	7.1	—	39.1	43.2	3.28	1.99	2670	249	9.05	274	44.5	2.90	6.83	279	68.2
101	7.7	—	42.7	36.2	3.27	2.17	2420	227	9.02	248	40.3	2.89	5.21	253	61.7
93	4.5	—	37.3	47.5	2.17	2.76	2070	192	8.70	92.9	22.1	1.84	6.03	221	34.7
83	5.0	—	41.6	38.1	2.15	3.07	1830	171	8.67	81.4	19.5	1.83	4.34	196	30.5
73	5.6	—	46.7	30.3	2.13	3.46	1600	151	8.64	70.6	17.0	1.81	3.02	172	26.6
68	6.0	—	49.1	27.4	2.12	3.73	1480	140	8.60	64.7	15.7	1.80	2.45	160	24.4
62	6.7	—	52.5	24.0	2.10	4.14	1330	127	8.54	57.5	13.9	1.77	1.83	144	21.7
57	5.0	—	52.0	24.4	1.64	4.94	1170	111	8.36	30.6	9.35	1.35	1.77	129	14.8
50	6.1	—	54.8	22.0	1.60	5.96	984	94.5	8.18	24.9	7.64	1.30	1.14	110	12.2
44	7.2	—	59.0	19.0	1.57	7.06	843	81.6	8.06	20.7	6.36	1.26	0.77	95.4	10.2

AMERICAN INSTITUTE OF STEEL CONSTRUCTION

Table S-1a Dimensions and physical properties of W shapes *(continued)*

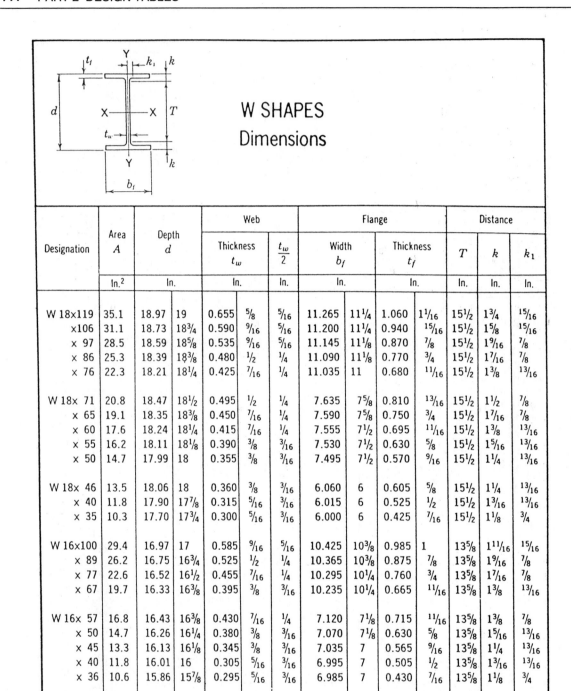

W SHAPES
Dimensions

Designation	Area A	Depth d		Web		Flange				Distance		
				Thickness t_w	$\dfrac{t_w}{2}$	Width b_f		Thickness t_f		T	k	k_1
	In.^{2}	In.		In.	In.	In.		In.		In.	In.	In.
W 18x119	35.1	18.97	19	0.655	5/8	5/16	11.265	11 1/4	1.060	1 1/16	15 1/2	1 3/4	15/16
x106	31.1	18.73	18 3/4	0.590	9/16	5/16	11.200	11 1/4	0.940	15/16	15 1/2	1 5/8	15/16
x 97	28.5	18.59	18 5/8	0.535	9/16	5/16	11.145	11 1/8	0.870	7/8	15 1/2	1 9/16	7/8
x 86	25.3	18.39	18 3/8	0.480	1/2	1/4	11.090	11 1/8	0.770	3/4	15 1/2	1 7/16	7/8
x 76	22.3	18.21	18 1/4	0.425	7/16	1/4	11.035	11	0.680	11/16	15 1/2	1 3/8	13/16
W 18x 71	20.8	18.47	18 1/2	0.495	1/2	1/4	7.635	7 5/8	0.810	13/16	15 1/2	1 1/2	7/8
x 65	19.1	18.35	18 3/8	0.450	7/16	1/4	7.590	7 5/8	0.750	3/4	15 1/2	1 7/16	7/8
x 60	17.6	18.24	18 1/4	0.415	7/16	1/4	7.555	7 1/2	0.695	11/16	15 1/2	1 3/8	13/16
x 55	16.2	18.11	18 1/8	0.390	3/8	3/16	7.530	7 1/2	0.630	5/8	15 1/2	1 5/16	13/16
x 50	14.7	17.99	18	0.355	3/8	3/16	7.495	7 1/2	0.570	9/16	15 1/2	1 1/4	13/16
W 18x 46	13.5	18.06	18	0.360	3/8	3/16	6.060	6	0.605	5/8	15 1/2	1 1/4	13/16
x 40	11.8	17.90	17 7/8	0.315	5/16	3/16	6.015	6	0.525	1/2	15 1/2	1 3/16	13/16
x 35	10.3	17.70	17 3/4	0.300	5/16	3/16	6.000	6	0.425	7/16	15 1/2	1 1/8	3/4
W 16x100	29.4	16.97	17	0.585	9/16	5/16	10.425	10 3/8	0.985	1	13 5/8	1 11/16	15/16
x 89	26.2	16.75	16 3/4	0.525	1/2	1/4	10.365	10 3/8	0.875	7/8	13 5/8	1 9/16	7/8
x 77	22.6	16.52	16 1/2	0.455	7/16	1/4	10.295	10 1/4	0.760	3/4	13 5/8	1 7/16	7/8
x 67	19.7	16.33	16 3/8	0.395	3/8	3/16	10.235	10 1/4	0.665	11/16	13 5/8	1 3/8	13/16
W 16x 57	16.8	16.43	16 3/8	0.430	7/16	1/4	7.120	7 1/8	0.715	11/16	13 5/8	1 3/8	7/8
x 50	14.7	16.26	16 1/4	0.380	3/8	3/16	7.070	7 1/8	0.630	5/8	13 5/8	1 5/16	13/16
x 45	13.3	16.13	16 1/8	0.345	3/8	3/16	7.035	7	0.565	9/16	13 5/8	1 1/4	13/16
x 40	11.8	16.01	16	0.305	5/16	3/16	6.995	7	0.505	1/2	13 5/8	1 3/16	13/16
x 36	10.6	15.86	15 7/8	0.295	5/16	3/16	6.985	7	0.430	7/16	13 5/8	1 1/8	3/4
W 16x 31	9.12	15.88	15 7/8	0.275	1/4	1/8	5.525	5 1/2	0.440	7/16	13 5/8	1 1/8	3/4
x 26	7.68	15.69	15 3/4	0.250	1/4	1/8	5.500	5 1/2	0.345	3/8	13 5/8	1 1/16	3/4

AMERICAN INSTITUTE OF STEEL CONSTRUCTION

Table S-1b Dimensions and physical properties of W shapes

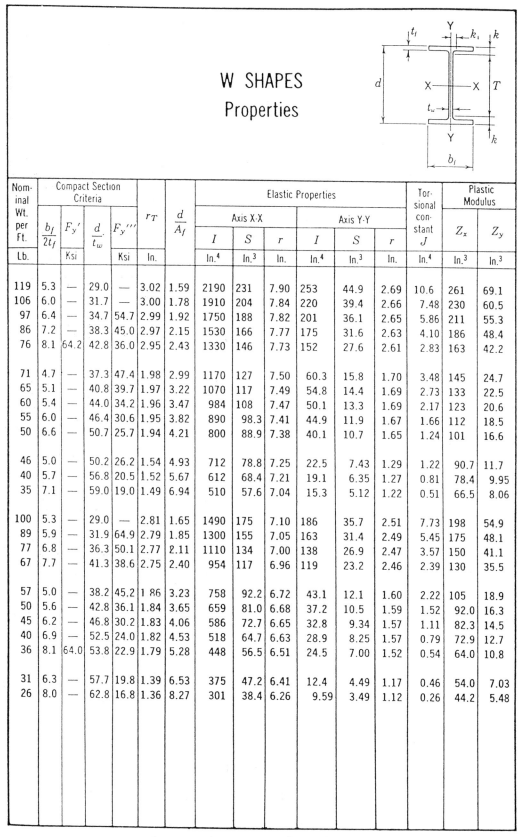

W SHAPES
Properties

Nominal Wt. per Ft.	Compact Section Criteria				r_T	$\dfrac{d}{A_f}$	Elastic Properties						Torsional constant	Plastic Modulus	
	$\dfrac{b_f}{2t_f}$	F_y'	$\dfrac{d}{t_w}$	F_y'''			Axis X-X			Axis Y-Y				Z_x	Z_y
							I	S	r	I	S	r	J		
Lb.		Ksi		Ksi	In.		In.⁴	In.³	In.	In.⁴	In.³	In.	In.⁴	In.³	In.³
119	5.3	—	29.0	—	3.02	1.59	2190	231	7.90	253	44.9	2.69	10.6	261	69.1
106	6.0	—	31.7	—	3.00	1.78	1910	204	7.84	220	39.4	2.66	7.48	230	60.5
97	6.4	—	34.7	54.7	2.99	1.92	1750	188	7.82	201	36.1	2.65	5.86	211	55.3
86	7.2	—	38.3	45.0	2.97	2.15	1530	166	7.77	175	31.6	2.63	4.10	186	48.4
76	8.1	64.2	42.8	36.0	2.95	2.43	1330	146	7.73	152	27.6	2.61	2.83	163	42.2
71	4.7	—	37.3	47.4	1.98	2.99	1170	127	7.50	60.3	15.8	1.70	3.48	145	24.7
65	5.1	—	40.8	39.7	1.97	3.22	1070	117	7.49	54.8	14.4	1.69	2.73	133	22.5
60	5.4	—	44.0	34.2	1.96	3.47	984	108	7.47	50.1	13.3	1.69	2.17	123	20.6
55	6.0	—	46.4	30.6	1.95	3.82	890	98.3	7.41	44.9	11.9	1.67	1.66	112	18.5
50	6.6	—	50.7	25.7	1.94	4.21	800	88.9	7.38	40.1	10.7	1.65	1.24	101	16.6
46	5.0	—	50.2	26.2	1.54	4.93	712	78.8	7.25	22.5	7.43	1.29	1.22	90.7	11.7
40	5.7	—	56.8	20.5	1.52	5.67	612	68.4	7.21	19.1	6.35	1.27	0.81	78.4	9.95
35	7.1	—	59.0	19.0	1.49	6.94	510	57.6	7.04	15.3	5.12	1.22	0.51	66.5	8.06
100	5.3	—	29.0	—	2.81	1.65	1490	175	7.10	186	35.7	2.51	7.73	198	54.9
89	5.9	—	31.9	64.9	2.79	1.85	1300	155	7.05	163	31.4	2.49	5.45	175	48.1
77	6.8	—	36.3	50.1	2.77	2.11	1110	134	7.00	138	26.9	2.47	3.57	150	41.1
67	7.7	—	41.3	38.6	2.75	2.40	954	117	6.96	119	23.2	2.46	2.39	130	35.5
57	5.0	—	38.2	45.2	1 86	3.23	758	92.2	6.72	43.1	12.1	1.60	2.22	105	18.9
50	5.6	—	42.8	36.1	1.84	3.65	659	81.0	6.68	37.2	10.5	1.59	1.52	92.0	16.3
45	6.2	—	46.8	30.2	1.83	4.06	586	72.7	6.65	32.8	9.34	1.57	1.11	82.3	14.5
40	6.9	—	52.5	24.0	1.82	4.53	518	64.7	6.63	28.9	8.25	1.57	0.79	72.9	12.7
36	8.1	64.0	53.8	22.9	1.79	5.28	448	56.5	6.51	24.5	7.00	1.52	0.54	64.0	10.8
31	6.3	—	57.7	19.8	1.39	6.53	375	47.2	6.41	12.4	4.49	1.17	0.46	54.0	7.03
26	8.0	—	62.8	16.8	1.36	8.27	301	38.4	6.26	9.59	3.49	1.12	0.26	44.2	5.48

AMERICAN INSTITUTE OF STEEL CONSTRUCTION

Table S-1b Dimensions and physical properties of W shapes *(continued)*

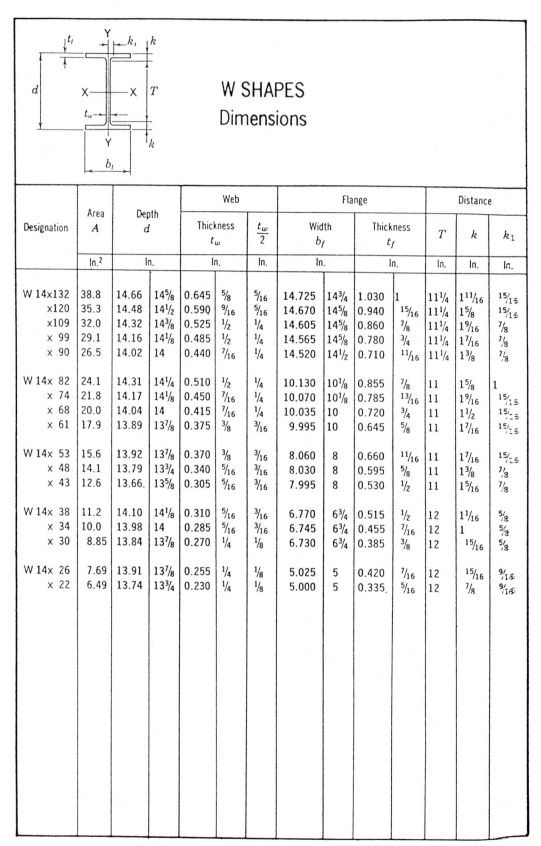

W SHAPES
Dimensions

Designation	Area A	Depth d		Web			Flange			Distance		
				Thickness t_w	$\dfrac{t_w}{2}$		Width b_f		Thickness t_f	T	k	k_1
	In.²	In.		In.	In.		In.		In.	In.	In.	In.
W 14x132	38.8	14.66	14⅝	0.645	⅝	⁵⁄₁₆	14.725	14¾	1.030	1	11¼	1¹¹⁄₁₆	¹⁵⁄₁₆
x120	35.3	14.48	14½	0.590	⁹⁄₁₆	⁵⁄₁₆	14.670	14⅝	0.940	¹⁵⁄₁₆	11¼	1⅝	¹⁵⁄₁₆
x109	32.0	14.32	14⅜	0.525	½	¼	14.605	14⅝	0.860	⅞	11¼	1⁹⁄₁₆	⅞
x 99	29.1	14.16	14⅛	0.485	½	¼	14.565	14⅝	0.780	¾	11¼	1⁷⁄₁₆	⅞
x 90	26.5	14.02	14	0.440	⁷⁄₁₆	¼	14.520	14½	0.710	¹¹⁄₁₆	11¼	1⅜	⅞
W 14x 82	24.1	14.31	14¼	0.510	½	¼	10.130	10⅛	0.855	⅞	11	1⅝	1
x 74	21.8	14.17	14⅛	0.450	⁷⁄₁₆	¼	10.070	10⅛	0.785	¹³⁄₁₆	11	1⁹⁄₁₆	¹⁵⁄₁₆
x 68	20.0	14.04	14	0.415	⁷⁄₁₆	¼	10.035	10	0.720	¾	11	1½	¹⁵⁄₁₆
x 61	17.9	13.89	13⅞	0.375	⅜	³⁄₁₆	9.995	10	0.645	⅝	11	1⁷⁄₁₆	¹⁵⁄₁₆
W 14x 53	15.6	13.92	13⅞	0.370	⅜	³⁄₁₆	8.060	8	0.660	¹¹⁄₁₆	11	1⁷⁄₁₆	¹⁵⁄₁₆
x 48	14.1	13.79	13¾	0.340	⁵⁄₁₆	³⁄₁₆	8.030	8	0.595	⅝	11	1⅜	⅞
x 43	12.6	13.66	13⅝	0.305	⁵⁄₁₆	³⁄₁₆	7.995	8	0.530	½	11	1⁵⁄₁₆	⅞
W 14x 38	11.2	14.10	14⅛	0.310	⁵⁄₁₆	³⁄₁₆	6.770	6¾	0.515	½	12	1¹⁄₁₆	⅝
x 34	10.0	13.98	14	0.285	⁵⁄₁₆	³⁄₁₆	6.745	6¾	0.455	⁷⁄₁₆	12	1	⅝
x 30	8.85	13.84	13⅞	0.270	¼	⅛	6.730	6¾	0.385	⅜	12	¹⁵⁄₁₆	⅝
W 14x 26	7.69	13.91	13⅞	0.255	¼	⅛	5.025	5	0.420	⁷⁄₁₆	12	¹⁵⁄₁₆	⁹⁄₁₆
x 22	6.49	13.74	13¾	0.230	¼	⅛	5.000	5	0.335	⁵⁄₁₆	12	⅞	⁹⁄₁₆

Table S-1c Dimensions and physical properties of W shapes

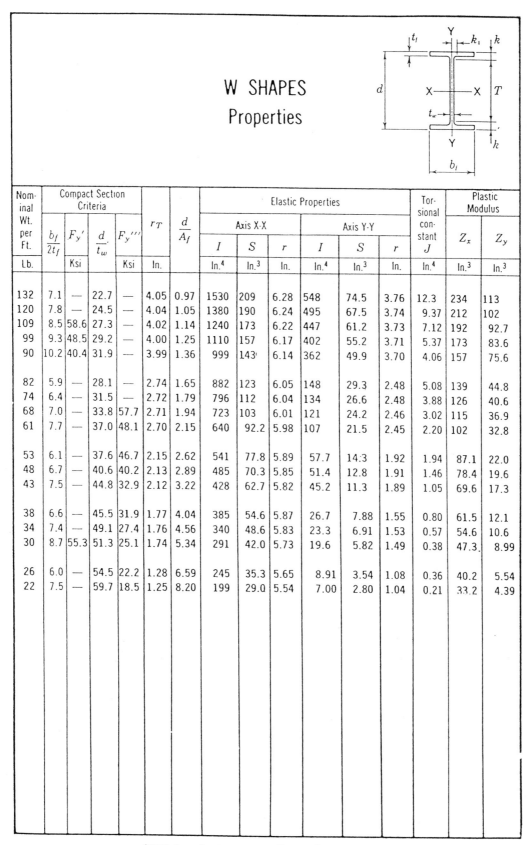

W SHAPES
Properties

Nominal Wt. per Ft.	Compact Section Criteria				r_T	$\dfrac{d}{A_f}$	Elastic Properties						Torsional constant	Plastic Modulus	
	$\dfrac{b_f}{2t_f}$	$F_y{}'$	$\dfrac{d}{t_w}$	$F_y{}'''$			Axis X-X			Axis Y-Y				Z_x	Z_y
							I	S	r	I	S	r	J		
Lb.		Ksi		Ksi	In.		In.⁴	In.³	In.	In.⁴	In.³	In.	In.⁴	In.³	In.³
132	7.1	—	22.7	—	4.05	0.97	1530	209	6.28	548	74.5	3.76	12.3	234	113
120	7.8	—	24.5	—	4.04	1.05	1380	190	6.24	495	67.5	3.74	9.37	212	102
109	8.5	58.6	27.3	—	4.02	1.14	1240	173	6.22	447	61.2	3.73	7.12	192	92.7
99	9.3	48.5	29.2	—	4.00	1.25	1110	157	6.17	402	55.2	3.71	5.37	173	83.6
90	10.2	40.4	31.9	—	3.99	1.36	999	143'	6.14	362	49.9	3.70	4.06	157	75.6
82	5.9	—	28.1	—	2.74	1.65	882	123	6.05	148	29.3	2.48	5.08	139	44.8
74	6.4	—	31.5	—	2.72	1.79	796	112	6.04	134	26.6	2.48	3.88	126	40.6
68	7.0	—	33.8	57.7	2.71	1.94	723	103	6.01	121	24.2	2.46	3.02	115	36.9
61	7.7	—	37.0	48.1	2.70	2.15	640	92.2	5.98	107	21.5	2.45	2.20	102	32.8
53	6.1	—	37.6	46.7	2.15	2.62	541	77.8	5.89	57.7	14.3	1.92	1.94	87.1	22.0
48	6.7	—	40.6	40.2	2.13	2.89	485	70.3	5.85	51.4	12.8	1.91	1.46	78.4	19.6
43	7.5	—	44.8	32.9	2.12	3.22	428	62.7	5.82	45.2	11.3	1.89	1.05	69.6	17.3
38	6.6	—	45.5	31.9	1.77	4.04	385	54.6	5.87	26.7	7.88	1.55	0.80	61.5	12.1
34	7.4	—	49.1	27.4	1.76	4.56	340	48.6	5.83	23.3	6.91	1.53	0.57	54.6	10.6
30	8.7	55.3	51.3	25.1	1.74	5.34	291	42.0	5.73	19.6	5.82	1.49	0.38	47.3	8.99
26	6.0	—	54.5	22.2	1.28	6.59	245	35.3	5.65	8.91	3.54	1.08	0.36	40.2	5.54
22	7.5	—	59.7	18.5	1.25	8.20	199	29.0	5.54	7.00	2.80	1.04	0.21	33.2	4.39

AMERICAN INSTITUTE OF STEEL CONSTRUCTION

Table S-1c Dimensions and physical properties of W shapes *(continued)*

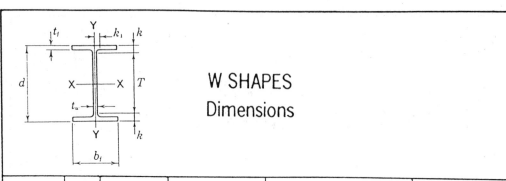

W SHAPES
Dimensions

Designation	Area A	Depth d		Web Thickness t_w		$\frac{t_w}{2}$	Flange Width b_f		Flange Thickness t_f		T	k	k_1
	In.²	In.		In.		In.	In.		In.		In.	In.	In.
W 12×336	98.8	16.82	$16\frac{7}{8}$	1.775	$1\frac{3}{4}$	$\frac{7}{8}$	13.385	$13\frac{3}{8}$	2.955	$2\frac{15}{16}$	$9\frac{1}{2}$	$3\frac{11}{16}$	$1\frac{1}{2}$
×305	89.6	16.32	$16\frac{3}{8}$	1.625	$1\frac{5}{8}$	$\frac{13}{16}$	13.235	$13\frac{1}{4}$	2.705	$2\frac{11}{16}$	$9\frac{1}{2}$	$3\frac{7}{16}$	$1\frac{7}{16}$
×279	81.9	15.85	$15\frac{7}{8}$	1.530	$1\frac{1}{2}$	$\frac{3}{4}$	13.140	$13\frac{1}{8}$	2.470	$2\frac{1}{2}$	$9\frac{1}{2}$	$3\frac{3}{16}$	$1\frac{3}{8}$
×252	74.1	15.41	$15\frac{3}{8}$	1.395	$1\frac{3}{8}$	$\frac{11}{16}$	13.005	13	2.250	$2\frac{1}{4}$	$9\frac{1}{2}$	$2\frac{15}{16}$	$1\frac{5}{16}$
×230	67.7	15.05	15	1.285	$1\frac{5}{16}$	$\frac{11}{16}$	12.895	$12\frac{7}{8}$	2.070	$2\frac{1}{16}$	$9\frac{1}{2}$	$2\frac{3}{4}$	$1\frac{1}{4}$
×210	61.8	14.71	$14\frac{3}{4}$	1.180	$1\frac{3}{16}$	$\frac{5}{8}$	12.790	$12\frac{3}{4}$	1.900	$1\frac{7}{8}$	$9\frac{1}{2}$	$2\frac{5}{8}$	$1\frac{1}{4}$
×190	55.8	14.38	$14\frac{3}{8}$	1.060	$1\frac{1}{16}$	$\frac{9}{16}$	12.670	$12\frac{5}{8}$	1.735	$1\frac{3}{4}$	$9\frac{1}{2}$	$2\frac{7}{16}$	$1\frac{3}{16}$
×170	50.0	14.03	14	0.960	$\frac{15}{16}$	$\frac{1}{2}$	12.570	$12\frac{5}{8}$	1.560	$1\frac{9}{16}$	$9\frac{1}{2}$	$2\frac{1}{4}$	$1\frac{1}{8}$
×152	44.7	13.71	$13\frac{3}{4}$	0.870	$\frac{7}{8}$	$\frac{7}{16}$	12.480	$12\frac{1}{2}$	1.400	$1\frac{3}{8}$	$9\frac{1}{2}$	$2\frac{1}{8}$	$1\frac{1}{16}$
×136	39.9	13.41	$13\frac{3}{8}$	0.790	$\frac{13}{16}$	$\frac{7}{16}$	12.400	$12\frac{3}{8}$	1.250	$1\frac{1}{4}$	$9\frac{1}{2}$	$1\frac{15}{16}$	1
×120	35.3	13.12	$13\frac{1}{8}$	0.710	$\frac{11}{16}$	$\frac{3}{8}$	12.320	$12\frac{3}{8}$	1.105	$1\frac{1}{8}$	$9\frac{1}{2}$	$1\frac{13}{16}$	1
×106	31.2	12.89	$12\frac{7}{8}$	0.610	$\frac{5}{8}$	$\frac{5}{16}$	12.220	$12\frac{1}{4}$	0.990	1	$9\frac{1}{2}$	$1\frac{11}{16}$	$\frac{15}{16}$
× 96	28.2	12.71	$12\frac{3}{4}$	0.550	$\frac{9}{16}$	$\frac{5}{16}$	12.160	$12\frac{1}{8}$	0.900	$\frac{7}{8}$	$9\frac{1}{2}$	$1\frac{5}{8}$	$\frac{7}{8}$
× 87	25.6	12.53	$12\frac{1}{2}$	0.515	$\frac{1}{2}$	$\frac{1}{4}$	12.125	$12\frac{1}{8}$	0.810	$\frac{13}{16}$	$9\frac{1}{2}$	$1\frac{1}{2}$	$\frac{7}{8}$
× 79	23.2	12.38	$12\frac{3}{8}$	0.470	$\frac{1}{2}$	$\frac{1}{4}$	12.080	$12\frac{1}{8}$	0.735	$\frac{3}{4}$	$9\frac{1}{2}$	$1\frac{7}{16}$	$\frac{7}{8}$
× 72	21.1	12.25	$12\frac{1}{4}$	0.430	$\frac{7}{16}$	$\frac{1}{4}$	12.040	12	0.670	$\frac{11}{16}$	$9\frac{1}{2}$	$1\frac{3}{8}$	$\frac{7}{8}$
× 65	19.1	12.12	$12\frac{1}{8}$	0.390	$\frac{3}{8}$	$\frac{3}{16}$	12.000	12	0.605	$\frac{5}{8}$	$9\frac{1}{2}$	$1\frac{5}{16}$	$\frac{13}{16}$
W 12× 58	17.0	12.19	$12\frac{1}{4}$	0.360	$\frac{3}{8}$	$\frac{3}{16}$	10.010	10	0.640	$\frac{5}{8}$	$9\frac{1}{2}$	$1\frac{3}{8}$	$\frac{13}{16}$
× 53	15.6	12.06	12	0.345	$\frac{3}{8}$	$\frac{3}{16}$	9.995	10	0.575	$\frac{9}{16}$	$9\frac{1}{2}$	$1\frac{1}{4}$	$\frac{13}{16}$
W 12× 50	14.7	12.19	$12\frac{1}{4}$	0.370	$\frac{3}{8}$	$\frac{3}{16}$	8.080	$8\frac{1}{8}$	0.640	$\frac{5}{8}$	$9\frac{1}{2}$	$1\frac{3}{8}$	$\frac{13}{16}$
× 45	13.2	12.06	12	0.335	$\frac{5}{16}$	$\frac{3}{16}$	8.045	8	0.575	$\frac{9}{16}$	$9\frac{1}{2}$	$1\frac{1}{4}$	$\frac{13}{16}$
× 40	11.8	11.94	12	0.295	$\frac{5}{16}$	$\frac{3}{16}$	8.005	8	0.515	$\frac{1}{2}$	$9\frac{1}{2}$	$1\frac{1}{4}$	$\frac{3}{4}$
W 12× 35	10.3	12.50	$12\frac{1}{2}$	0.300	$\frac{5}{16}$	$\frac{3}{16}$	6.560	$6\frac{1}{2}$	0.520	$\frac{1}{2}$	$10\frac{1}{2}$	1	$\frac{9}{16}$
× 30	8.79	12.34	$12\frac{3}{8}$	0.260	$\frac{1}{4}$	$\frac{1}{8}$	6.520	$6\frac{1}{2}$	0.440	$\frac{7}{16}$	$10\frac{1}{2}$	$\frac{15}{16}$	$\frac{1}{2}$
× 26	7.65	12.22	$12\frac{1}{4}$	0.230	$\frac{1}{4}$	$\frac{1}{8}$	6.490	$6\frac{1}{2}$	0.380	$\frac{3}{8}$	$10\frac{1}{2}$	$\frac{7}{8}$	$\frac{1}{2}$
W 12× 22	6.48	12.31	$12\frac{1}{4}$	0.260	$\frac{1}{4}$	$\frac{1}{8}$	4.030	4	0.425	$\frac{7}{16}$	$10\frac{1}{2}$	$\frac{7}{8}$	$\frac{1}{2}$
× 19	5.57	12.16	$12\frac{1}{8}$	0.235	$\frac{1}{4}$	$\frac{1}{8}$	4.005	4	0.350	$\frac{3}{8}$	$10\frac{1}{2}$	$\frac{13}{16}$	$\frac{1}{2}$
× 16	4.71	11.99	12	0.220	$\frac{1}{4}$	$\frac{1}{8}$	3.990	4	0.265	$\frac{1}{4}$	$10\frac{1}{2}$	$\frac{3}{4}$	$\frac{1}{2}$
× 14	4.16	11.91	$11\frac{7}{8}$	0.200	$\frac{3}{16}$	$\frac{1}{8}$	3.970	4	0.225	$\frac{1}{4}$	$10\frac{1}{2}$	$\frac{11}{16}$	$\frac{1}{2}$

AMERICAN INSTITUTE OF STEEL CONSTRUCTION

Table S-1d Dimensions and physical properties of W shapes

W SHAPES
Properties

Nominal Wt. per Ft.	Compact Section Criteria				r_T	$\dfrac{d}{A_f}$	Elastic Properties						Torsional constant	Plastic Modulus	
							Axis X-X			Axis Y-Y					
	$\dfrac{b_f}{2t_f}$	$F_y{}'$	$\dfrac{d}{t_w}$	$F_y{}'''$			I	S	r	I	S	r	J	Z_x	Z_y
Lb.		Ksi		Ksi	In.		In.^{4}	In.^{3}	In.	In.^{4}	In.^{3}	In.	In.^{4}	In.^{3}	In.^{3}
336	2.3	—	9.5	—	3.71	0.43	4060	483	6.41	1190	177	3.47	243	603	274
305	2.4	—	10.0	—	3.67	0.46	3550	435	6.29	1050	159	3.42	185	537	244
279	2.7	—	10.4	—	3.64	0.49	3110	393	6.16	937	143	3.38	143	481	220
252	2.9	—	11.0	—	3.59	0.53	2720	353	6.06	828	127	3.34	108	428	196
230	3.1	—	11.7	—	3.56	0.56	2420	321	5.97	742	115	3.31	83.8	386	177
210	3.4	—	12.5	—	3.53	0.61	2140	292	5.89	664	104	3.28	64.7	348	159
190	3.7	—	13.6	—	3.50	0.65	1890	263	5.82	589	93.0	3.25	48.8	311	143
170	4.0	—	14.6	—	3.47	0.72	1650	235	5.74	517	82.3	3.22	35.6	275	126
152	4.5	—	15.8	—	3.44	0.79	1430	209	5.66	454	72.8	3.19	25.8	243	111
136	5.0	—	17.0	—	3.41	0.87	1240	186	5.58	398	64.2	3.16	18.5	214	98.0
120	5.6	—	18.5	—	3.38	0.96	1070	163	5.51	345	56.0	3.13	12.9	186	85.4
106	6.2	—	21.1	—	3.36	1.07	933	145	5.47	301	49.3	3.11	9.13	164	75.1
96	6.8	—	23.1	—	3.34	1.16	833	131	5.44	270	44.4	3.09	6.86	147	67.5
87	7.5	—	24.3	—	3.32	1.28	740	118	5.38	241	39.7	3.07	5.10	132	60.4
79	8.2	62.6	26.3	—	3.31	1.39	662	107	5.34	216	35.8	3.05	3.84	119	54.3
72	9.0	52.3	28.5	—	3.29	1.52	597	97.4	5.31	195	32.4	3.04	2.93	108	49.2
65	9.9	43.0	31.1	—	3.28	1.67	533	87.9	5.28	174	29.1	3.02	2.18	96.8	44.1
58	7.8	—	33.9	57.6	2.72	1.90	475	78.0	5.28	107	21.4	2.51	2.10	86.4	32.5
53	8.7	55.9	35.0	54.1	2.71	2.10	425	70.6	5.23	95.8	19.2	2.48	1.58	77.9	29.1
50	6.3	—	32.9	60.9	2.17	2.36	394	64.7	5.18	56.3	13.9	1.96	1.78	72.4	21.4
45	7.0	—	36.0	51.0	2.15	2.61	350	58.1	5.15	50.0	12.4	1.94	1.31	64.7	19.0
40	7.8	—	40.5	40.3	2.14	2.90	310	51.9	5.13	44.1	11.0	1.93	0.95	57.5	16.8
35	6.3	—	41.7	38.0	1.74	3.66	285	45.6	5.25	24.5	7.47	1.54	0.74	51.2	11.5
30	7.4	—	47.5	29.3	1.73	4.30	238	38.6	5.21	20.3	6.24	1.52	0.46	43.1	9.56
26	8.5	57.9	53.1	23.4	1.72	4.95	204	33.4	5.17	17.3	5.34	1.51	0.30	37.2	8.17
22	4.7	—	47.3	29.5	1.02	7.19	156	25.4	4.91	4.66	2.31	0.847	0.29	29.3	3.66
19	5.7	—	51.7	24.7	1.00	8.67	130	21.3	4.82	3.76	1.88	0.822	0.18	24.7	2.98
16	7.5	—	54.5	22.2	0.96	11.3	103	17.1	4.67	2.82	1.41	0.773	0.10	20.1	2.26
14	8.8	54.3	59.6	18.6	0.95	13.3	88.6	14.9	4.62	2.36	1.19	0.753	0.07	17.4	1.90

AMERICAN INSTITUTE OF STEEL CONSTRUCTION

Table S-1d Dimensions and physical properties of W shapes *(continued)*

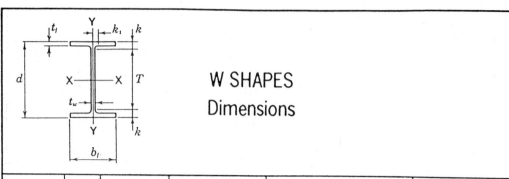

W SHAPES
Dimensions

Designation	Area A	Depth d		Web Thickness t_w		$\dfrac{t_w}{2}$	Flange Width b_f		Flange Thickness t_f		T	k	k_1
	In.²	In.		In.		In.	In.		In.		In.	In.	In.
W 10x112	32.9	11.36	11³⁄₈	0.755	³⁄₄	³⁄₈	10.415	10³⁄₈	1.250	1¹⁄₄	7⁵⁄₈	1⁷⁄₈	¹⁵⁄₁₆
x100	29.4	11.10	11¹⁄₈	0.680	¹¹⁄₁₆	³⁄₈	10.340	10³⁄₈	1.120	1¹⁄₈	7⁵⁄₈	1³⁄₄	⁷⁄₈
x 88	25.9	10.84	10⁷⁄₈	0.605	⁵⁄₈	⁵⁄₁₆	10.265	10¹⁄₄	0.990	1	7⁵⁄₈	1⁵⁄₈	¹³⁄₁₆
x 77	22.6	10.60	10⁵⁄₈	0.530	¹⁄₂	¹⁄₄	10.190	10¹⁄₄	0.870	⁷⁄₈	7⁵⁄₈	1¹⁄₂	¹³⁄₁₆
x 68	20.0	10.40	10³⁄₈	0.470	¹⁄₂	¹⁄₄	10.130	10¹⁄₈	0.770	³⁄₄	7⁵⁄₈	1³⁄₈	³⁄₄
x 60	17.6	10.22	10¹⁄₄	0.420	⁷⁄₁₆	¹⁄₄	10.080	10¹⁄₈	0.680	¹¹⁄₁₆	7⁵⁄₈	1⁵⁄₁₆	³⁄₄
x 54	15.8	10.09	10¹⁄₈	0.370	³⁄₈	³⁄₁₆	10.030	10	0.615	⁵⁄₈	7⁵⁄₈	1¹⁄₄	¹¹⁄₁₆
x 49	14.4	9.98	10	0.340	⁵⁄₁₆	³⁄₁₆	10.000	10	0.560	⁹⁄₁₆	7⁵⁄₈	1³⁄₁₆	¹¹⁄₁₆
W 10x 45	13.3	10.10	10¹⁄₈	0.350	³⁄₈	³⁄₁₆	8.020	8	0.620	⁵⁄₈	7⁵⁄₈	1¹⁄₄	¹¹⁄₁₆
x 39	11.5	9.92	9⁷⁄₈	0.315	⁵⁄₁₆	³⁄₁₆	7.985	8	0.530	¹⁄₂	7⁵⁄₈	1¹⁄₈	¹¹⁄₁₆
x 33	9.71	9.73	9³⁄₄	0.290	⁵⁄₁₆	³⁄₁₆	7.960	8	0.435	⁷⁄₁₆	7⁵⁄₈	1¹⁄₁₆	¹¹⁄₁₆
W 10x 30	8.84	10.47	10¹⁄₂	0.300	⁵⁄₁₆	³⁄₁₆	5.810	5³⁄₄	0.510	¹⁄₂	8⁵⁄₈	¹⁵⁄₁₆	¹⁄₂
x 26	7.61	10.33	10³⁄₈	0.260	¹⁄₄	¹⁄₈	5.770	5³⁄₄	0.440	⁷⁄₁₆	8⁵⁄₈	⁷⁄₈	¹⁄₂
x 22	6.49	10.17	10¹⁄₈	0.240	¹⁄₄	¹⁄₈	5.750	5³⁄₄	0.360	³⁄₈	8⁵⁄₈	³⁄₄	¹⁄₂
W 10x 19	5.62	10.24	10¹⁄₄	0.250	¹⁄₄	¹⁄₈	4.020	4	0.395	³⁄₈	8⁵⁄₈	¹³⁄₁₆	¹⁄₂
x 17	4.99	10.11	10¹⁄₈	0.240	¹⁄₄	¹⁄₈	4.010	4	0.330	⁵⁄₁₆	8⁵⁄₈	³⁄₄	¹⁄₂
x 15	4.41	9.99	10	0.230	¹⁄₄	¹⁄₈	4.000	4	0.270	¹⁄₄	8⁵⁄₈	¹¹⁄₁₆	⁷⁄₁₆
x 12	3.54	9.87	9⁷⁄₈	0.190	³⁄₁₆	¹⁄₈	3.960	4	0.210	³⁄₁₆	8⁵⁄₈	⁵⁄₈	⁷⁄₁₆

AMERICAN INSTITUTE OF STEEL CONSTRUCTION

Table S-1e Dimensions and physical properties of W shapes

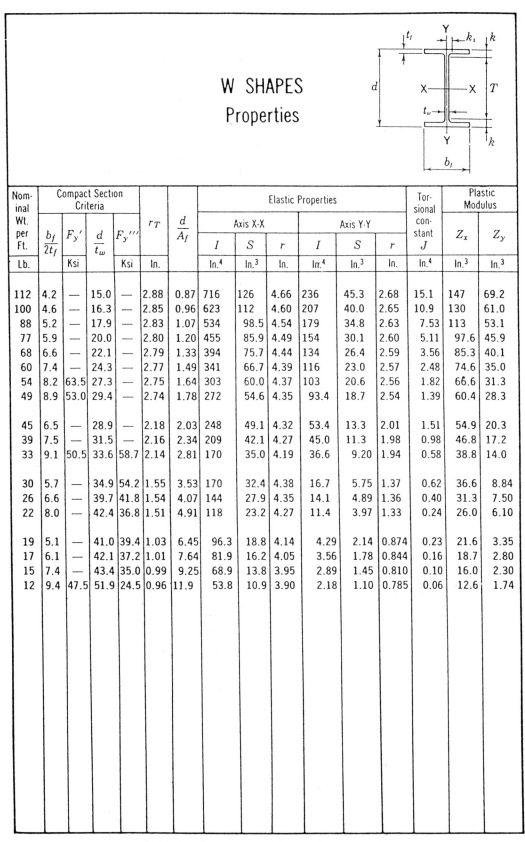

W SHAPES
Properties

Nominal Wt. per Ft.	Compact Section Criteria				r_T	$\dfrac{d}{A_f}$	Elastic Properties						Torsional constant	Plastic Modulus	
							Axis X-X			Axis Y-Y					
	$\dfrac{b_f}{2t_f}$	F_y'	$\dfrac{d}{t_w}$	F_y'''			I	S	r	I	S	r	J	Z_x	Z_y
Lb.		Ksi		Ksi	In.		In.⁴	In.³	In.	In.⁴	In.³	In.	In.⁴	In.³	In.³
112	4.2	—	15.0	—	2.88	0.87	716	126	4.66	236	45.3	2.68	15.1	147	69.2
100	4.6	—	16.3	—	2.85	0.96	623	112	4.60	207	40.0	2.65	10.9	130	61.0
88	5.2	—	17.9	—	2.83	1.07	534	98.5	4.54	179	34.8	2.63	7.53	113	53.1
77	5.9	—	20.0	—	2.80	1.20	455	85.9	4.49	154	30.1	2.60	5.11	97.6	45.9
68	6.6	—	22.1	—	2.79	1.33	394	75.7	4.44	134	26.4	2.59	3.56	85.3	40.1
60	7.4	—	24.3	—	2.77	1.49	341	66.7	4.39	116	23.0	2.57	2.48	74.6	35.0
54	8.2	63.5	27.3	—	2.75	1.64	303	60.0	4.37	103	20.6	2.56	1.82	66.6	31.3
49	8.9	53.0	29.4	—	2.74	1.78	272	54.6	4.35	93.4	18.7	2.54	1.39	60.4	28.3
45	6.5	—	28.9	—	2.18	2.03	248	49.1	4.32	53.4	13.3	2.01	1.51	54.9	20.3
39	7.5	—	31.5	—	2.16	2.34	209	42.1	4.27	45.0	11.3	1.98	0.98	46.8	17.2
33	9.1	50.5	33.6	58.7	2.14	2.81	170	35.0	4.19	36.6	9.20	1.94	0.58	38.8	14.0
30	5.7	—	34.9	54.2	1.55	3.53	170	32.4	4.38	16.7	5.75	1.37	0.62	36.6	8.84
26	6.6	—	39.7	41.8	1.54	4.07	144	27.9	4.35	14.1	4.89	1.36	0.40	31.3	7.50
22	8.0	—	42.4	36.8	1.51	4.91	118	23.2	4.27	11.4	3.97	1.33	0.24	26.0	6.10
19	5.1	—	41.0	39.4	1.03	6.45	96.3	18.8	4.14	4.29	2.14	0.874	0.23	21.6	3.35
17	6.1	—	42.1	37.2	1.01	7.64	81.9	16.2	4.05	3.56	1.78	0.844	0.16	18.7	2.80
15	7.4	—	43.4	35.0	0.99	9.25	68.9	13.8	3.95	2.89	1.45	0.810	0.10	16.0	2.30
12	9.4	47.5	51.9	24.5	0.96	11.9	53.8	10.9	3.90	2.18	1.10	0.785	0.06	12.6	1.74

AMERICAN INSTITUTE OF STEEL CONSTRUCTION

Table S-1e Dimensions and physical properties of W shapes *(continued)*

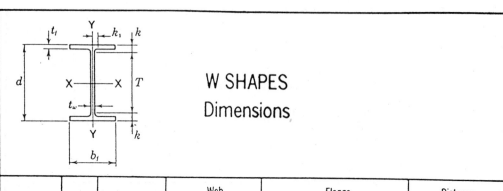

W SHAPES
Dimensions

Designation	Area A	Depth d		Web			Flange				Distance		
				Thickness t_w		$\frac{t_w}{2}$	Width b_f		Thickness t_f		T	k	k_1
	In.²	In.		In.		In.	In.		In.		In.	In.	In.
W 8×67	19.7	9.00	9	0.570	9/16	5/16	8.280	8¼	0.935	15/16	6⅛	1 7/16	11/16
×58	17.1	8.75	8¾	0.510	½	¼	8.220	8¼	0.810	13/16	6⅛	1 5/16	11/16
×48	14.1	8.50	8½	0.400	⅜	3/16	8.110	8⅛	0.685	11/16	6⅛	1 3/16	⅝
×40	11.7	8.25	8¼	0.360	⅜	3/16	8.070	8⅛	0.560	9/16	6⅛	1 1/16	⅝
×35	10.3	8.12	8⅛	0.310	5/16	3/16	8.020	8	0.495	½	6⅛	1	9/16
×31	9.13	8.00	8	0.285	5/16	3/16	7.995	8	0.435	7/16	6⅛	15/16	9/16
W 8×28	8.25	8.06	8	0.285	5/16	3/16	6.535	6½	0.465	7/16	6⅛	15/16	9/16
×24	7.08	7.93	7⅞	0.245	¼	⅛	6.495	6½	0.400	⅜	6⅛	⅞	9/16
W 8×21	6.16	8.28	8¼	0.250	¼	⅛	5.270	5¼	0.400	⅜	6⅝	13/16	½
×18	5.26	8.14	8⅛	0.230	¼	⅛	5.250	5¼	0.330	5/16	6⅝	¾	7/16
W 8×15	4.44	8.11	8⅛	0.245	¼	⅛	4.015	4	0.315	5/16	6⅝	¾	½
×13	3.84	7.99	8	0.230	¼	⅛	4.000	4	0.255	¼	6⅝	11/16	7/16
×10	2.96	7.89	7⅞	0.170	3/16	⅛	3.940	4	0.205	3/16	6⅝	⅝	7/16
W 6×25	7.34	6.38	6⅜	0.320	5/16	3/16	6.080	6⅛	0.455	7/16	4¾	13/16	7/16
×20	5.87	6.20	6¼	0.260	¼	⅛	6.020	6	0.365	⅜	4¾	¾	7/16
×15	4.43	5.99	6	0.230	¼	⅛	5.990	6	0.260	¼	4¾	⅝	⅜
W 6×16	4.74	6.28	6¼	0.260	¼	⅛	4.030	4	0.405	⅜	4¾	¾	7/16
×12	3.55	6.03	6	0.230	¼	⅛	4.000	4	0.280	¼	4¾	⅝	⅜
× 9	2.68	5.90	5⅞	0.170	3/16	⅛	3.940	4	0.215	3/16	4¾	9/16	⅜
W 5×19	5.54	5.15	5⅛	0.270	¼	⅛	5.030	5	0.430	7/16	3½	13/16	7/16
×16	4.68	5.01	5	0.240	¼	⅛	5.000	5	0.360	⅜	3½	¾	7/16
W 4×13	3.83	4.16	4⅛	0.280	¼	⅛	4.060	4	0.345	⅜	2¾	11/16	7/16

AMERICAN INSTITUTE OF STEEL CONSTRUCTION

Table S-1f Dimensions and physical properties of W shapes

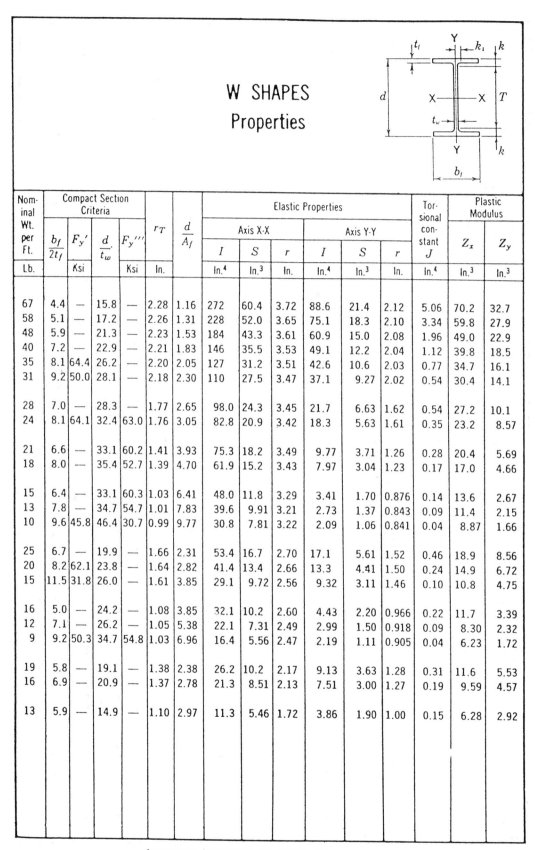

W SHAPES
Properties

Nom- inal Wt. per Ft.	Compact Section Criteria				r_T	$\dfrac{d}{A_f}$	Elastic Properties						Tor- sional con- stant J	Plastic Modulus	
	$\dfrac{b_f}{2t_f}$	F_y'	$\dfrac{d}{t_w}$	F_y'''			Axis X-X			Axis Y-Y				Z_x	Z_y
							I	S	r	I	S	r			
Lb.		Ksi		Ksi	In.		In.⁴	In.³	In.	In.⁴	In.³	In.	In.⁴	In.³	In.³
67	4.4	—	15.8	—	2.28	1.16	272	60.4	3.72	88.6	21.4	2.12	5.06	70.2	32.7
58	5.1	—	17.2	—	2.26	1.31	228	52.0	3.65	75.1	18.3	2.10	3.34	59.8	27.9
48	5.9	—	21.3	—	2.23	1.53	184	43.3	3.61	60.9	15.0	2.08	1.96	49.0	22.9
40	7.2	—	22.9	—	2.21	1.83	146	35.5	3.53	49.1	12.2	2.04	1.12	39.8	18.5
35	8.1	64.4	26.2	—	2.20	2.05	127	31.2	3.51	42.6	10.6	2.03	0.77	34.7	16.1
31	9.2	50.0	28.1	—	2.18	2.30	110	27.5	3.47	37.1	9.27	2.02	0.54	30.4	14.1
28	7.0	—	28.3	—	1.77	2.65	98.0	24.3	3.45	21.7	6.63	1.62	0.54	27.2	10.1
24	8.1	64.1	32.4	63.0	1.76	3.05	82.8	20.9	3.42	18.3	5.63	1.61	0.35	23.2	8.57
21	6.6	—	33.1	60.2	1.41	3.93	75.3	18.2	3.49	9.77	3.71	1.26	0.28	20.4	5.69
18	8.0	—	35.4	52.7	1.39	4.70	61.9	15.2	3.43	7.97	3.04	1.23	0.17	17.0	4.66
15	6.4	—	33.1	60.3	1.03	6.41	48.0	11.8	3.29	3.41	1.70	0.876	0.14	13.6	2.67
13	7.8	—	34.7	54.7	1.01	7.83	39.6	9.91	3.21	2.73	1.37	0.843	0.09	11.4	2.15
10	9.6	45.8	46.4	30.7	0.99	9.77	30.8	7.81	3.22	2.09	1.06	0.841	0.04	8.87	1.66
25	6.7	—	19.9	—	1.66	2.31	53.4	16.7	2.70	17.1	5.61	1.52	0.46	18.9	8.56
20	8.2	62.1	23.8	—	1.64	2.82	41.4	13.4	2.66	13.3	4.41	1.50	0.24	14.9	6.72
15	11.5	31.8	26.0	—	1.61	3.85	29.1	9.72	2.56	9.32	3.11	1.46	0.10	10.8	4.75
16	5.0	—	24.2	—	1.08	3.85	32.1	10.2	2.60	4.43	2.20	0.966	0.22	11.7	3.39
12	7.1	—	26.2	—	1.05	5.38	22.1	7.31	2.49	2.99	1.50	0.918	0.09	8.30	2.32
9	9.2	50.3	34.7	54.8	1.03	6.96	16.4	5.56	2.47	2.19	1.11	0.905	0.04	6.23	1.72
19	5.8	—	19.1	—	1.38	2.38	26.2	10.2	2.17	9.13	3.63	1.28	0.31	11.6	5.53
16	6.9	—	20.9	—	1.37	2.78	21.3	8.51	2.13	7.51	3.00	1.27	0.19	9.59	4.57
13	5.9	—	14.9	—	1.10	2.97	11.3	5.46	1.72	3.86	1.90	1.00	0.15	6.28	2.92

AMERICAN INSTITUTE OF STEEL CONSTRUCTION

Table S-1f Dimensions and physical properties of W shapes *(continued)*

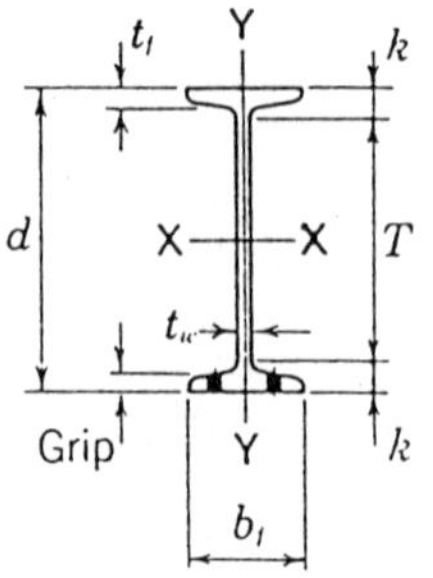

S SHAPES
Dimensions

Designation	Area A	Depth d		Web Thickness t_w		Web $\frac{t_w}{2}$	Flange Width b_f		Flange Thickness t_f		Distance T	Distance k	Grip	Max. Flge. Fas-ten-er
	In.²	In.		In.		In.	In.		In.		In.	In.	In.	In.
S 24x121	35.6	24.50	24 1/2	0.800	13/16	7/16	8.050	8	1.090	1 1/16	20 1/2	2	1 1/8	1
x106	31.2	24.50	24 1/2	0.620	5/8	5/16	7.870	7 7/8	1.090	1 1/16	20 1/2	2	1 1/8	1
S 24x100	29.3	24.00	24	0.745	3/4	3/8	7.245	7 1/4	0.870	7/8	20 1/2	1 3/4	7/8	1
x90	26.5	24.00	24	0.625	5/8	5/16	7.125	7 1/8	0.870	7/8	20 1/2	1 3/4	7/8	1
x80	23.5	24.00	24	0.500	1/2	1/4	7.000	7	0.870	7/8	20 1/2	1 3/4	7/8	1
S 20x96	28.2	20.30	20 1/4	0.800	13/16	7/16	7.200	7 1/4	0.920	15/16	16 3/4	1 3/4	15/16	1
x86	25.3	20.30	20 1/4	0.660	11/16	3/8	7.060	7	0.920	15/16	16 3/4	1 3/4	15/16	1
S 20x75	22.0	20.00	20	0.635	5/8	5/16	6.385	6 3/8	0.795	13/16	16 3/4	1 5/8	13/16	7/8
x66	19.4	20.00	20	0.505	1/2	1/4	6.255	6 1/4	0.795	13/16	16 3/4	1 5/8	13/16	7/8
S 18x70	20.6	18.00	18	0.711	11/16	3/8	6.251	6 1/4	0.691	11/16	15	1 1/2	11/16	7/8
x54.7	16.1	18.00	18	0.461	7/16	1/4	6.001	6	0.691	11/16	15	1 1/2	11/16	7/8
S 15x50	14.7	15.00	15	0.550	9/16	5/16	5.640	5 5/8	0.622	5/8	12 1/4	1 3/8	9/16	3/4
x42.9	12.6	15.00	15	0.411	7/16	1/4	5.501	5 1/2	0.622	5/8	12 1/4	1 3/8	9/16	3/4
S 12x50	14.7	12.00	12	0.687	11/16	3/8	5.477	5 1/2	0.659	11/16	9 1/8	1 7/16	11/16	3/4
x40.8	12.0	12.00	12	0.462	7/16	1/4	5.252	5 1/4	0.659	11/16	9 1/8	1 7/16	5/8	3/4
S 12x35	10.3	12.00	12	0.428	7/16	1/4	5.078	5 1/8	0.544	9/16	9 5/8	1 3/16	1/2	3/4
x31.8	9.35	12.00	12	0.350	3/8	3/16	5.000	5	0.544	9/16	9 5/8	1 3/16	1/2	3/4
S 10x35	10.3	10.00	10	0.594	5/8	5/16	4.944	5	0.491	1/2	7 3/4	1 1/8	1/2	3/4
x25.4	7.46	10.00	10	0.311	5/16	3/16	4.661	4 5/8	0.491	1/2	7 3/4	1 1/8	1/2	3/4
S 8x23	6.77	8.00	8	0.441	7/16	1/4	4.171	4 1/8	0.426	7/16	6	1	7/16	3/4
x18.4	5.41	8.00	8	0.271	1/4	1/8	4.001	4	0.426	7/16	6	1	7/16	3/4
S 7x20	5.88	7.00	7	0.450	7/16	1/4	3.860	3 7/8	0.392	3/8	5 1/8	15/16	3/8	5/8
x15.3	4.50	7.00	7	0.252	1/4	1/8	3.662	3 5/8	0.392	3/8	5 1/8	15/16	3/8	5/8
S 6x17.25	5.07	6.00	6	0.465	7/16	1/4	3.565	3 5/8	0.359	3/8	4 1/4	7/8	3/8	5/8
x12.5	3.67	6.00	6	0.232	1/4	1/8	3.332	3 3/8	0.359	3/8	4 1/4	7/8	3/8	—
S 5x14.75	4.34	5.00	5	0.494	1/2	1/4	3.284	3 1/4	0.326	5/16	3 3/8	13/16	5/16	—
x10	2.94	5.00	5	0.214	3/16	1/8	3.004	3	0.326	5/16	3 3/8	13/16	5/16	—
S 4x9.5	2.79	4.00	4	0.326	5/16	3/16	2.796	2 3/4	0.293	5/16	2 1/2	3/4	5/16	—
x7.7	2.26	4.00	4	0.193	3/16	1/8	2.663	2 5/8	0.293	5/16	2 1/2	3/4	5/16	—
S 3x7.5	2.21	3.00	3	0.349	3/8	3/16	2.509	2 1/2	0.260	1/4	1 5/8	11/16	1/4	—
x5.7	1.67	3.00	3	0.170	3/16	1/8	2.330	2 3/8	0.260	1/4	1 5/8	11/16	1/4	—

AMERICAN INSTITUTE OF STEEL CONSTRUCTION

Table S-2 Dimensions and physical properties of S shapes

S SHAPES
Properties

Nom-inal Wt. per Ft	Compact Section Criteria				r_T	$\dfrac{d}{A_f}$	Elastic Properties						Tor-sional con-stant J	Plastic Modulus	
	$\dfrac{b_f}{2t_f}$	$F_y{}'$	$\dfrac{d}{t_w}$	$F_y{}'''$			Axis X-X			Axis Y-Y				Z_x	Z_y
							I	S	r	I	S	r			
Lb.		Ksi		Ksi	In.		In.⁴	In.³	In.	In.⁴	In.³	In.	In.⁴	In.³	In.³
121	3.7	—	30.6	—	1.86	2.79	3160	258	9.43	83.3	20.7	1.53	12.8	306	36.2
106	3.6	—	39.5	42.3	1.86	2.86	2940	240	9.71	77.1	19.6	1.57	10.1	279	33.2
100	4.2	—	32.2	63.6	1.59	3.81	2390	199	9.02	47.7	13.2	1.27	7.58	240	23.9
90	4.1	—	38.4	44.8	1.60	3.87	2250	187	9.21	44.9	12.6	1.30	6.04	222	22.3
80	4.0	—	48.0	28.7	1.61	3.94	2100	175	9.47	42.2	12.1	1.34	4.88	204	20.7
96	3.9	—	25.4	—	1.63	3.06	1670	165	7.71	50.2	13.9	1.33	8.39	198	24.9
86	3.8	—	30.8	—	1.63	3.13	1580	155	7.89	46.8	13.3	1.36	6.64	183	23.0
75	4.0	—	31.5	—	1.43	3.94	1280	128	7.62	29.8	9.32	1.16	4.59	153	16.7
66	3.9	—	39.6	42.1	1.44	4.02	1190	119	7.83	27.7	8.85	1.19	3.58	140	15.3
70	4.5	—	25.3	—	1.36	4.17	926	103	6.71	24.1	7.72	1.08	4.15	125	14.4
54.7	4.3	—	39.0	43.3	1.37	4.34	804	89.4	7.07	20.8	6.94	1.14	2.37	105	12.1
50	4.5	—	27.3	—	1.26	4.28	486	64.8	5.75	15.7	5.57	1.03	2.12	77.1	9.97
42.9	4.4	—	36.5	49.6	1.26	4.38	447	59.6	5.95	14.4	5.23	1.07	1.54	69.3	9.02
50	4.2	—	17.5	—	1.25	3.32	305	50.8	4.55	15.7	5.74	1.03	2.82	61.2	10.3
40.8	4.0	—	26.0	—	1.24	3.46	272	45.4	4.77	13.6	5.16	1.06	1.76	53.1	8.85
35	4.7	—	28.0	—	1.16	4.34	229	38.2	4.72	9.87	3.89	0.980	1.08	44.8	6.79
31.8	4.6	—	34.3	56.2	1.16	4.41	218	36.4	4.83	9.36	3.74	1.00	0.90	42.0	6.40
35	5.0	—	16.8	—	1.10	4.12	147	29.4	3.78	8.36	3.38	0.901	1.29	35.4	6.22
25.4	4.7	—	32.2	63.9	1.09	4.37	124	24.7	4.07	6.79	2.91	0.954	0.60	28.4	4.96
23	4.9	—	18.1	—	0.95	4.51	64.9	16.2	3.10	4.31	2.07	0.798	0.55	19.3	3.68
18.4	4.7	—	29.5	—	0.94	4.70	57.6	14.4	3.26	3.73	1.86	0.831	0.34	16.5	3.16
20	4.9	—	15.6	—	0.88	4.63	42.4	12.1	2.69	3.17	1.64	0.734	0.45	14.5	2.96
15.3	4.7	—	27.8	—	0.87	4.88	36.7	10.5	2.86	2.64	1.44	0.766	0.24	12.1	2.44
17.25	5.0	—	12.9	—	0.81	4.69	26.3	8.77	2.28	2.31	1.30	0.675	0.37	10.6	2.36
12.5	4.6	—	25.9	—	0.79	5.02	22.1	7.37	2.45	1.82	1.09	0.705	0.17	8.47	1.85
14.75	5.0	—	10.1	—	0.74	4.66	15.2	6.09	1.87	1.67	1.01	0.620	0.32	7.42	1.88
10	4.6	—	23.4	—	0.72	5.10	12.3	4.92	2.05	1.22	0.809	0.643	0.11	5.67	1.37
9.5	4.8	—	12.3	—	0.65	4.88	6.79	3.39	1.56	0.903	0.646	0.569	0.12	4.04	1.13
7.7	4.5	—	20.7	—	0.64	5.13	6.08	3.04	1.64	0.764	0.574	0.581	0.07	3.51	0.964
7.5	4.8	—	8.6	—	0.59	4.60	2.93	1.95	1.15	0.586	0.468	0.516	0.09	2.36	0.826
5.7	4.5	—	17.6	—	0.57	4.95	2.52	1.68	1.23	0.455	0.390	0.522	0.04	1.95	0.653

AMERICAN INSTITUTE OF STEEL CONSTRUCTION

Table S-2 Dimensions and physical properties of S shapes *(continued)*

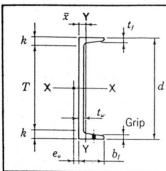

CHANNELS
AMERICAN STANDARD
Dimensions

Designation	Area A	Depth d	Web		Flange		Distance		Grip	Max. Flge. Fastener
			Thickness t_w	$\frac{t_w}{2}$	Width b_f	Average thickness t_f	T	k		
	In.²	In.	In.	In.	In.	In.	In.	In.	In.	In.
C 15x50	14.7	15.00	0.716	11/16 3/8	3.716 3 3/4	0.650 5/8	12 1/8	1 7/16	5/8	1
x40	11.8	15.00	0.520	1/2 1/4	3.520 3 1/2	0.650 5/8	12 1/8	1 7/16	5/8	1
x33.9	9.96	15.00	0.400	3/8 3/16	3.400 3 3/8	0.650 5/8	12 1/8	1 7/16	5/8	1
C 12x30	8.82	12.00	0.510	1/2 1/4	3.170 3 1/8	0.501 1/2	9 3/4	1 1/8	1/2	7/8
x25	7.35	12.00	0.387	3/8 3/16	3.047 3	0.501 1/2	9 3/4	1 1/8	1/2	7/8
x20.7	6.09	12.00	0.282	5/16 1/8	2.942 3	0.501 1/2	9 3/4	1 1/8	1/2	7/8
C 10x30	8.82	10.00	0.673	11/16 5/16	3.033 3	0.436 7/16	8	1	7/16	3/4
x25	7.35	10.00	0.526	1/2 1/4	2.886 2 7/8	0.436 7/16	8	1	7/16	3/4
x20	5.88	10.00	0.379	3/8 3/16	2.739 2 3/4	0.436 7/16	8	1	7/16	3/4
x15.3	4.49	10.00	0.240	1/4 1/8	2.600 2 5/8	0.436 7/16	8	1	7/16	3/4
C 9x20	5.88	9.00	0.448	7/16 1/4	2.648 2 5/8	0.413 7/16	7 1/8	15/16	7/16	3/4
x15	4.41	9.00	0.285	5/16 1/8	2.485 2 1/2	0.413 7/16	7 1/8	15/16	7/16	3/4
x13.4	3.94	9.00	0.233	1/4 1/8	2.433 2 3/8	0.413 7/16	7 1/8	15/16	7/16	3/4
C 8x18.75	5.51	8.00	0.487	1/2 1/4	2.527 2 1/2	0.390 3/8	6 1/8	15/16	3/8	3/4
x13.75	4.04	8.00	0.303	5/16 1/8	2.343 2 3/8	0.390 3/8	6 1/8	15/16	3/8	3/4
x11.5	3.38	8.00	0.220	1/4 1/8	2.260 2 1/4	0.390 3/8	6 1/8	15/16	3/8	3/4
C 7x14.75	4.33	7.00	0.419	7/16 3/16	2.299 2 1/4	0.366 3/8	5 1/4	7/8	3/8	5/8
x12.25	3.60	7.00	0.314	5/16 3/16	2.194 2 1/4	0.366 3/8	5 1/4	7/8	3/8	5/8
x 9.8	2.87	7.00	0.210	3/16 1/8	2.090 2 1/8	0.366 3/8	5 1/4	7/8	3/8	5/8
C 6x13	3.83	6.00	0.437	7/16 3/16	2.157 2 1/8	0.343 5/16	4 3/8	13/16	5/16	5/8
x10.5	3.09	6.00	0.314	5/16 3/16	2.034 2	0.343 5/16	4 3/8	13/16	3/8	5/8
x 8.2	2.40	6.00	0.200	3/16 1/8	1.920 1 7/8	0.343 5/16	4 3/8	13/16	5/16	5/8
C 5x 9	2.64	5.00	0.325	5/16 3/16	1.885 1 7/8	0.320 5/16	3 1/2	3/4	5/16	5/8
x 6.7	1.97	5.00	0.190	3/16 1/8	1.750 1 3/4	0.320 5/16	3 1/2	3/4	—	—
C 4x 7.25	2.13	4.00	0.321	5/16 3/16	1.721 1 3/4	0.296 5/16	2 5/8	11/16	5/16	5/8
x 5.4	1.59	4.00	0.184	3/16 1/16	1.584 1 5/8	0.296 5/16	2 5/8	11/16	—	—
C 3x 6	1.76	3.00	0.356	3/8 3/16	1.596 1 5/8	0.273 1/4	1 5/8	11/16	—	—
x 5	1.47	3.00	0.258	1/4 1/8	1.498 1 1/2	0.273 1/4	1 5/8	11/16	—	—
x 4.1	1.21	3.00	0.170	3/16 1/16	1.410 1 3/8	0.273 1/4	1 5/8	11/16	—	—

AMERICAN INSTITUTE OF STEEL CONSTRUCTION

Table S-3 Dimensions and physical properties of channels

CHANNELS
AMERICAN STANDARD
Properties

Nominal Weight per Ft.	$\bar{x}$	Shear Center Location e_o	$\frac{d}{A_f}$	Axis X-X			Axis Y-Y		
				I	S	r	I	S	r
	In.	In.		In.⁴	In.³	In.	In.⁴	In.³	In.
50	0.798	0.583	6.21	404	53.8	5.24	11.0	3.78	0.867
40	0.777	0.767	6.56	349	46.5	5.44	9.23	3.37	0.886
33.9	0.787	0.896	6.79	315	42.0	5.62	8.13	3.11	0.904
30	0.674	0.618	7.55	162	27.0	4.29	5.14	2.06	0.763
25	0.674	0.746	7.85	144	24.1	4.43	4.47	1.88	0.780
20.7	0.698	0.870	8.13	129	21.5	4.61	3.88	1.73	0.799
30	0.649	0.369	7.55	103	20.7	3.42	3.94	1.65	0.669
25	0.617	0.494	7.94	91.2	18.2	3.52	3.36	1.48	0.676
20	0.606	0.637	8.36	78.9	15.8	3.66	2.81	1.32	0.692
15.3	0.634	0.796	8.81	67.4	13.5	3.87	2.28	1.16	0.713
20	0.583	0.515	8.22	60.9	13.5	3.22	2.42	1.17	0.642
15	0.586	0.682	8.76	51.0	11.3	3.40	1.93	1.01	0.661
13.4	0.601	0.743	8.95	47.9	10.6	3.48	1.76	0.962	0.669
18.75	0.565	0.431	8.12	44.0	11.0	2.82	1.98	1.01	0.599
13.75	0.553	0.604	8.75	36.1	9.03	2.99	1.53	0.854	0.615
11.5	0.571	0.697	9.08	32.6	8.14	3.11	1.32	0.781	0.625
14.75	0.532	0.441	8.31	27.2	7.78	2.51	1.38	0.779	0.564
12.25	0.525	0.538	8.71	24.2	6.93	2.60	1.17	0.703	0.571
9.8	0.540	0.647	9.14	21.3	6.08	2.72	0.968	0.625	0.581
13	0.514	0.380	8.10	17.4	5.80	2.13	1.05	0.642	0.525
10.5	0.499	0.486	8.59	15.2	5.06	2.22	0.866	0.564	0.529
8.2	0.511	0.599	9.10	13.1	4.38	2.34	0.693	0.492	0.537
9	0.478	0.427	8.29	8.90	3.56	1.83	0.632	0.450	0.489
6.7	0.484	0.552	8.93	7.49	3.00	1.95	0.479	0.378	0.493
7.25	0.459	0.386	7.84	4.59	2.29	1.47	0.433	0.343	0:450
5.4	0.457	0.502	8.52	3.85	1.93	1.56	0.319	0.283	0.449
6	0.455	0.322	6.87	2.07	1.38	1.08	0.305	0.268	0.416
5	0.438	0.392	7.32	1.85	1.24	1.12	0.247	0.233	0.410
4.1	0.436	0.461	7.78	1.66	1.10	1.17	0.197	0.202	0.404

AMERICAN INSTITUTE OF STEEL CONSTRUCTION

Table S-3 Dimensions and physical properties of channels *(continued)*

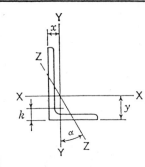

ANGLES
Equal legs and unequal legs
Properties for designing

Size and Thickness	k	Weight per Foot	Area	AXIS X-X.				AXIS Y-Y				AXIS Z-Z	
				I	S	r	y	I	S	r	x	r	Tan α
In.	In.	Lb.	In.^{2}	In.^{4}	In.^{3}	In.	In.	In.^{4}	In.^{3}	In.	In.	In.	α
L 9 x4 x $\frac{5}{8}$	$1\frac{1}{8}$	26.3	7.73	64.9	11.5	2.90	3.36	8.32	2.65	1.04	0.858	.847	0.216
$\frac{9}{16}$	$1\frac{1}{16}$	23.8	7.00	59.1	10.4	2.91	3.33	7.63	2.41	1.04	0.834	.850	0.218
$\frac{1}{2}$	1	21.3	6.25	53.2	9.34	2.92	3.31	6.92	2.17	1.05	0.810	.854	0.220
L 8 x8 x$1\frac{1}{8}$	$1\frac{3}{4}$	56.9	16.7	98.0	17.5	2.42	2.41	98.0	17.5	2.42	2.41	1.56	1.000
1	$1\frac{5}{8}$	51.0	15.0	89.0	15.8	2.44	2.37	89.0	15.8	2.44	2.37	1.56	1.000
$\frac{7}{8}$	$1\frac{1}{2}$	45.0	13.2	79.6	14.0	2.45	2.32	79.6	14.0	2.45	2.32	1.57	1.000
$\frac{3}{4}$	$1\frac{3}{8}$	38.9	11.4	69.7	12.2	2.47	2.28	69.7	12.2	2.47	2.28	1.58	1.000
$\frac{5}{8}$	$1\frac{1}{4}$	32.7	9.61	59.4	10.3	2.49	2.23	59.4	10.3	2.49	2.23	1.58	1.000
$\frac{9}{16}$	$1\frac{3}{16}$	29.6	8.68	54.1	9.34	2.50	2.21	54.1	9.34	2.50	2.21	1.59	1.000
$\frac{1}{2}$	$1\frac{1}{8}$	26.4	7.75	48.6	8.36	2.50	2.19	48.6	8.36	2.50	2.19	1.59	1.000
L 8 x6 x1	$1\frac{1}{2}$	44.2	13.0	80.8	15.1	2.49	2.65	38.8	8.92	1.73	1.65	1.28	0.543
$\frac{7}{8}$	$1\frac{3}{8}$	39.1	11.5	72.3	13.4	2.51	2.61	34.9	7.94	1.74	1.61	1.28	0.547
$\frac{3}{4}$	$1\frac{1}{4}$	33.8	9.94	63.4	11.7	2.53	2.56	30.7	6.92	1.76	1.56	1.29	0.551
$\frac{5}{8}$	$1\frac{1}{8}$	28.5	8.36	54.1	9.87	2.54	2.52	26.3	5.88	1.77	1.52	1.29	0.554
$\frac{9}{16}$	$1\frac{1}{16}$	25.7	7.56	49.3	8.95	2.55	2.50	24.0	5.34	1.78	1.50	1.30	0.556
$\frac{1}{2}$	1	23.0	6.75	44.3	8.02	2.56	2.47	21.7	4.79	1.79	1.47	1.30	0.558
$\frac{7}{16}$	$\frac{15}{16}$	20.2	5.93	39.2	7.07	2.57	2.45	19.3	4.23	1.80	1.45	1.31	0.560
L 8 x4 x1	$1\frac{1}{2}$	37.4	11.0	69.6	14.1	2.52	3.05	11.6	3.94	1.03	1.05	0.846	0.247
$\frac{3}{4}$	$1\frac{1}{4}$	28.7	8.44	54.9	10.9	2.55	2.95	9.36	3.07	1.05	0.953	0.852	0.258
$\frac{9}{16}$	$1\frac{1}{16}$	21.9	6.43	42.8	8.35	2.58	2.88	7.43	2.38	1.07	0.882	0.861	0.265
$\frac{1}{2}$	1	19.6	5.75	38.5	7.49	2.59	2.86	6.74	2.15	1.08	0.859	0.865	0.267
L 7 x4 x $\frac{3}{4}$	$1\frac{1}{4}$	26.2	7.69	37.8	8.42	2.22	2.51	9.05	3.03	1.09	1.01	0.860	0.324
$\frac{5}{8}$	$1\frac{1}{8}$	22.1	6.48	32.4	7.14	2.24	2.46	7.84	2.58	1.10	0.963	0.865	0.329
$\frac{1}{2}$	1	17.9	5.25	26.7	5.81	2.25	2.42	6.53	2.12	1.11	0.917	0.872	0.335
$\frac{3}{8}$	$\frac{7}{8}$	13.6	3.98	20.6	4.44	2.27	2.37	5.10	1.63	1.13	0.870	0.880	0.340

Angles in shaded rows may not be readily available. Availability is subject to rolling accumulation and geographical location, and should be checked with material suppliers.

AMERICAN INSTITUTE OF STEEL CONSTRUCTION

Table S-4a Dimensions and physical properties of angles

ANGLES
Equal legs and unequal legs
Properties for designing

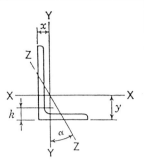

Size and Thickness	k	Weight per Foot	Area	AXIS X-X				AXIS Y-Y				AXIS Z-Z	
				I	S	r	y	I	S	r	x	r	Tan
In.	In.	Lb.	In.²	In.⁴	In.³	In.	In.	In.⁴	In.³	In.	In.	In.	α
L 6 x6 x1	1½	37.4	11.0	35.5	8.57	1.80	1.86	35.5	8.57	1.80	1.86	1.17	1.000
⅞	1⅜	33.1	9.73	31.9	7.63	1.81	1.82	31.9	7.63	1.81	1.82	1.17	1.000
¾	1¼	28.7	8.44	28.2	6.66	1.83	1.78	28.2	6.66	1.83	1.78	1.17	1.000
⅝	1⅛	24.2	7.11	24.2	5.66	1.84	1.73	24.2	5.66	1.84	1.73	1.18	1.000
9/16	1 1/16	21.9	6.43	22.1	5.14	1.85	1.71	22.1	5.14	1.85	1.71	1.18	1.000
½	1	19.6	5.75	19.9	4.61	1.86	1.68	19.9	4.61	1.86	1.68	1.18	1.000
7/16	15/16	17.2	5.06	17.7	4.08	1.87	1.66	17.7	4.08	1.87	1.66	1.19	1.000
⅜	⅞	14.9	4.36	15.4	3.53	1.88	1.64	15.4	3.53	1.88	1.64	1.19	1.000
5/16	13/16	12.4	3.65	13.0	2.97	1.89	1.62	13.0	2.97	1.89	1.62	1.20	1.000
L 6 x4 x ⅞	1⅜	27.2	7.98	27.7	7.15	1.86	2.12	9.75	3.39	1.11	1.12	0.857	0.421
¾	1¼	23.6	6.94	24.5	6.25	1.88	2.08	8.68	2.97	1.12	1.08	0.860	0.428
⅝	1⅛	20.0	5.86	21.1	5.31	1.90	2.03	7.52	2.54	1.13	1.03	0.864	0.435
9/16	1 1/16	18.1	5.31	19.3	4.83	1.90	2.01	6.91	2.31	1.14	1.01	0.866	0.438
½	1	16.2	4.75	17.4	4.33	1.91	1.99	6.27	2.08	1.15	0.987	0.870	0.440
7/16	15/16	14.3	4.18	15.5	3.83	1.92	1.96	5.60	1.85	1.16	0.964	0.873	0.443
⅜	⅞	12.3	3.61	13.5	3.32	1.93	1.94	4.90	1.60	1.17	0.941	0.877	0.446
5/16	13/16	10.3	3.03	11.4	2.79	1.94	1.92	4.18	1.35	1.17	0.918	0.882	0.448
L 6 x3½x ½	1	15.3	4.50	16.6	4.24	1.92	2.08	4.25	1.59	0.972	0.833	0.759	0.344
⅜	⅞	11.7	3.42	12.9	3.24	1.94	2.04	3.34	1.23	0.988	0.787	0.767	0.350
5/16	13/16	9.8	2.87	10.9	2.73	1.95	2.01	2.85	1.04	0.996	0.763	0.772	0.352
L 5 x5 x ⅞	1⅜	27.2	7.98	17.8	5.17	1.49	1.57	17.8	5.17	1.49	1.57	0.973	1.000
¾	1¼	23.6	6.94	15.7	4.53	1.51	1.52	15.7	4.53	1.51	1.52	0.975	1.000
⅝	1⅛	20.0	5.86	13.6	3.86	1.52	1.48	13.6	3.86	1.52	1.48	0.978	1.000
½	1	16.2	4.75	11.3	3.16	1.54	1.43	11.3	3.16	1.54	1.43	0.983	1.000
7/16	15/16	14.3	4.18	10.0	2.79	1.55	1.41	10.0	2.79	1.55	1.41	0.986	1.000
⅜	⅞	12.3	3.61	8.74	2.42	1.56	1.39	8.74	2.42	1.56	1.39	0.990	1.000
5/16	13/16	10.3	3.03	7.42	2.04	1.57	1.37	7.42	2.04	1.57	1.37	0.994	1.000

Angles in shaded rows may not be readily available. Availability is subject to rolling accumulation and geographical location, and should be checked with material suppliers.

AMERICAN INSTITUTE OF STEEL CONSTRUCTION

Table S-4a Dimensions and physical properties of angles *(continued)*

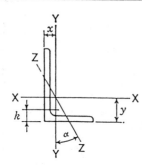

ANGLES
Equal legs and unequal legs
Properties for designing

Size and Thickness		k	Weight per Foot	Area	AXIS X-X				AXIS Y-Y				AXIS Z-Z	
					I	S	r	y	I	S	r	x	r	Tan
In.		In.	Lb.	In.²	In.⁴	In.³	In.	In.	In.⁴	In.³	In.	In.	In.	α
L 5 ×3½× ¾	1¼	19.8	5.81	13.9	4.28	1.55	1.75	5.55	2.22	0.977	0.996	0.748	0.464
	⅝	1⅛	16.8	4.92	12.0	3.65	1.56	1.70	4.83	1.90	0.991	0.951	0.751	0.472
	½	1	13.6	4.00	9.99	2.99	1.58	1.66	4.05	1.56	1.01	0.906	0.755	0.479
	⁷⁄₁₆	¹⁵⁄₁₆	12.0	3.53	8.90	2.64	1.59	1.63	3.63	1.39	1.01	0.883	0.758	0.482
	⅜	⅞	10.4	3.05	7.78	2.29	1.60	1.61	3.18	1.21	1.02	0.861	0.762	0.486
	⁵⁄₁₆	¹³⁄₁₆	8.7	2.56	6.60	1.94	1.61	1.59	2.72	1.02	1.03	0.838	0.766	0.489
	¼	¾	7.0	2.06	5.39	1.57	1.62	1.56	2.23	0.830	1.04	0.814	0.770	0.492
L 5 ×3 × ⅝	1	15.7	4.61	11.4	3.55	1.57	1.80	3.06	1.39	0.815	0.796	0.644	0.349
	½	1	12.8	3.75	9.45	2.91	1.59	1.75	2.58	1.15	0.829	0.750	0.648	0.357
	⁷⁄₁₆	¹⁵⁄₁₆	11.3	3.31	8.43	2.58	1.60	1.73	2.32	1.02	0.837	0.727	0.651	0.361
	⅜	⅞	9.8	2.86	7.37	2.24	1.61	1.70	2.04	0.888	0.845	0.704	0.654	0.364
	⁵⁄₁₆	¹³⁄₁₆	8.2	2.40	6.26	1.89	1.61	1.68	1.75	0.753	0.853	0.681	0.658	0.368
	¼	¾	6.6	1.94	5.11	1.53	1.62	1.66	1.44	0.614	0.861	0.657	0.663	0.371
L 4 ×4 × ¾	1⅛	18.5	5.44	7.67	2.81	1.19	1.27	7.67	2.81	1.19	1.27	0.778	1.000
	⅝	1	15.7	4.61	6.66	2.40	1.20	1.23	6.66	2.40	1.20	1.23	0.779	1.000
	½	⅞	12.8	3.75	5.56	1.97	1.22	1.18	5.56	1.97	1.22	1.18	0.782	1.000
	⁷⁄₁₆	¹³⁄₁₆	11.3	3.31	4.97	1.75	1.23	1.16	4.97	1.75	1.23	1.16	0.785	1.000
	⅜	¾	9.8	2.86	4.36	1.52	1.23	1.14	4.36	1.52	1.23	1.14	0.788	1.000
	⁵⁄₁₆	¹¹⁄₁₆	8.2	2.40	3.71	1.29	1.24	1.12	3.71	1.29	1.24	1.12	0.791	1.000
	¼	⅝	6.6	1.94	3.04	1.05	1.25	1.09	3.04	1.05	1.25	1.09	0.795	1.000
L 4 ×3½× ⅝	1¹⁄₁₆	14.7	4.30	6.37	2.35	1.22	1.29	4.52	1.84	1.03	1.04	0.719	0.745
	½	¹⁵⁄₁₆	11.9	3.50	5.32	1.94	1.23	1.25	3.79	1.52	1.04	1.00	0.722	0.750
	⁷⁄₁₆	⅞	10.6	3.09	4.76	1.72	1.24	1.23	3.40	1.35	1.05	0.978	0.724	0.753
	⅜	¹³⁄₁₆	9.1	2.67	4.18	1.49	1.25	1.21	2.95	1.17	1.06	0.955	0.727	0.755
	⁵⁄₁₆	¾	7.7	2.25	3.56	1.26	1.26	1.18	2.55	0.994	1.07	0.932	0.730	0.757
	¼	¹¹⁄₁₆	6.2	1.81	2.91	1.03	1.27	1.16	2.09	0.808	1.07	0.909	0.734	0.759

Angles in shaded rows may not be readily available. Availability is subject to rolling accumulation and geographical location, and should be checked with material suppliers.

AMERICAN INSTITUTE OF STEEL CONSTRUCTION

Table S-4b Dimensions and physical properties of angles

ANGLES
Equal legs and unequal legs
Properties for designing

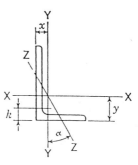

Size and Thickness	k	Weight per Foot	Area	AXIS X-X				AXIS Y-Y				AXIS Z-Z	
				I	S	r	y	I	S	r	x	r	Tan
In.	In.	Lb.	In.²	In.⁴	In.³	In.	In.	In.⁴	In.³	In.	In.	In.	α
L 4 x3 x 5/8	1 1/16	13.6	3.98	6.03	2.30	1.23	1.37	2.87	1.35	0.849	0.871	0.637	0.534
1/2	15/16	11.1	3.25	5.05	1.89	1.25	1.33	2.42	1.12	0.864	0.827	0.639	0.543
7/16	7/8	9.8	2.87	4.52	1.68	1.25	1.30	2.18	0.992	0.871	0.804	0.641	0.547
3/8	13/16	8.5	2.48	3.96	1.46	1.26	1.28	1.92	0.866	0.879	0.782	0.644	0.551
5/16	3/4	7.2	2.09	3.38	1.23	1.27	1.26	1.65	0.734	0.887	0.759	0.647	0.554
1/4	11/16	5.8	1.69	2.77	1.00	1.28	1.24	1.36	0.599	0.896	0.736	0.651	0.558
L 3½x3½x 1/2	7/8	11.1	3.25	3.64	1.49	1.06	1.06	3.64	1.49	1.06	1.06	0.683	1.000
7/16	13/16	9.8	2.87	3.26	1.32	1.07	1.04	3.26	1.32	1.07	1.04	0.684	1.000
3/8	3/4	8.5	2.48	2.87	1.15	1.07	1.01	2.87	1.15	1.07	1.01	0.687	1.000
5/16	11/16	7.2	2.09	2.45	0.976	1.08	0.990	2.45	0.976	1.08	0.990	0.690	1.000
1/4	5/8	5.8	1.69	2.01	0.794	1.09	0.968	2.01	0.794	1.09	0.968	0.694	1.000
L 3½x3 x 1/2	15/16	10.2	3.00	3.45	1.45	1.07	1.13	2.33	1.10	0.881	0.875	0.621	0.714
7/16	7/8	9.1	2.65	3.10	1.29	1.08	1.10	2.09	0.975	0.889	0.853	0.622	0.718
3/8	13/16	7.9	2.30	2.72	1.13	1.09	1.08	1.85	0.851	0.897	0.830	0.625	0.721
5/16	3/4	6.6	1.93	2.33	0.954	1.10	1.06	1.58	0.722	0.905	0.808	0.627	0.724
1/4	11/16	5.4	1.56	1.91	0.776	1.11	1.04	1.30	0.589	0.914	0.785	0.631	0.727
L 3½x2½x 1/2	15/16	9.4	2.75	3.24	1.41	1.09	1.20	1.36	0.760	0.704	0.705	0.534	0.486
7/16	7/8	8.3	2.43	2.91	1.26	1.09	1.18	1.23	0.677	0.711	0.682	0.535	0.491
3/8	13/16	7.2	2.11	2.56	1.09	1.10	1.16	1.09	0.592	0.719	0.660	0.537	0.496
5/16	3/4	6.1	1.78	2.19	0.927	1.11	1.14	0.939	0.504	0.727	0.637	0.540	0.501
1/4	11/16	4.9	1.44	1.80	0.755	1.12	1.11	0.777	0.412	0.735	0.614	0.544	0.506
L 3 x3 x 1/2	13/16	9.4	2.75	2.22	1.07	0.898	0.932	2.22	1.07	0.898	0.932	0.584	1.000
7/16	3/4	8.3	2.43	1.99	0.954	0.905	0.910	1.99	0.954	0.905	0.910	0.585	1.000
3/8	11/16	7.2	2.11	1.76	0.833	0.913	0.888	1.76	0.833	0.913	0.888	0.587	1.000
5/16	5/8	6.1	1.78	1.51	0.707	0.922	0.865	1.51	0.707	0.922	0.865	0.589	1.000
1/4	9/16	4.9	1.44	1.24	0.577	0.930	0.842	1.24	0.577	0.930	0.842	0.592	1.000
3/16	1/2	3.71	1.09	0.962	0.441	0.939	0.820	0.962	0.441	0.939	0.820	0.596	1.000

Angles in shaded rows may not be readily available. Availability is subject to rolling accumulation and geographical location, and should be checked with material suppliers.

AMERICAN INSTITUTE OF STEEL CONSTRUCTION

Table S-4b Dimensions and physical properties of angles *(continued)*

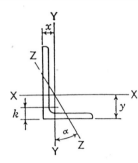

ANGLES
Equal legs and unequal legs
Properties for designing

Size and Thickness	k	Weight per Foot	Area	AXIS X-X				AXIS Y-Y				AXIS Z-Z	
				I	S	r	y	I	S	r	x	r	Tan
In.	In.	Lb.	In.²	In.⁴	In.³	In.	In.	In.⁴	In.³	In.	In.	In.	α
L 3 x2½x ½	⁷⁄₈	8.5	2.50	2.08	1.04	0.913	1.00	1.30	0.744	0.722	0.750	0.520	0.667
⁷⁄₁₆	¹³⁄₁₆	7.6	2.21	1.88	0.928	0.920	0.978	1.18	0.664	0.729	0.728	0.521	0.672
³⁄₈	³⁄₄	6.6	1.92	1.66	0.810	0.928	0.956	1.04	0.581	0.736	0.706	0.522	0.676
⁵⁄₁₆	¹¹⁄₁₆	5.6	1.62	1.42	0.688	0.937	0.933	0.898	0.494	0.744	0.683	0.525	0.680
¼	⁵⁄₈	4.5	1.31	1.17	0.561	0.945	0.911	0.743	0.404	0.753	0.661	0.528	0.684
³⁄₁₆	⁹⁄₁₆	3.39	0.996	0.907	0.430	0.954	0.888	0.577	0.310	0.761	0.638	0.533	0.688
L 3 x2 x ½	¹³⁄₁₆	7.7	2.25	1.92	1.00	0.924	1.08	0.672	0.474	0.546	0.583	0.428	0.414
⁷⁄₁₆	³⁄₄	6.8	2.00	1.73	0.894	0.932	1.06	0.609	0.424	0.553	0.561	0.429	0.421
³⁄₈	¹¹⁄₁₆	5.9	1.73	1.53	0.781	0.940	1.04	0.543	0.371	0.559	0.539	0.430	0.428
⁵⁄₁₆	⁵⁄₈	5.0	1.46	1.32	0.664	0.948	1.02	0.470	0.317	0.567	0.516	0.432	0.435
¼	⁹⁄₁₆	4.1	1.19	1.09	0.542	0.957	0.993	0.392	0.260	0.574	0.493	0.435	0.440
³⁄₁₆	½	3.07	0.902	0.842	0.415	0.966	0.970	0.307	0.200	0.583	0.470	0.439	0.446
L 2½x2½x ½	¹³⁄₁₆	7.7	2.25	1.23	0.724	0.739	0.806	1.23	0.724	0.739	0.806	0.487	1.000
³⁄₈	¹¹⁄₁₆	5.9	1.73	0.984	0.566	0.753	0.762	0.984	0.566	0.753	0.762	0.487	1.000
⁵⁄₁₆	⁵⁄₈	5.0	1.46	0.849	0.482	0.761	0.740	0.849	0.482	0.761	0.740	0.489	1.000
¼	⁹⁄₁₆	4.1	1.19	0.703	0.394	0.769	0.717	0.703	0.394	0.769	0.717	0.491	1.000
³⁄₁₆	½	3.07	0.902	0.547	0.303	0.778	0.694	0.547	0.303	0.778	0.694	0.495	1.000
L 2½x2 x ³⁄₈	¹¹⁄₁₆	5.3	1.55	0.912	0.547	0.768	0.831	0.514	0.363	0.577	0.581	0.420	0.614
⁵⁄₁₆	⁵⁄₈	4.5	1.31	0.788	0.466	0.776	0.809	0.446	0.310	0.584	0.559	0.422	0.620
¼	⁹⁄₁₆	3.62	1.06	0.654	0.381	0.784	0.787	0.372	0.254	0.592	0.537	0.424	0.626
³⁄₁₆	½	2.75	0.809	0.509	0.293	0.793	0.764	0.291	0.196	0.600	0.514	0.427	0.631
L 2 x2 x ³⁄₈	¹¹⁄₁₆	4.7	1.36	0.479	0.351	0.594	0.636	0.479	0.351	0.594	0.636	0.389	1.000
⁵⁄₁₆	⁵⁄₈	3.92	1.15	0.416	0.300	0.601	0.614	0.416	0.300	0.601	0.614	0.390	1.000
¼	⁹⁄₁₆	3.19	0.938	0.348	0.247	0.609	0.592	0.348	0.247	0.609	0.592	0.391	1.000
³⁄₁₆	½	2.44	0.715	0.272	0.190	0.617	0.569	0.272	0.190	0.617	0.569	0.394	1.000
⅛	⁷⁄₁₆	1.65	0.484	0.190	0.131	0.626	0.546	0.190	0.131	0.626	0.546	0.398	1.000

Angles in shaded rows may not be readily available. Availability is subject to rolling accumulation and geographical location, and should be checked with material suppliers.

AMERICAN INSTITUTE OF STEEL CONSTRUCTION

Table S-4c Dimensions and physical properties of angles

Structural Tees

The following list of structural tees are cut from wide flange W shapes and American Standard S shapes. Structural tees are made by cutting W shapes or S shapes in half. In many cases, the thin web as a projecting element disqualifies the section as a compact section; such tees are classified as "slender element" sections.

Those tees classified as a slender element section in the tables can be identified as such by looking at the columns headed Q_s or C_s' at the right hand side of the table. If a value is entered, the section is classified as a slender element section. If no value is entered in these columns, the section retains its compact or noncompact classification.

The values of Q_s an C_s' are computed for each slender element section from the appropriate equations of ASD Appendix B. Q_s is the capacity reduction factor and C_s' is the value of KL/r above which the section acts as an Euler column. See Table S-11 for a tabulation of the allowable axial load.

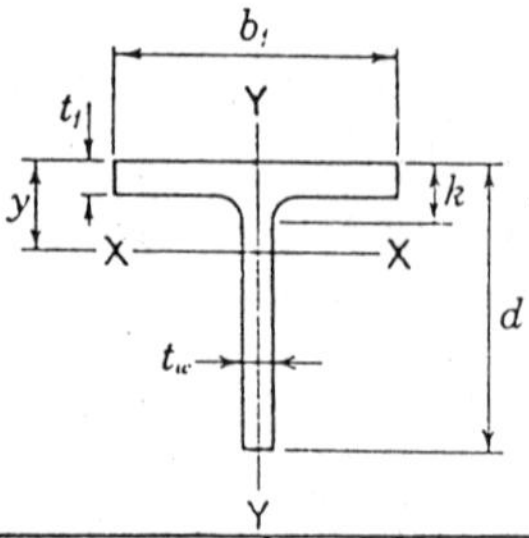

STRUCTURAL TEES
Cut from W shapes
Dimensions

Desig-nation	Area	Depth of Tee d		Stem Thickness t_w		$\dfrac{t_w}{2}$	Area of Stem	Flange Width b_f		Flange Thickness t_f		Dis-tance k
	In.²	In.		In.		In.	In.²	In.		In.		In.
WT 13.5x89	26.1	13.905	13⅞	0.725	¾	⅜	10.1	14.085	14⅛	1.190	1³⁄₁₆	1⅞
x80.5	23.7	13.795	13¾	0.660	¹¹⁄₁₆	⅜	9.10	14.020	14	1.080	1¹⁄₁₆	1¹³⁄₁₆
x73	21.5	13.690	13¾	0.605	⅝	⁵⁄₁₆	8.28	13.965	14	0.975	1	1¹¹⁄₁₆
WT 13.5x57	16.8	13.645	13⅝	0.570	⁹⁄₁₆	⁵⁄₁₆	7.78	10.070	10⅛	0.930	¹⁵⁄₁₆	1⅝
x51	15.0	13.545	13½	0.515	½	¼	6.98	10.015	10	0.830	¹³⁄₁₆	1⁹⁄₁₆
x47	13.8	13.460	13½	0.490	½	¼	6.60	9.990	10	0.745	¾	1⁷⁄₁₆
x42	12.4	13.355	13⅜	0.460	⁷⁄₁₆	¼	6.14	9.960	10	0.640	⅝	1⅜
WT 12 x81	23.9	12.500	12½	0.705	¹¹⁄₁₆	⅜	8.81	12.955	13	1.220	1¼	2
x73	21.5	12.370	12⅜	0.650	⅝	⁵⁄₁₆	8.04	12.900	12⅞	1.090	1¹⁄₁₆	1⅞
x65.5	19.3	12.240	12¼	0.605	⅝	⁵⁄₁₆	7.41	12.855	12⅞	0.960	¹⁵⁄₁₆	1¾
x58.5	17.2	12.130	12⅛	0.550	⁹⁄₁₆	⁵⁄₁₆	6.67	12.800	12¾	0.850	⅞	1⅝
x52	15.3	12.030	12	0.500	½	¼	6.01	12.750	12¾	0.750	¾	1½
WT 12 x47	13.8	12.155	12⅛	0.515	½	¼	6.26	9.065	9⅛	0.875	⅞	1⅝
x42	12.4	12.050	12	0.470	½	¼	5.66	9.020	9	0.770	¾	1⁹⁄₁₆
x38	11.2	11.960	12	0.440	⁷⁄₁₆	¼	5.26	8.990	9	0.680	¹¹⁄₁₆	1⁷⁄₁₆
x34	10.0	11.865	11⅞	0.415	⁷⁄₁₆	¼	4.92	8.965	9	0.585	⁹⁄₁₆	1⅜
WT 12 x31	9.11	11.870	11⅞	0.430	⁷⁄₁₆	¼	5.10	7.040	7	0.590	⁹⁄₁₆	1⅜
x27.5	8.10	11.785	11¾	0.395	⅜	³⁄₁₆	4.66	7.005	7	0.505	½	1⁵⁄₁₆
WT 10.5x73.5	21.6	11.030	11	0.720	¾	⅜	7.94	12.510	12½	1.150	1⅛	1⅞
x66	19.4	10.915	10⅞	0.650	⅝	⁵⁄₁₆	7.09	12.440	12½	1.035	1¹⁄₁₆	1¹³⁄₁₆
x61	17.9	10.840	10⅞	0.600	⅝	⁵⁄₁₆	6.50	12.390	12⅜	0.960	¹⁵⁄₁₆	1¹¹⁄₁₆
x55.5	16.3	10.755	10¾	0.550	⁹⁄₁₆	⁵⁄₁₆	5.92	12.340	12⅜	0.875	⅞	1⅝
x50.5	14.9	10.680	10⅝	0.500	½	¼	5.34	12.290	12¼	0.800	¹³⁄₁₆	1⁹⁄₁₆
WT 10.5x46.5	13.7	10.810	10¾	0.580	⁹⁄₁₆	⁵⁄₁₆	6.27	8.420	8⅜	0.930	¹⁵⁄₁₆	1¹¹⁄₁₆
x41.5	12.2	10.715	10¾	0.515	½	¼	5.52	8.355	8⅜	0.835	¹³⁄₁₆	1⁹⁄₁₆
x36.5	10.7	10.620	10⅝	0.455	⁷⁄₁₆	¼	4.83	8.295	8¼	0.740	¾	1½
x34	10.0	10.565	10⅝	0.430	⁷⁄₁₆	¼	4.54	8.270	8¼	0.685	¹¹⁄₁₆	1⁷⁄₁₆
x31	9.13	10.495	10½	0.400	⅜	³⁄₁₆	4.20	8.240	8¼	0.615	⅝	1⅜
WT 10.5x28.5	8.37	10.530	10½	0.405	⅜	³⁄₁₆	4.26	6.555	6½	0.650	⅝	1⅜
x25	7.36	10.415	10⅜	0.380	⅜	³⁄₁₆	3.96	6.530	6½	0.535	⁹⁄₁₆	1⁵⁄₁₆
x22	6.49	10.330	10⅜	0.350	⅜	³⁄₁₆	3.62	6.500	6½	0.450	⁷⁄₁₆	1³⁄₁₆

AMERICAN INSTITUTE OF STEEL CONSTRUCTION

Table S-5a Dimensions and physical properties of WT shapes

STRUCTURAL TEES
Cut from W shapes
Properties

$$C_c' = \sqrt{\frac{2\pi^2 E}{Q_s Q_a F_y}}, \quad Q_a = 1.0$$

Nominal Weight per Ft.	$\frac{d}{t_w}$	AXIS X-X				AXIS Y-Y			$F_y = 36$ ksi		$F_y = 50$ ksi	
		I	S	r	y	I	S	r	Q_s	C_c'	Q_s	C_c'
Lb.		In.⁴	In.³	In.	In.	In.⁴	In.³	In.				
89	19.2	414	38.2	3.98	3.05	278	39.4	3.26	—	—	0.937	111
80.5	20.9	372	34.4	3.96	2.99	248	35.4	3.24	—	—	0.851	116
73	22.6	336	31.2	3.95	2.95	222	31.7	3.21	0.938	130	0.765	122
57	23.9	289	28.3	4.15	3.42	79.4	15.8	2.18	0.883	134	0.700	128
51	26.3	258	25.3	4.14	3.37	69.6	13.9	2.15	0.780	143	0.578	141
47	27.5	239	23.8	4.16	3.41	62.0	12.4	2.12	0.728	148	0.529	147
42	29.0	216	21.9	4.18	3.48	52.8	10.6	2.07	0.664	155	0.476	155
81	17.7	293	29.9	3.50	2.70	221	34.2	3.05	—	—	—	—
73	19.0	264	27.2	3.50	2.66	195	30.3	3.01	—	—	0.947	110
65.5	20.2	238	24.8	3.52	2.65	170	26.5	2.97	—	—	0.887	114
58.5	22.1	212	22.3	3.51	2.62	149	23.2	2.94	0.960	129	0.791	120
52	24.1	189	20.0	3.51	2.59	130	20.3	2.91	0.874	135	0.690	129
47	23.6	186	20.3	3.67	2.99	54.5	12.0	1.98	0.896	133	0.715	127
42	25.6	166	18.3	3.67	2.97	47.2	10.5	1.95	0.810	140	0.610	137
38	27.2	151	16.9	3.68	3.00	41.3	9.18	1.92	0.741	146	0.541	146
34	28.6	137	15.6	3.70	3.06	35.2	7.85	1.87	0.681	153	0.489	153
31	27.6	131	15.6	3.79	3.46	17.2	4.90	1.38	0.724	148	0.525	148
27.5	29.8	117	14.1	3.80	3.50	14.5	4.15	1.34	0.626	159	0.450	159
73.5	15.3	204	23.7	3.08	2.39	188	30.0	2.95	—	—	—	—
66	16.8	181	21.1	3.06	2.33	166	26.7	2.93	—	—	—	—
61	18.1	166	19.3	3.04	2.28	152	24.6	2.92	—	—	0.993	107
55.5	19.6	150	17.5	3.03	2.23	137	22.2	2.90	—	—	0.917	112
50.5	21.4	135	15.8	3.01	2.18	124	20.2	2.89	0.990	127	0.826	118
46.5	18.6	144	17.9	3.25	2.74	46.4	11.0	1.84	—	—	0.968	109
41.5	20.8	127	15.7	3.22	2.66	40.7	9.75	1.83	—	—	0.856	116
36.5	23.3	110	13.8	3.21	2.60	35.3	8.51	1.81	0.908	132	0.730	125
34	24.6	103	12.9	3.20	2.59	32.4	7.83	1.80	0.853	137	0.664	131
31	26.2	93.8	11.9	3.21	2.58	28.7	6.97	1.77	0.784	142	0.583	140
28.5	26.0	90.4	11.8	3.29	2.85	15.3	4.67	1.35	0.793	142	0.592	139
25	27.4	80.3	10.7	3.30	2.93	12.5	3.82	1.30	0.733	147	0.533	147
22	29.5	71.1	9.68	3.31	2.98	10.3	3.18	1.26	0.638	158	0.460	158

Where no value of C_c' or Q_s is shown, the Tee complies with Specification Sect. 1.9.1.2.

AMERICAN INSTITUTE OF STEEL CONSTRUCTION

Table S-5a Dimensions and physical properties of WT shapes *(continued)*

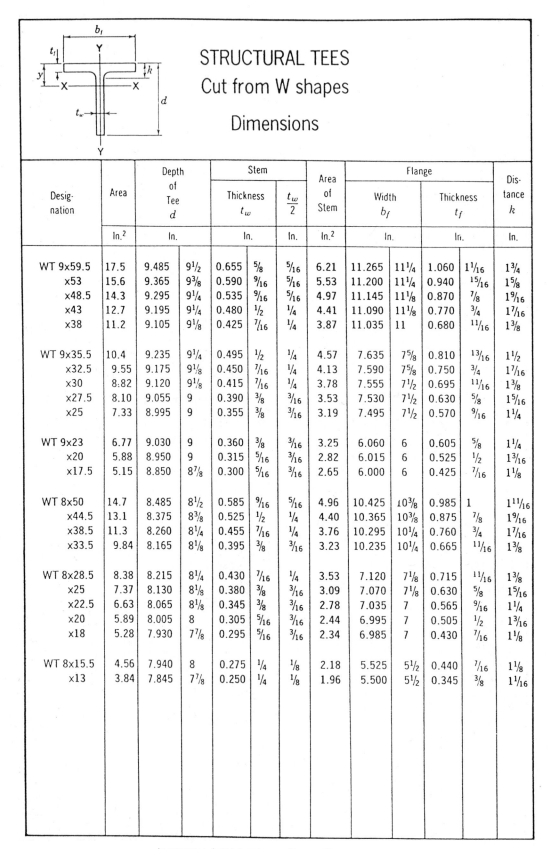

STRUCTURAL TEES
Cut from W shapes
Dimensions

Desig-nation	Area	Depth of Tee d		Stem			Area of Stem	Flange				Dis-tance k
				Thickness t_w		$\frac{t_w}{2}$		Width b_f		Thickness t_f		
	In.²	In.		In.		In.	In.²	In.		In.		In.
WT 9x59.5	17.5	9.485	9½	0.655	⅝	5/16	6.21	11.265	11¼	1.060	1 1/16	1¾
x53	15.6	9.365	9⅜	0.590	9/16	5/16	5.53	11.200	11¼	0.940	15/16	1⅝
x48.5	14.3	9.295	9¼	0.535	9/16	5/16	4.97	11.145	11⅛	0.870	⅞	1 9/16
x43	12.7	9.195	9¼	0.480	½	¼	4.41	11.090	11⅛	0.770	¾	1 7/16
x38	11.2	9.105	9⅛	0.425	7/16	¼	3.87	11.035	11	0.680	11/16	1⅜
WT 9x35.5	10.4	9.235	9¼	0.495	½	¼	4.57	7.635	7⅝	0.810	13/16	1½
x32.5	9.55	9.175	9⅛	0.450	7/16	¼	4.13	7.590	7⅝	0.750	¾	1 7/16
x30	8.82	9.120	9⅛	0.415	7/16	¼	3.78	7.555	7½	0.695	11/16	1⅜
x27.5	8.10	9.055	9	0.390	⅜	3/16	3.53	7.530	7½	0.630	⅝	1 5/16
x25	7.33	8.995	9	0.355	⅜	3/16	3.19	7.495	7½	0.570	9/16	1¼
WT 9x23	6.77	9.030	9	0.360	⅜	3/16	3.25	6.060	6	0.605	⅝	1¼
x20	5.88	8.950	9	0.315	5/16	3/16	2.82	6.015	6	0.525	½	1 3/16
x17.5	5.15	8.850	8⅞	0.300	5/16	3/16	2.65	6.000	6	0.425	7/16	1⅛
WT 8x50	14.7	8.485	8½	0.585	9/16	5/16	4.96	10.425	10⅜	0.985	1	1 11/16
x44.5	13.1	8.375	8⅜	0.525	½	¼	4.40	10.365	10⅜	0.875	⅞	1 9/16
x38.5	11.3	8.260	8¼	0.455	7/16	¼	3.76	10.295	10¼	0.760	¾	1 7/16
x33.5	9.84	8.165	8⅛	0.395	⅜	3/16	3.23	10.235	10¼	0.665	11/16	1⅜
WT 8x28.5	8.38	8.215	8¼	0.430	7/16	¼	3.53	7.120	7⅛	0.715	11/16	1⅜
x25	7.37	8.130	8⅛	0.380	⅜	3/16	3.09	7.070	7⅛	0.630	⅝	1 5/16
x22.5	6.63	8.065	8⅛	0.345	⅜	3/16	2.78	7.035	7	0.565	9/16	1¼
x20	5.89	8.005	8	0.305	5/16	3/16	2.44	6.995	7	0.505	½	1 3/16
x18	5.28	7.930	7⅞	0.295	5/16	3/16	2.34	6.985	7	0.430	7/16	1⅛
WT 8x15.5	4.56	7.940	8	0.275	¼	⅛	2.18	5.525	5½	0.440	7/16	1⅛
x13	3.84	7.845	7⅞	0.250	¼	⅛	1.96	5.500	5½	0.345	⅜	1 1/16

AMERICAN INSTITUTE OF STEEL CONSTRUCTION

Table S-5b Dimensions and physical properties of WT shapes

STRUCTURAL TEES
Cut from W shapes
Properties

$$C_c' = \sqrt{\frac{2\pi^2 E}{Q_s Q_a F_y}}, \quad Q_a = 1.0$$

Nominal Weight per Ft.	$\dfrac{d}{t_w}$	AXIS X-X				AXIS Y-Y			$F_y = 36$ ksi		$F_y = 50$ ksi	
		I	S	r	y	I	S	r	Q_s	C_c'	Q_s	C_c'
Lb.		In.⁴	In.³	In.	In.	In.⁴	In.³	In.				
59.5	14.5	119	15.9	2.60	2.03	126	22.5	2.69	—	—	—	—
53	15.9	104	14.1	2.59	1.97	110	19.7	2.66	—	—	—	—
48.5	17.4	93.8	12.7	2.56	1.91	100	18.0	2.65	—	—	—	—
43	19.2	82.4	11.2	2.55	1.86	87.6	15.8	2.63	—	—	0.937	111
38	21.4	71.8	9.83	2.54	1.80	76.2	13.8	2.61	0.990	127	0.826	118
35.5	18.7	78.2	11.2	2.74	2.26	30.1	7.89	1.70	—	—	0.963	109
32.5	20.4	70.7	10.1	2.72	2.20	27.4	7.22	1.69	—	—	0.877	114
30	22.0	64.7	9.29	2.71	2.16	25.0	6.63	1.69	0.964	128	0.796	120
27.5	23.2	59.5	8.63	2.71	2.16	22.5	5.97	1.67	0.913	132	0.735	125
25	25.3	53.5	7.79	2.70	2.12	20.0	5.35	1.65	0.823	139	0.625	135
23	25.1	52.1	7.77	2.77	2.33	11.3	3.72	1.29	0.831	138	0.635	134
20	28.4	44.8	6.73	2.76	2.29	9.55	3.17	1.27	0.690	152	0.496	152
17.5	29.5	40.1	6.21	2.79	2.39	7.67	2.56	1.22	0.638	158	0.460	158
50	14.5	76.8	11.4	2.28	1.76	93.1	17.9	2.51	—	—	—	—
44.5	16.0	67.2	10.1	2.27	1.70	81.3	15.7	2.49	—	—	—	—
38.5	18.2	56.9	8.59	2.24	1.63	69.2	13.4	2.47	—	—	0.988	108
33.5	20.7	48.6	7.36	2.22	1.56	59.5	11.6	2.46	—	—	0.861	115
28.5	19.1	48.7	7.77	2.41	1.94	21.6	6.06	1.60	—	—	0.942	110
25	21.4	42.3	6.78	2.40	1.89	18.6	5.26	1.59	0.990	127	0.826	118
22.5	23.4	37.8	6.10	2.39	1.86	16.4	4.67	1.57	0.904	133	0.725	126
20	26.2	33.1	5.35	2.37	1.81	14.4	4.12	1.57	0.784	142	0.583	140
18	26.9	30.6	5.05	2.41	1.88	12.2	3.50	1.52	0.754	145	0.553	144
15.5	28.9	27.4	4.64	2.45	2.02	6.20	2.24	1.17	0.668	154	0.479	155
13	31.4	23.5	4.09	2.47	2.09	4.80	1.74	1.12	0.563	168	0.406	168

Where no value of C_c' or Q_s is shown, the Tee complies with Specification Sect. 1.9.1.2.

AMERICAN INSTITUTE OF STEEL CONSTRUCTION

Table S-5b Dimensions and physical properties of WT shapes *(continued)*

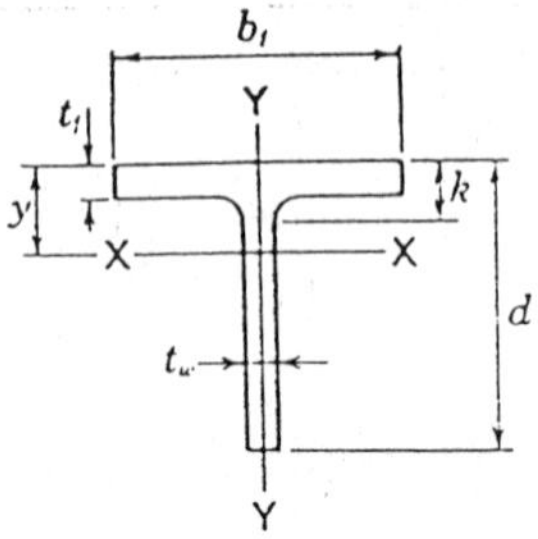

STRUCTURAL TEES
Cut from W shapes
Dimensions

Desig nation	Area	Depth of Tee d	Stem Thickness t_w		$\dfrac{t_w}{2}$	Area of Stem	Flange Width b_f		Thickness t_f		Distance k
	In.²	In.	In.		In.	In.²	In.		In.		In.
WT 9x59.5	17.5	9.485	9½	0.655	⅝	5/16	6.21	11.265	11¼	1.060	1 1/16	1¾
x53	15.6	9.365	9⅜	0.590	9/16	5/16	5.53	11.200	11¼	0.940	15/16	1⅝
x48.5	14.3	9.295	9¼	0.535	9/16	5/16	4.97	11.145	11⅛	0.870	⅞	1 9/16
x43	12.7	9.195	9¼	0.480	½	¼	4.41	11.090	11⅛	0.770	¾	1 7/16
x38	11.2	9.105	9⅛	0.425	7/16	¼	3.87	11.035	11	0.680	11/16	1⅜
WT 9x35.5	10.4	9.235	9¼	0.495	½	¼	4.57	7.635	7⅝	0.810	13/16	1½
x32.5	9.55	9.175	9⅛	0.450	7/16	¼	4.13	7.590	7⅝	0.750	¾	1 7/16
x30	8.82	9.120	9⅛	0.415	7/16	¼	3.78	7.555	7½	0.695	11/16	1⅜
x27.5	8.10	9.055	9	0.390	⅜	3/16	3.53	7.530	7½	0.630	⅝	1 5/16
x25	7.33	8.995	9	0.355	⅜	3/16	3.19	7.495	7½	0.570	9/16	1¼
WT 9x23	6.77	9.030	9	0.360	⅜	3/16	3.25	6.060	6	0.605	⅝	1¼
x20	5.88	8.950	9	0.315	5/16	3/16	2.82	6.015	6	0.525	½	1 3/16
x17.5	5.15	8.850	8⅞	0.300	5/16	3/16	2.65	6.000	6	0.425	7/16	1⅛
WT 8x50	14.7	8.485	8½	0.585	9/16	5/16	4.96	10.425	10⅜	0.985	1	1 11/16
x44.5	13.1	8.375	8⅜	0.525	½	¼	4.40	10.365	10⅜	0.875	⅞	1 9/16
x38.5	11.3.	8.260	8¼	0.455	7/16	¼	3.76	10.295	10¼	0.760	¾	1 7/16
x33.5	9.84	8.165	8⅛	0.395	⅜	3/16	3.23	10.235	10¼	0.665	11/16	1⅜
WT 8x28.5	8.38	8.215	8¼	0.430	7/16	¼	3.53	7.120	7⅛	0.715	11/16	1⅜
x25	7.37	8.130	8⅛	0.380	⅜	3/16	3.09	7.070	7⅛	0.630	⅝	1 5/16
x22.5	6.63	8.065	8⅛	0.345	⅜	3/16	2.78	7.035	7	0.565	9/16	1¼
x20	5.89	8.005	8	0.305	5/16	3/16	2.44	6.995	7	0.505	½	1 3/16
x18	5.28	7.930	7⅞	0.295	5/16	3/16	2.34	6.985	7	0.430	7/16	1⅛
WT 8x15.5	4.56	7.940	8	0.275	¼	⅛	2.18	5.525	5½	0.440	7/16	1⅛
x13	3.84	7.845	7⅞	0.250	¼	⅛	1.96	5.500	5½	0.345	⅜	1 1/16

AMERICAN INSTITUTE OF STEEL CONSTRUCTION

Table S-5c Dimensions and physical properties of WT shapes

STRUCTURAL TEES
Cut from W shapes
Properties

$$C_c' = \sqrt{\frac{2\pi^2 E}{Q_s Q_a F_y}}, \quad Q_a = 1.0$$

Nominal Weight per Ft.	$\frac{d}{t_w}$	AXIS X-X				AXIS Y-Y			$F_y = 36$ ksi		$F_y = 50$ ksi	
		I	S	r	y	I	S	r	Q_s	C_c'	Q_s	C_c'
Lb.		In.⁴	In.³	In.	In.	In.⁴	In.³	In.	Q_s	C_c'	Q_s	C_c'
66	11.4	57.8	9.57	1.73	1.29	274	37.2	3.76	—	—	—	—
60	12.3	51.7	8.61	1.71	1.24	247	33.7	3.74	—	—	—	—
54.5	13.6	45.3	7.56	1.68	1.17	223	30.6	3.73	—	—	—	—
49.5	14.6	40.9	6.88	1.67	1.14	201	27.6	3.71	—	—	—	—
45	15.9	36.4	6.16	1.66	1.09	181	25.0	3.70	—	—	—	—
41	14.0	41.2	7.14	1.85	1.39	74.2	14.6	2.48	—	—	—	—
37	15.7	36.0	6.25	1.82	1.32	66.9	13.3	2.48	—	—	—	—
34	16.9	32.6	5.69	1.81	1.29	60.7	12.1	2.46	—	—	—	—
30.5	18.5	28.9	5.07	1.80	1.25	53.7	10.7	2.45	—	—	0.973	108
26.5	18.8	27.6	4.94	1.88	1.38	28.8	7.16	1.92	—	—	0.958	109
24	20.3	24.9	4.48	1.87	1.35	25.7	6.40	1.91	—	—	0.882	114
21.5	22.4	21.9	3.98	1.86	1.31	22.6	5.65	1.89	0.947	130	0.775	122
19	22.7	23.3	4.22	2.04	1.54	13.3	3.94	1.55	0.934	130	0.760	123
17	24.5	20.9	3.83	2.04	1.53	11.7	3.45	1.53	0.857	136	0.669	131
15	25.6	19.0	3.55	2.07	1.58	9.79	2.91	1.49	0.810	140	0.610	137
13	27.3	17.3	3.31	2.12	1.72	4.45	1.77	1.08	0.737	147	0.537	146
11	29.9	14.8	2.91	2.14	1.76	3.50	1.40	1.04	0.621	160	0.447	160

Where no value of C_c' or Q_s is shown, the Tee complies with Specification Sect. 1.9.1.2.

AMERICAN INSTITUTE OF STEEL CONSTRUCTION

Table S-5c Dimensions and physical properties of WT shapes *(continued)*

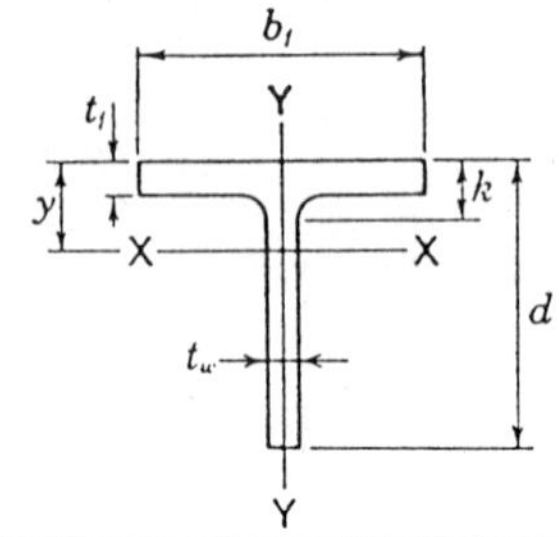

STRUCTURAL TEES
Cut from W shapes
Dimensions

Desig-nation	Area	Depth of Tee d		Stem Thickness t_w		$\dfrac{t_w}{2}$	Area of Stem	Flange Width b_f		Flange Thickness t_f		Dis-tance k
	In.²	In.		In.		In.	In.²	In.		In.		In.
WT 6x168	49.4	8.410	8⅜	1.775	1¾	⅞	14.9	13.385	13⅜	2.955	2¹⁵⁄₁₆	3¹¹⁄₁₆
x152.5	44.8	8.160	8⅛	1.625	1⅝	¹³⁄₁₆	13.3	13.235	13¼	2.705	2¹¹⁄₁₆	3⁷⁄₁₆
x139.5	41.0	7.925	7⅞	1.530	1½	¾	12.1	13.140	13⅛	2.470	2½	3³⁄₁₆
x126	37.0	7.705	7¾	1.395	1⅜	¹¹⁄₁₆	10.7	13.005	13	2.250	2¼	2¹⁵⁄₁₆
x115	33.9	7.525	7½	1.285	1⁵⁄₁₆	¹¹⁄₁₆	9.67	12.895	12⅞	2.070	2¹⁄₁₆	2¾
x105	30.9	7.355	7⅜	1.180	1³⁄₁₆	⅝	8.68	12.790	12¾	1.900	1⅞	2⅝
x 95	27.9	7.190	7¼	1.060	1¹⁄₁₆	⁹⁄₁₆	7.62	12.670	12⅝	1.735	1¾	2⁷⁄₁₆
x 85	25.0	7.015	7	0.960	¹⁵⁄₁₆	½	6.73	12.570	12⅝	1.560	1⁹⁄₁₆	2¼
x 76	22.4	6.855	6⅞	0.870	⅞	⁷⁄₁₆	5.96	12.480	12½	1.400	1⅜	2⅛
x 68	20.0	6.705	6¾	0.790	¹³⁄₁₆	⁷⁄₁₆	5.30	12.400	12⅜	1.250	1¼	1¹⁵⁄₁₆
x 60	17.6	6.560	6½	0.710	¹¹⁄₁₆	⅜	4.66	12.320	12⅜	1.105	1⅛	1¹³⁄₁₆
x 53	15.6	6.445	6½	0.610	⅝	⁵⁄₁₆	3.93	12.220	12¼	0.990	1	1¹¹⁄₁₆
x 48	14.1	6.355	6⅜	0.550	⁹⁄₁₆	⁵⁄₁₆	3.50	12.160	12⅛	0.900	⅞	1⅝
x 43.5	12.8	6.265	6¼	0.515	½	¼	3.23	12.125	12⅛	0.810	¹³⁄₁₆	1½
x 39.5	11.6	6.190	6¼	0.470	½	¼	2.91	12.080	12⅛	0.735	¾	1⁷⁄₁₆
x 36	10.6	6.125	6⅛	0.430	⁷⁄₁₆	¼	2.63	12.040	12	0.670	¹¹⁄₁₆	1⅜
x 32.5	9.54	6.060	6	0.390	⅜	³⁄₁₆	2.36	12.000	12	0.605	⅝	1⁵⁄₁₆
WT 6x 29	8.52	6.095	6⅛	0.360	⅜	³⁄₁₆	2.19	10.010	10	0.640	⅝	1⅜
x 26.5	7.78	6.030	6	0.345	⅜	³⁄₁₆	2.08	9.995	10	0.575	⁹⁄₁₆	1¼
WT 6x 25	7.34	6.095	6⅛	0.370	⅜	³⁄₁₆	2.26	8.080	8⅛	0.640	⅝	1⅜
x 22.5	6.61	6.030	6	0.335	⁵⁄₁₆	³⁄₁₆	2.02	8.045	8	0.575	⁹⁄₁₆	1¼
x 20	5.89	5.970	6	0.295	⁵⁄₁₆	³⁄₁₆	1.76	8.005	8	0.515	½	1¼
WT 6x 17.5	5.17	6.250	6¼	0.300	⁵⁄₁₆	³⁄₁₆	1.88	6.560	6½	0.520	½	1
x 15	4.40	6.170	6⅛	0.260	¼	⅛	1.60	6.520	6½	0.440	⁷⁄₁₆	¹⁵⁄₁₆
x 13	3.82	6.110	6⅛	0.230	¼	⅛	1.41	6.490	6½	0.380	⅜	⅞
WT 6x 11	3.24	6.155	6⅛	0.260	¼	⅛	1.60	4.030	4	0.425	⁷⁄₁₆	⅞
x 9.5	2.79	6.080	6⅛	0.235	¼	⅛	1.43	4.005	4	0.350	⅜	¹³⁄₁₆
x 8	2.36	5.995	6	0.220	¼	⅛	1.32	3.990	4	0.265	¼	¾
x 7	2.08	5.955	6	0.200	³⁄₁₆	⅛	1.19	3.970	4	0.225	¼	¹¹⁄₁₆

AMERICAN INSTITUTE OF STEEL CONSTRUCTION

Table S-5d Dimensions and physical properties of WT shapes

STRUCTURAL TEES
Cut from W shapes
Properties

$$C_c' = \sqrt{\frac{2\pi^2 E}{Q_s Q_a F_y}}, \quad Q_a = 1.0$$

Nominal Weight per Ft.	$\dfrac{d}{t_w}$	AXIS X-X				AXIS Y-Y			$F_y = 36$ ksi		$F_y = 50$ ksi	
		I	S	r	y	I	S	r	Q_s	C_c'	Q_s	C_c'
Lb.		In.⁴	In.³	In.	In.	In.⁴	In.³	In.	Q_s	C_c'	Q_s	C_c'
168	4.7	190	31.2	1.96	2.31	593	88.6	3.47	—	—	—	—
152.5	5.0	162	27.0	1.90	2.16	525	79.3	3.42	—	—	—	—
139.5	5.2	141	24.1	1.86	2.05	469	71.3	3.38	—	—	—	—
126	5.5	121	20.9	1.81	1.92	414	63.6	3.34	—	—	—	—
115	5.9	106	18.5	1.77	1.82	371	57.5	3.31	—	—	—	—
105	6.2	92.1	16.4	1.73	1.72	332	51.9	3.28	—	—	—	—
95	6.8	79.0	14.2	1.68	1.62	295	46.5	3.25	—	—	—	—
85	7.3	67.8	12.3	1.65	1.52	259	41.2	3.22	—	—	—	—
76	7.9	58.5	10.8	1.62	1.43	227	36.4	3.19	—	—	—	—
68	8.5	50.6	9.46	1.59	1.35	199	32.1	3.16	—	—	—	—
60	9.2	43.4	8.22	1.57	1.28	172	28.0	3.13	—	—	—	—
53	10.6	36.3	6.91	1.53	1.19	151	24.7	3.11	—	—	—	—
48	11.6	32.0	6.12	1.51	1.13	135	22.2	3.09	—	—	—	—
43.5	12.2	28.9	5.60	1.50	1.10	120	19.9	3.07	—	—	—	—
39.5	13.2	25.8	5.03	1.49	1.06	108	17.9	3.05	—	—	—	—
36	14.2	23.2	4.54	1.48	1.02	97.5	16.2	3.04	—	—	—	—
32.5	15.5	20.6	4.06	1.47	0.985	87.2	14.5	3.02	—	—	—	—
29	16.9	19.1	3.76	1.50	1.03	53.5	10.7	2.51	—	—	—	—
26.5	17.5	17.7	3.54	1.51	1.02	47.9	9.58	2.48	—	—	—	—
25	16.5	18.7	3.79	1.60	1.17	28.2	6.97	1.96	—	—	—	—
22.5	18.0	16.6	3.39	1.58	1.13	25.0	6.21	1.94	—	—	0.998	107
20	20.2	14.4	2.95	1.57	1.08	22.0	5.51	1.93	—	—	0.887	114
17.5	20.8	16.0	3.23	1.76	1.30	12.2	3.73	1.54	—	—	0.856	116
15	23.7	13.5	2.75	1.75	1.27	10.2	3.12	1.52	0.891	134	0.710	127
13	26.6	11.7	2.40	1.75	1.25	8.66	2.67	1.51	0.767	144	0.565	142
11	23.7	11.7	2.59	1.90	1.63	2.33	1.16	0.847	0.891	134	0.710	127
9.5	25.9	10.1	2.28	1.90	1.65	1.88	0.939	0.822	0.797	141	0.596	139
8	27.2	8.70	2.04	1.92	1.74	1.41	0.706	0.773	0.741	146	0.541	146
7	29.8	7.67	1.83	1.92	1.76	1.18	0.594	0.753	0.626	159	0.450	159

Where no value of C_c' or Q_s is shown, the Tee complies with Specification Sect. 1.9.1.2.

AMERICAN INSTITUTE OF STEEL CONSTRUCTION

Table S-5d Dimensions and physical properties of WT shapes *(continued)*

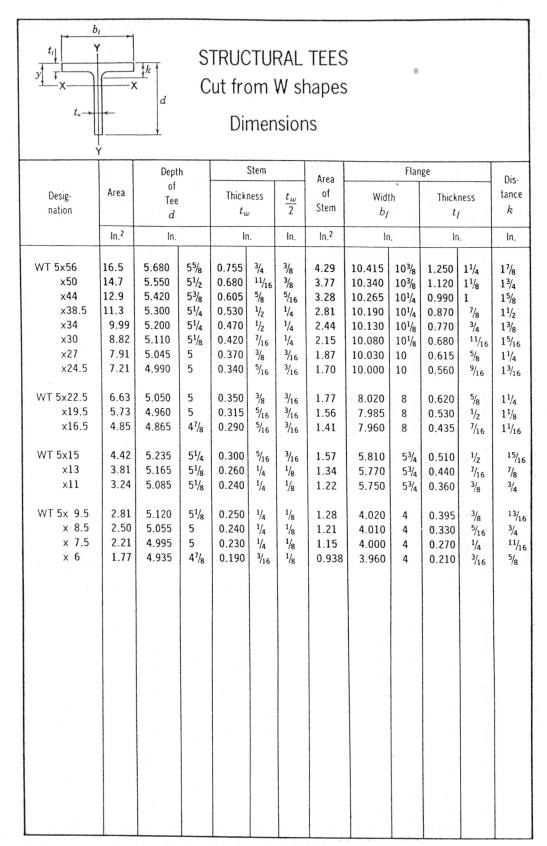

STRUCTURAL TEES
Cut from W shapes
Dimensions

Desig-nation	Area	Depth of Tee d		Stem		Area of Stem	Flange				Dis-tance k
				Thickness t_w	$\frac{t_w}{2}$		Width b_f		Thickness t_f		
	In.²	In.		In.	In.	In.²	In.		In.		In.
WT 5×56	16.5	5.680	5⅝	0.755	¾	⅜	4.29	10.415	10⅜	1.250	1¼	1⅞
×50	14.7	5.550	5½	0.680	11/16	⅜	3.77	10.340	10⅜	1.120	1⅛	1¾
×44	12.9	5.420	5⅜	0.605	⅝	5/16	3.28	10.265	10¼	0.990	1	1⅝
×38.5	11.3	5.300	5¼	0.530	½	¼	2.81	10.190	10¼	0.870	⅞	1½
×34	9.99	5.200	5¼	0.470	½	¼	2.44	10.130	10⅛	0.770	¾	1⅜
×30	8.82	5.110	5⅛	0.420	7/16	¼	2.15	10.080	10⅛	0.680	11/16	1 5/16
×27	7.91	5.045	5	0.370	⅜	3/16	1.87	10.030	10	0.615	⅝	1¼
×24.5	7.21	4.990	5	0.340	5/16	3/16	1.70	10.000	10	0.560	9/16	1 3/16
WT 5×22.5	6.63	5.050	5	0.350	⅜	3/16	1.77	8.020	8	0.620	⅝	1¼
×19.5	5.73	4.960	5	0.315	5/16	3/16	1.56	7.985	8	0.530	½	1⅛
×16.5	4.85	4.865	4⅞	0.290	5/16	3/16	1.41	7.960	8	0.435	7/16	1 1/16
WT 5×15	4.42	5.235	5¼	0.300	5/16	3/16	1.57	5.810	5¾	0.510	½	15/16
×13	3.81	5.165	5⅛	0.260	¼	⅛	1.34	5.770	5¾	0.440	7/16	⅞
×11	3.24	5.085	5⅛	0.240	¼	⅛	1.22	5.750	5¾	0.360	⅜	¾
WT 5× 9.5	2.81	5.120	5⅛	0.250	¼	⅛	1.28	4.020	4	0.395	⅜	13/16
× 8.5	2.50	5.055	5	0.240	¼	⅛	1.21	4.010	4	0.330	5/16	¾
× 7.5	2.21	4.995	5	0.230	¼	⅛	1.15	4.000	4	0.270	¼	11/16
× 6	1.77	4.935	4⅞	0.190	3/16	⅛	0.938	3.960	4	0.210	3/16	⅝

AMERICAN INSTITUTE OF STEEL CONSTRUCTION

Table S-5e Dimensions and physical properties of WT shapes

STRUCTURAL TEES
Cut from W shapes
Properties

$$C_c' = \sqrt{\frac{2\pi^2 E}{Q_s Q_a F_y}}, \quad Q_a = 1.0$$

Nominal Weight per Ft.	$\dfrac{d}{t_w}$	AXIS X-X				AXIS Y-Y			$F_y = 36$ ksi		$F_y = 50$ ksi	
		I	S	r	y	I	S	r	Q_s	C_c'	Q_s	C_c'
Lb.		In.⁴	In.³	In.	In.	In.⁴	In.³	In.				
56	7.5	28.6	6.40	1.32	1.21	118	22.6	2.68	—	—	—	—
50	8.2	24.5	5.56	1.29	1.13	103	20.0	2.65	—	—	—	—
44	9.0	20.8	4.77	1.27	1.06	89.3	17.4	2.63	—	—	—	—
38.5	10.0	17.4	4.04	1.24	0.990	76.8	15.1	2.60	—	—	—	—
34	11.1	14.9	3.49	1.22	0.932	66.8	13.2	2.59	—	—	—	—
30	12.2	12.9	3.04	1.21	0.884	58.1	11.5	2.57	—	—	—	—
27	13.6	11.1	2.64	1.19	0.836	51.7	10.3	2.56	—	—	—	—
24.5	14.7	10.0	2.39	1.18	0.807	46.7	9.34	2.54	—	—	—	—
22.5	14.4	10.2	2.47	1.24	0.907	26.7	6.65	2.01	—	—	—	—
19.5	15.7	8.84	2.16	1.24	0.876	22.5	5.64	1.98	—	—	—	—
16.5	16.8	7.71	1.93	1.26	0.869	18.3	4.60	1.94	—	—	—	—
15	17.4	9.28	2.24	1.45	1.10	8.35	2.87	1.37	—	—	—	—
13	19.9	7.86	1.91	1.44	1.06	7.05	2.44	1.36	—	—	0.902	113
11	21.2	6.88	1.72	1.46	1.07	5.71	1.99	1.33	0.999	126	0.836	117
9.5	20.5	6.68	1.74	1.54	1.28	2.15	1.07	0.874	—	—	0.872	115
8.5	21.1	6.06	1.62	1.56	1.32	1.78	0.888	0.844	—	—	0.841	117
7.5	21.7	5.45	1.50	1.57	1.37	1.45	0.723	0.810	0.977	128	0.811	119
6	26.0	4.35	1.22	1.57	1.36	1.09	0.551	0.785	0.793	142	0.592	139

Where no value of C_c' or Q_s is shown, the Tee complies with Specification Sect. 1.9.1.2.

AMERICAN INSTITUTE OF STEEL CONSTRUCTION

Table S-5e Dimensions and physical properties of WT shapes *(continued)*

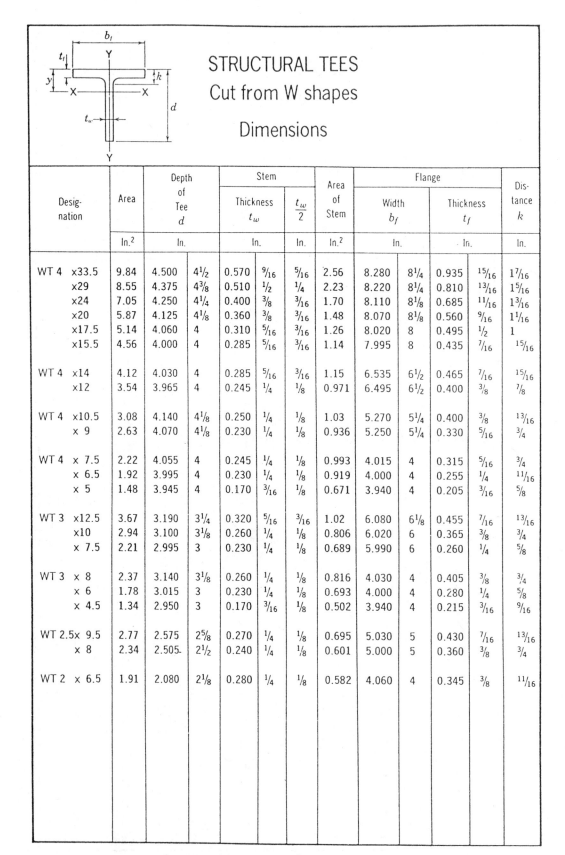

STRUCTURAL TEES
Cut from W shapes
Dimensions

Desig-nation	Area In.²	Depth of Tee d In.		Stem Thickness t_w In.		$\dfrac{t_w}{2}$ In.	Area of Stem In.²	Flange Width b_f In.		Flange Thickness t_f In.		Distance k In.
WT 4 x33.5	9.84	4.500	4½	0.570	9/16	5/16	2.56	8.280	8¼	0.935	15/16	1 7/16
x29	8.55	4.375	4⅜	0.510	½	¼	2.23	8.220	8¼	0.810	13/16	1 5/16
x24	7.05	4.250	4¼	0.400	⅜	3/16	1.70	8.110	8⅛	0.685	11/16	1 3/16
x20	5.87	4.125	4⅛	0.360	⅜	3/16	1.48	8.070	8⅛	0.560	9/16	1 1/16
x17.5	5.14	4.060	4	0.310	5/16	3/16	1.26	8.020	8	0.495	½	1
x15.5	4.56	4.000	4	0.285	5/16	3/16	1.14	7.995	8	0.435	7/16	15/16
WT 4 x14	4.12	4.030	4	0.285	5/16	3/16	1.15	6.535	6½	0.465	7/16	15/16
x12	3.54	3.965	4	0.245	¼	⅛	0.971	6.495	6½	0.400	⅜	⅞
WT 4 x10.5	3.08	4.140	4⅛	0.250	¼	⅛	1.03	5.270	5¼	0.400	⅜	13/16
x 9	2.63	4.070	4⅛	0.230	¼	⅛	0.936	5.250	5¼	0.330	5/16	¾
WT 4 x 7.5	2.22	4.055	4	0.245	¼	⅛	0.993	4.015	4	0.315	5/16	¾
x 6.5	1.92	3.995	4	0.230	¼	⅛	0.919	4.000	4	0.255	¼	11/16
x 5	1.48	3.945	4	0.170	3/16	⅛	0.671	3.940	4	0.205	3/16	⅝
WT 3 x12.5	3.67	3.190	3¼	0.320	5/16	3/16	1.02	6.080	6⅛	0.455	7/16	13/16
x10	2.94	3.100	3⅛	0.260	¼	⅛	0.806	6.020	6	0.365	⅜	¾
x 7.5	2.21	2.995	3	0.230	¼	⅛	0.689	5.990	6	0.260	¼	⅝
WT 3 x 8	2.37	3.140	3⅛	0.260	¼	⅛	0.816	4.030	4	0.405	⅜	¾
x 6	1.78	3.015	3	0.230	¼	⅛	0.693	4.000	4	0.280	¼	⅝
x 4.5	1.34	2.950	3	0.170	3/16	⅛	0.502	3.940	4	0.215	3/16	9/16
WT 2.5x 9.5	2.77	2.575	2⅝	0.270	¼	⅛	0.695	5.030	5	0.430	7/16	13/16
x 8	2.34	2.505	2½	0.240	¼	⅛	0.601	5.000	5	0.360	⅜	¾
WT 2 x 6.5	1.91	2.080	2⅛	0.280	¼	⅛	0.582	4.060	4	0.345	⅜	11/16

Table S-5f Dimensions and physical properties of WT shapes

STRUCTURAL TEES
Cut from W shapes
Properties

Nominal Weight per Ft.	$\dfrac{d}{t_w}$	AXIS X-X				AXIS Y-Y			$C_c' = \sqrt{\dfrac{2\pi^2 E}{Q_s Q_a F_y}}$, $Q_a = 1.0$			
									$F_y = 36$ ksi		$F_y = 50$ ksi	
		I	S	r	y	I	S	r	Q_s	C_c'	Q_s	C_c'
Lb.		In.⁴	In.³	In.	In.	In.⁴	In.³	In.	Q_s	C_c'	Q_s	C_c'
33.5	7.9	10.9	3.05	1.05	0.936	44.3	10.7	2.12	—	—	—	—
29	8.6	9.12	2.61	1.03	0.874	37.5	9.13	2.10	—	—	—	—
24	10.6	6.85	1.97	0.986	0.777	30.5	7.52	2.08	—	—	—	—
20	11.5	5.73	1.69	0.988	0.735	24.5	6.08	2.04	—	—	—	—
17.5	13.1	4.81	1.43	0.967	0.688	21.3	5.31	2.03	—	—	—	—
15.5	14.0	4.28	1.28	0.968	0.667	18.5	4.64	2.02	—	—	—	—
14	14.1	4.22	1.28	1.01	0.734	10.8	3.31	1.62	—	—	—	—
12	16.2	3.53	1.08	0.999	0.695	9.14	2.81	1.61	—	—	—	—
10.5	16.6	3.90	1.18	1.12	0.831	4.89	1.85	1.26	—	—	—	—
9	17.7	3.41	1.05	1.14	0.834	3.98	1.52	1.23	—	—	—	—
7.5	16.6	3.28	1.07	1.22	0.998	1.70	0.849	0.876	—	—	—	—
6.5	17.4	2.89	0.974	1.23	1.03	1.37	0.683	0.843	—	—	—	—
5	23.2	2.15	0.717	1.20	0.953	1.05	0.532	0.841	0.913	132	0.735	125
12.5	10.0	2.28	0.886	0.789	0.610	8.53	2.81	1.52	—	—	—	—
10	11.9	1.76	0.693	0.774	0.560	6.64	2.21	1.50	—	—	—	—
7.5	13.0	1.41	0.577	0.797	0.558	4.66	1.56	1.45	—	—	—	—
8	12.1	1.69	0.685	0.844	0.676	2.21	1.10	0.966	—	—	—	—
6	13.1	1.32	0.564	0.861	0.677	1.50	0.748	0.918	—	—	—	—
4.5	17.4	0.950	0.408	0.842	0.623	1.10	0.557	0.905	—	—	—	—
9.5	9.5	1.01	0.485	0.605	0.487	4.56	1.82	1.28	—	—	—	—
8	10.4	0.845	0.413	0.601	0.458	3.75	1.50	1.27	—	—	—	—
6.5	7.4	0.526	0.321	0.524	0.440	1.93	0.950	1.00	—	—	—	—

Where no value of C_c' or Q_s is shown, the Tee complies with Specification Sect. 1.9.1.2.

AMERICAN INSTITUTE OF STEEL CONSTRUCTION

Table S-5f Dimensions and physical properties of WT shapes *(continued)*

STRUCTURAL TEES
Cut from S shapes
Dimensions

Designation	Area	Depth of Tee d		Stem Thickness t_w		$\dfrac{t_w}{2}$	Area of Stem	Flange Width b_f		Flange Thickness t_f		Distance k	Grip	Max. Flge. Fastener
	In.²	In.		In.		In.	In.²	In.		In.		In.	In.	In.
ST 12 x 60.5	17.8	12.250	12¼	0.800	13/16	7/16	9.80	8.050	8	1.090	1 1/16	2	1 1/8	1
12 x 53	15.6	12.250	12¼	0.620	5/8	5/16	7.59	7.870	7 7/8	1.090	1 1/16	2	1 1/8	1
ST 12 x 50	14.7	12.000	12	0.745	3/4	3/8	8.94	7.245	7 1/4	0.870	7/8	1 3/4	7/8	1
12 x 45	13.2	12.000	12	0.625	5/8	5/16	7.50	7.125	7 1/8	0.870	7/8	1 3/4	7/8	1
12 x 40	11.7	12.000	12	0.500	1/2	1/4	6.00	7.000	7	0.870	7/8	1 3/4	7/8	1
ST 10 x 48	14.1	10.150	10 1/8	0.800	13/16	7/16	8.12	7.200	7 1/4	0.920	15/16	1 3/4	15/16	1
10 x 43	12.7	10.150	10 1/8	0.660	11/16	3/8	6.70	7.060	7	0.920	15/16	1 3/4	15/16	1
ST 10 x 37.5	11.0	10.000	10	0.635	5/8	5/16	6.35	6.385	6 3/8	0.795	13/16	1 5/8	13/16	7/8
10 x 33	9.70	10.000	10	0.505	1/2	1/4	5.05	6.255	6 1/4	0.795	13/16	1 5/8	13/16	7/8
ST 9 x 35	10.3	9.000	9	0.711	11/16	3/8	6.40	6.251	6 1/4	0.691	11/16	1 1/2	11/16	7/8
9 x 27.35	8.04	9.000	9	0.461	7/16	1/4	4.15	6.001	6	0.691	11/16	1 1/2	11/16	7/8
ST 7.5 x 25	7.35	7.500	7½	0.550	9/16	5/16	4.13	5.640	5 5/8	0.622	5/8	1 3/8	9/16	3/4
7.5 x 21.45	6.31	7.500	7½	0.411	7/16	1/4	3.08	5.501	5 1/2	0.622	5/8	1 3/8	9/16	3/4
ST 6 x 25	7.35	6.000	6	0.687	11/16	3/8	4.12	5.477	5 1/2	0.659	11/16	1 7/16	11/16	3/4
6 x 20.4	6.00	6.000	6	0.462	7/16	1/4	2.77	5.252	5 1/4	0.659	11/16	1 7/16	5/8	3/4
ST 6 x 17.5	5.15	6.000	6	0.428	7/16	1/4	2.57	5.078	5 1/8	0.545	9/16	1 3/16	1/2	3/4
6 x 15.9	4.68	6.000	6	0.350	3/8	3/16	2.10	5.000	5	0.544	9/16	1 3/16	1/2	3/4
ST 5 x 17.5	5.15	5.000	5	0.594	5/8	5/16	2.97	4.944	5	0.491	1/2	1 1/8	1/2	3/4
5 x 12.7	3.73	5.000	5	0.311	5/16	3/16	1.55	4.661	4 5/8	0.491	1/2	1 1/8	1/2	3/4
ST 4 x 11.5	3.38	4.000	4	0.441	7/16	1/4	1.76	4.171	4 1/8	0.425	7/16	1	7/16	3/4
4 x 9.2	2.70	4.000	4	0.271	1/4	1/8	1.08	4.001	4	0.425	7/16	1	7/16	3/4
ST 3.5 x 10	2.94	3.500	3½	0.450	7/16	1/4	1.57	3.860	3 7/8	0.392	3/8	15/16	3/8	5/8
3.5 x 7.65	2.25	3.500	3½	0.252	1/4	1/8	0.882	3.662	3 5/8	0.392	3/8	15/16	3/8	5/8
ST 3 x 8.625	2.53	3.000	3	0.465	7/16	1/4	1.39	3.565	3 5/8	0.359	3/8	7/8	3/8	5/8
3 x 6.25	1.83	3.000	3	0.232	1/4	1/8	0.696	3.332	3 3/8	0.359	3/8	7/8	3/8	—
ST 2.5 x 7.375	2.17	2.500	2½	0.494	1/2	1/4	1.23	3.284	3 1/4	0.326	5/16	13/16	5/16	—
2.5 x 5	1.47	2.500	2½	0.214	3/16	1/8	0.535	3.004	3	0.326	5/16	13/16	5/16	—
ST 2 x 4.75	1.40	2.000	2	0.326	5/16	3/16	0.652	2.796	2 3/4	0.293	5/16	3/4	5/16	—
2 x 3.85	1.13	2.000	2	0.193	3/16	1/8	0.386	2.663	2 5/8	0.293	5/16	3/4	5/16	—
ST 1.5 x 3.75	1.10	1.500	1½	0.349	3/8	3/16	0.523	2.509	2 1/2	0.260	1/4	11/16	1/4	—
1.5 x 2.85	0.835	1.500	1½	0.170	3/16	1/8	0.255	2.330	2 3/8	0.260	1/4	11/16	1/4	—

AMERICAN INSTITUTE OF STEEL CONSTRUCTION

Table S-6 Dimensions and physical properties of ST shapes

STRUCTURAL TEES
Cut from S shapes
Properties

$$C_c' = \sqrt{\frac{2\pi^2 E}{Q_s Q_a F_y}}, Q_a = 1.0$$

Nominal Wt. per Ft.	$\frac{d}{t_w}$	AXIS X-X				AXIS Y-Y			$F_y = 36$ ksi		$F_y = 50$ ksi	
		I	S	r	y	I	S	r	Q_s	C_c'	Q_s	C_c'
Lb.		In.⁴	In.³	In.	In.	In.⁴	In.³	In.				
60.5	15.3	259	30.1	3.82	3.63	41.7	10.4	1.53	—	—	—	—
53	19.8	216	24.1	3.72	3.28	38.5	9.80	1.57	—	—	0.907	112
50	16.1	215	26.3	3.83	3.84	23.8	6.58	1.27	—	—	—	—
45	19.2	190	22.6	3.79	3.60	22.5	6.31	1.30	—	—	0.937	111
40	24.0	162	18.7	3.72	3.29	21.1	6.04	1.34	0.878	135	0.695	128
48	12.7	143	20.3	3.18	3.13	25.1	6.97	1.33	—	—	—	—
43	15.4	125	17.2	3.14	2.91	23.4	6.63	1.36	—	—	—	—
37.5	15.7	109	15.8	3.15	3.07	14.9	4.66	1.16	—	—	—	—
33	19.8	93.1	12.9	3.10	2.81	13.8	4.43	1.19	—	—	0.907	112
35	12.7	84.7	14.0	2.87	2.94	12.1	3.86	1.08	—	—	—	—
27.35	19.5	62.4	9.61	2.79	2.50	10.4	3.47	1.14	—	—	0.922	111
25	13.6	40.6	7.73	2.35	2.25	7.85	2.78	1.03	—	—	—	—
21.45	18.2	33.0	6.00	2.29	2.01	7.19	2.61	1.07	—	—	0.988	108
25	8.7	25.2	6.05	1.85	1.84	7.85	2.87	1.03	—	—	—	—
20.4	13.0	18.9	4.28	1.78	1.58	6.78	2.58	1.06	—	—	—	—
17.5	14.0	17.2	3.95	1.83	1.64	4.94	1.95	0.980	—	—	—	—
15.9	17.1	14.9	3.31	1.78	1.51	4.68	1.87	1.00	—	—	—	—
17.5	8.4	12.5	3.63	1.56	1.56	4.18	1.69	0.901	—	—	—	—
12.7	16.1	7.83	2.06	1.45	1.20	3.39	1.46	0.954	—	—	—	—
11.5	9.1	5.03	1.77	1.22	1.15	2.15	1.03	0.798	—	—	—	—
9.2	14.8	3.51	1.15	1.14	0.941	1.86	0.932	0.831	—	—	—	—
10	7.8	3.36	1.36	1.07	1.04	1.59	0.821	0.734	—	—	—	—
7.65	13.9	2.19	0.816	0.987	0.817	1.32	0.720	0.766	—	—	—	—
8.625	6.5	2.13	1.02	0.917	0.914	1.15	0.648	0.675	—	—	—	—
6.25	12.9	1.27	0.552	0.833	0.691	0.911	0.547	0.705	—	—	—	—
7.375	5.1	1.27	0.740	0.764	0.789	0.833	0.507	0.620	—	—	—	—
5	11.7	0.681	0.353	0.681	0.569	0.608	0.405	0.643	—	—	—	—
4.75	6.1	0.470	0.325	0.580	0.553	0.451	0.323	0.569	—	—	—	—
3.85	10.4	0.316	0.203	0.528	0.448	0.382	0.287	0.581	—	—	—	—
3.75	4.3	0.204	0.191	0.430	0.432	0.293	0.234	0.516	—	—	—	—
2.85	8.8	0.118	0.101	0.376	0.329	0.227	0.195	0.522	—	—	—	—

Where no value of C_c' or Q_s is shown, the Tee complies with Specification Sect. 1.9.1.2.

AMERICAN INSTITUTE OF STEEL CONSTRUCTION

Table S-6 Dimensions and physical properties of ST shapes *(continued)*

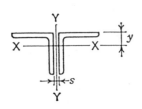

DOUBLE ANGLES
Two equal leg angles
Properties of sections

Designation	Wt. per Ft. 2 Angles Lb.	Area of 2 Angles In.²	AXIS X — X I In.⁴	S In.³	r In.	y In.	AXIS Y — Y Radii of Gyration Back to Back of Angles, Inches 0	³⁄₈	³⁄₄	Q_s* Angles in Contact $F_y =$ 36 ksi	$F_y =$ 50 ksi	Angles Separated $F_y =$ 36 ksi	$F_y =$ 50 ksi
L 8 x8 x1¹⁄₈	113.8	33.5	195.0	35.1	2.42	2.41	3.42	3.55	3.69	—	—	—	—
1	102.0	30.0	177.0	31.6	2.44	2.37	3.40	3.53	3.67	—	—	—	—
⁷⁄₈	90.0	26.5	159.0	28.0	2.45	2.32	3.38	3.51	3.64	—	—	—	—
³⁄₄	77.8	22.9	139.0	24.4	2.47	2.28	3.36	3.49	3.62	—	—	—	—
⁵⁄₈	65.4	19.2	118.0	20.6	2.49	2.23	3.34	3.47	3.60	—	—	.997	.935
¹⁄₂	52.8	15.5	97.3	16.7	2.50	2.19	3.32	3.45	3.58	.995	.921	.911	.834
L 6 x6 x1	74.8	22.0	70.9	17.1	1.80	1.86	2.59	2.73	2.87	—	—	—	—
⁷⁄₈	66.2	19.5	63.8	15.3	1.81	1.82	2.57	2.70	2.85	—	—	—	—
³⁄₄	57.4	16.9	56.3	13.3	1.83	1.78	2.55	2.68	2.82	—	—	—	—
⁵⁄₈	48.4	14.2	48.3	11.3	1.84	1.73	2.53	2.66	2.80	—	—	—	—
¹⁄₂	39.2	11.5	39.8	9.23	1.86	1.68	2.51	2.64	2.78	—	—	—	.961
³⁄₈	29.8	8.72	30.8	7.06	1.88	1.64	2.49	2.62	2.75	.995	.921	.911	.834
L 5 x5 x ⁷⁄₈	54.4	16.0	35.5	10.3	1.49	1.57	2.16	2.30	2.45	—	—	—	—
³⁄₄	47.2	13.9	31.5	9.06	1.51	1.52	2.14	2.28	2.42	—	—	—	—
¹⁄₂	32.4	9.50	22.5	6.31	1.54	1.43	2.10	2.24	2.38	—	—	—	—
³⁄₈	24.6	7.22	17.5	4.84	1.56	1.39	2.09	2.22	2.35	—	—	.982	.919
⁵⁄₁₆	20.6	6.05	14.8	4.08	1.57	1.37	2.08	2.21	2.34	.995	.921	.911	.834
L 4 x4 x ³⁄₄	37.0	10.9	15.3	5.62	1.19	1.27	1.74	1.88	2.83	—	—	—	—
⁵⁄₈	31.4	9.22	13.3	4.80	1.20	1.23	1.72	1.86	2.00	—	—	—	—
¹⁄₂	25.6	7.50	11.1	3.95	1.22	1.18	1.70	1.83	1.98	—	—	—	—
³⁄₈	19.6	5.72	8.72	3.05	1.23	1.14	1.68	1.81	1.95	—	—	—	—
⁵⁄₁₆	16.4	4.80	7.43	2.58	1.24	1.12	1.67	1.80	1.94	—	—	.997	.935
¹⁄₄	13.2	3.88	6.08	2.09	1.25	1.09	1.66	1.79	1.93	.995	.921	.911	.834

* Where no value of Q_s is shown, the angles comply with Specification Sect. 1.9.1.2 and may be considered fully effective.

For $F_y = 36$ ksi: $C'_c = 126.1/\sqrt{Q_s}$
For $F_y = 50$ ksi: $C'_c = 107.0/\sqrt{Q_s}$

AMERICAN INSTITUTE OF STEEL CONSTRUCTION

Table S-7a Dimensions and physical properties of double angles

DOUBLE ANGLES
Two equal leg angles
Properties of sections

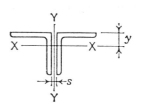

Designation	Wt. per Ft. 2 Angles	Area of 2 Angles	AXIS X — X				AXIS Y — Y Radii of Gyration Back to Back of Angles, Inches			Q_s* Angles in Contact		Angles Separated	
			I	S	r	y	0	3/8	3/4	$F_y =$ 36 ksi	$F_y =$ 50 ksi	$F_y =$ 36 ksi	$F_y =$ 50 ksi
	Lb.	In.²	In.⁴	In.³	In.	In.							
L 3½x3½x 3/8	17.0	4.97	5.73	2.30	1.07	1.01	1.48	1.61	1.75	—	—	—	—
5/16	14.4	4.18	4.90	1.95	1.08	.990	1.47	1.60	1.74	—	—	—	.986
1/4	11.6	3.38	4.02	1.59	1.09	.968	1.46	1.59	1.73	—	.982	.965	.897
L 3 x3 x 1/2	18.8	5.50	4.43	2.14	.898	.932	1.29	1.43	1.59	—	—	—	—
3/8	14.4	4.22	3.52	1.67	.913	.888	1.27	1.41	1.56	—	—	—	—
5/16	12.2	3.55	3.02	1.41	.922	.865	1.26	1.40	1.55	—	—	—	—
1/4	9.8	2.88	2.49	1.15	.930	.842	1.26	1.39	1.53	—	—	—	.961
3/16	7.42	2.18	1.92	.882	.939	.820	1.25	1.38	1.52	.995	.921	.911	.834
L 2½x2½x 3/8	11.8	3.47	1.97	1.13	.753	.762	1.07	1.21	1.36	—	—	—	—
5/16	10.0	2.93	1.70	.964	.761	.740	1.06	1.20	1.35	—	—	—	—
1/4	8.2	2.38	1.41	.789	.769	.717	1.05	1.19	1.34	—	—	—	—
3/16	6.14	1.80	1.09	.685	.778	.694	1.04	1.18	1.32	—	—	.982	.919
L 2 x2 x 3/8	9.4	2.72	.958	.702	.594	.636	.870	1.01	1.17	—	—	—	—
5/16	7.84	2.30	.832	.681	.601	.614	.859	1.00	1.16	—	—	—	—
1/4	6.38	1.88	.695	.494	.609	.592	.849	.989	1.14	—	—	—	—
3/16	4.88	1.43	.545	.381	.617	.569	.840	.977	1.13	—	—	—	—
1/8	3.30	.960	.380	.261	.626	.546	.831	.965	1.11	.995	.921	.911	.834

* Where no value of Q_s is shown, the angles comply with Specification Sect. 1.9.1.2 and may be considered fully effective.

For $F_y = 36$ ksi: $C'_c = 126.1/\sqrt{Q_s}$

For $F_y = 50$ ksi: $C'_c = 107.0/\sqrt{Q_s}$

AMERICAN INSTITUTE OF STEEL CONSTRUCTION

Table S-7a Dimensions and physical properties of double angles *(continued)*

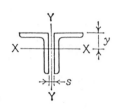

DOUBLE ANGLES
Two unequal leg angles
Properties of sections

Long legs back to back

Designation	Wt. per Ft. 2 Angles	Area of 2 Angles	AXIS X — X				AXIS Y — Y				Q_s*			
								Radii of Gyration Back to Back of Angles, Inches			Angles in Contact		Angles Separated	
			I	S	r	y	0	$3/8$	$3/4$	$F_y =$ 36ksi	$F_y =$ 50ksi	$F_y =$ 36ksi	$F_y =$ 50ksi
	Lb.	In.²	In.⁴	In.³	In.	In.							
L 8 x6 x1	88.4	26.0	161.0	30.2	2.49	2.65	2.39	2.52	2.66	—	—	—	—
¾	67.6	19.9	126.0	23.3	2.53	2.56	2.35	2.48	2.62	—	—	—	—
½	46.0	13.5	88.6	16.0	2.56	2.47	2.32	2.44	2.57	—	—	.911	.834
L 8 x4 x1	74.8	22.0	139.0	28.1	2.52	3.05	1.47	1.61	1.75	—	—	—	—
¾	57.4	16.9	109.0	21.8	2.55	2.95	1.42	1.55	1.69	—	—	—	—
½	39.2	11.5	77.0	15.0	2.59	2.86	1.38	1.51	1.64	—	—	.911	.834
L 7 x4 x ¾	52.4	15.4	75.6	16.8	2.22	2.51	1.48	1.62	1.76	—	—	—	—
½	35.8	10.5	53.3	11.6	2.25	2.42	1.44	1.57	1.71	—	—	.965	.897
⅜	27.2	7.97	41.1	8.88	2.27	2.37	1.43	1.55	1.68	—	—	.839	.750
L 6 x4 x ¾	47.2	13.9	49.0	12.5	1.88	2.08	1.55	1.69	1.83	—	—	—	—
⅝	40.0	11.7	42.1	10.6	1.90	2.03	1.53	1.67	1.81	—	—	—	—
½	32.4	9.50	34.8	8.67	1.91	1.99	1.51	1.64	1.78	—	—	—	.961
⅜	24.6	7.22	26.9	6.64	1.93	1.94	1.50	1.62	1.76	—	—	.911	.834
L 6 x3½x ⅜	23.4	6.84	25.7	6.49	1.94	2.04	1.26	1.39	1.53	—	—	.911	.834
5/16	19.6	5.74	21.8	5.47	1.95	2.01	1.26	1.38	1.51	—	—	.825	.733
L 5 x3½x ¾	39.6	11.6	27.8	8.55	1.55	1.75	1.40	1.53	1.68	—	—	—	—
½	27.2	8.00	20.0	5.97	1.58	1.66	1.35	1.49	1.63	—	—	—	—
⅜	20.8	6.09	15.6	4.59	1.60	1.61	1.34	1.46	1.60	—	—	.982	.919
5/16	17.4	5.12	13.2	3.87	1.61	1.59	1.33	1.45	1.59	—	—	.911	.834
L 5 x3 x ½	25.6	7.50	18.9	5.82	1.59	1.75	1.12	1.25	1.40	—	—	—	—
⅜	19.6	5.72	14.7	4.47	1.61	1.70	1.10	1.23	1.37	—	—	.982	.919
5/16	16.4	4.80	12.5	3.77	1.61	1.68	1.09	1.22	1.36	—	—	.911	.834
¼	13.2	3.88	10.2	3.06	1.62	1.66	1.08	1.21	1.34	—	—	.804	.708

* Where no value of Q_s is shown, the angles comply with Specification Sect. 1.9.1.2 and may be considered fully effective.

For F_y = 36 ksi: $C'_c = 126.1/\sqrt{Q_s}$

For F_y = 50 ksi: $C'_c = 107.0/\sqrt{Q_s}$

AMERICAN INSTITUTE OF STEEL CONSTRUCTION

Table S-7b Dimensions and physical properties of double angles

DOUBLE ANGLES

Two unequal leg angles

Properties of sections

Long legs back to back

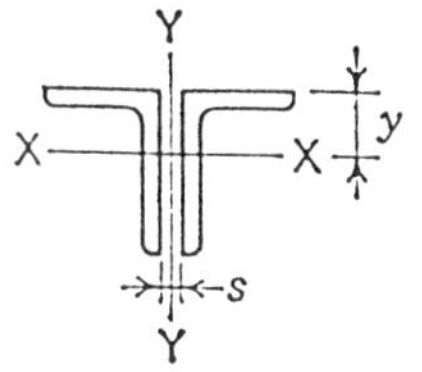

Designation	Wt. per Ft. 2 Angles	Area of 2 Angles	AXIS X — X				AXIS Y — Y: Radii of Gyration, Back to Back of Angles, Inches			Q_s*: Angles in Contact		Angles Separated	
			I	S	r	y				F_y = 36 ksi	F_y = 50 ksi	F_y = 36 ksi	F_y = 50 ksi
	Lb.	In.^{2}	In.^{4}	In.^{3}	In.	In.	0	3/8	3/4				
L 4 x3½x ½	23.8	7.00	10.6	3.87	1.23	1.25	1.44	1.58	1.72	—	—	—	—
3/8	18.2	5.34	8.35	2.99	1.25	1.21	1.42	1.56	1.70	—	—	—	—
5/16	15.4	4.49	7.12	2.53	1.26	1.18	1.42	1.55	1.69	—	—	.997	.935
1/4	12.4	3.63	5.83	2.05	1.27	1.16	1.41	1.54	1.67	—	.982	.911	.834
L 4 x3 x ½	22.2	6.50	10.1	3.78	1.25	1.33	1.20	1.33	1.48	—	—	—	—
3/8	17.0	4.97	7.93	2.92	1.26	1.28	1.18	1.31	1.45	—	—	—	—
5/16	14.4	4.18	6.76	2.47	1.27	1.26	1.17	1.30	1.44	—	—	.997	.935
1/4	11.6	3.38	5.54	2.00	1.28	1.24	1.16	1.29	1.43	—	—	.911	.834
L 3½x3 x 3/8	15.8	4.59	5.45	2.25	1.09	1.08	1.22	1.36	1.50	—	—	—	—
5/16	13.2	3.87	4.66	1.91	1.10	1.06	1.21	1.35	1.49	—	—	—	.986
1/4	10.8	3.13	3.83	1.55	1.11	1.04	1.20	1.33	1.48	—	—	.965	.897
L 3½x2½x 3/8	14.4	4.22	5.12	2.19	1.10	1.16	.976	1.11	1.26	—	—	—	—
5/16	12.2	3.55	4.38	1.85	1.11	1.14	.966	1.10	1.25	—	—	—	.986
1/4	9.8	2.88	3.60	1.51	1.12	1.11	.958	1.09	1.23	—	—	.965	.897
L 3 x2½x 3/8	13.2	3.84	3.31	1.62	.928	.956	1.02	1.16	1.31	—	—	—	—
1/4	9.0	2.63	2.35	1.12	.945	.911	1.00	1.13	1.28	—	—	—	.961
3/16	6.77	1.99	1.81	.859	.954	.888	.993	1.12	1.27	—	—	.911	.834
L 3 x2 x 3/8	11.8	3.47	3.06	1.56	.940	1.04	.777	.917	1.07	—	—	—	—
5/16	10.0	2.93	2.63	1.33	.948	1.02	.767	.903	1.06		—	—	—
1/4	8.2	2.38	2.17	1.08	.957	.993	.757	.891	1.04	—	—	—	.961
3/16	6.1	1.80	1.68	.830	.966	.970	.749	.879	1.03	—	—	.911	.834
L 2½x2 x 3/8	10.6	3.09	1.82	1.09	.768	.831	.819	.961	1.12	—	—	—	—
5/16	9.0	2.62	1.58	.932	.776	.809	.809	.948	1.10	—	—	—	—
1/4	7.2	2.13	1.31	.763	.784	.787	.799	.935	1.09	—	—	—	—
3/16	5.5	1.62	1.02	.586	.793	.764	.790	.923	1.07	—	—	.982	.919

* Where no value of Q_s is shown, the angles comply with Specification Sect. 1.9.1.2 and may be considered fully effective.

For F_y = 36 ksi: $C'_c = 126.1/\sqrt{Q_s}$

For F_y = 50 ksi: $C'_c = 107.0/\sqrt{Q_s}$

AMERICAN INSTITUTE OF STEEL CONSTRUCTION

Table S-7b Dimensions and physical properties of double angles *(continued)*

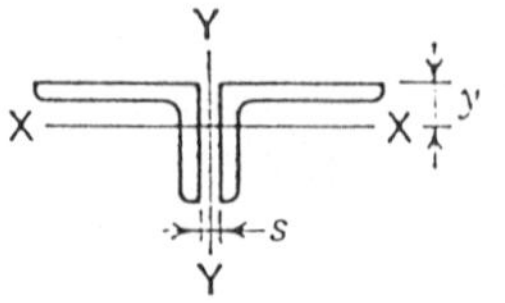

DOUBLE ANGLES

Two unequal leg angles

Properties of sections

Short legs back to back

Designation	Wt. per Ft. 2 Angles	Area of 2 Angles	AXIS X – X				AXIS Y – Y Radii of Gyration Back to Back of Angles, Inches			Q_s* Angles in Contact		Angles Separated	
			I	S	r	y				F_y = 36 ksi	F_y = 50 ksi	F_y = 36 ksi	F_y = 50 ksi
	Lb.	In.^{2}	In.^{4}	In.^{3}	In.	In.	0	3/8	3/4				
L 8 x6 x1	88.4	26.0	77.6	17.8	1.73	1.65	3.64	3.78	3.92	—	—	—	—
3/4	67.6	19.9	61.4	13.8	1.76	1.56	3.60	3.74	3.88	—	—	—	—
1/2	46.0	13.5	43.4	9.58	1.79	1.47	3.56	3.69	3.83	.995	.921	.911	.834
L 8 x4 x1	74.8	22.0	23.3	7.88	1.03	1.05	3.95	4.10	4.25	—	—	—	—
3/4	57.4	16.9	18.7	6.14	1.05	.953	3.90	4.05	4.19	—	—	—	—
1/2	39.2	11.5	13.5	4.29	1.08	.859	3.86	4.00	4.14	.995	.921	.911	.834
L 7 x4 x 3/4	52.4	15.4	16.1	6.05	1.09	1.01	3.35	3.49	3.64	—	—	—	—
1/2	35.8	10.5	13.1	4.23	1.11	.917	3.30	3.44	3.59	—	.982	.965	.897
3/8	27.2	7.97	10.2	3.26	1.13	.870	3.28	3.42	3.56	.926	.838	.839	.750
L 6 x4 x 3/4	47.2	13.9	17.4	5.94	1.12	1.08	2.80	2.94	3.09	—	—	—	—
5/8	40.0	11.7	15.0	5.07	1.13	1.03	2.78	2.92	3.06	—	—	—	—
1/2	32.4	9.50	12.5	4.16	1.15	.987	2.76	2.90	3.04	—	—	—	.961
3/8	24.6	7.22	9.81	3.21	1.17	.941	2.74	2.87	3.02	.995	.921	.911	.834
L 6 x3½x 3/8	23.4	6.84	6.68	2.46	.988	.787	2.81	2.95	3.09	.995	.921	.911	.834
5/16	19.6	5.74	5.70	2.08	.996	.763	2.80	2.94	3.08	.912	.822	.825	.733
L 5 x3½x 3/4	39.6	11.6	11.1	4.43	.977	.996	2.33	2.48	2.63	—	—	—	—
1/2	27.2	8.00	8.10	3.12	1.01	.906	2.29	2.43	2.57	—	—	—	—
3/8	20.8	6.09	6.37	2.41	1.02	.861	2.27	2.41	2.55	—	—	.982	.919
5/16	17.4	5.12	5.44	2.04	1.03	.838	2.26	2.39	2.54	.995	.921	.911	.834
L 5 x3 x 1/2	25.6	7.50	5.16	2.29	.829	.750	2.36	2.50	2.65	—	—	—	—
3/8	19.6	5.72	4.08	1.78	.845	.704	2.34	2.48	2.63	—	—	.982	.919
5/16	16.4	4.80	3.49	1.51	.853	.681	2.33	2.47	2.61	.995	.921	.911	.834
1/4	13.2	3.88	2.88	1.23	.861	.657	2.32	2.46	2.60	.891	.797	.804	.708

* Where no value of Q_s is shown, the angles comply with Specification Sect. 1.9.1.2 and may be considered fully effective.

For F_y = 36 ksi: $C'_c = 126.1/\sqrt{Q_s}$

For F_y = 50 ksi: $C'_c = 107.0/\sqrt{Q_s}$

AMERICAN INSTITUTE OF STEEL CONSTRUCTION

Table S-7c Dimensions and physical properties of double angles

DOUBLE ANGLES
Two unequal leg angles
Properties of sections

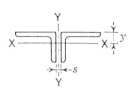

Short legs back to back

Designation	Wt. per Ft. 2 Angles	Area of 2 Angles	AXIS X — X				AXIS Y — Y				Q_s*			
			I	S	r	y	Radii of Gyration Back to Back of Angles, Inches			Angles in Contact		Angles Separated	
							0	$\frac{3}{8}$	$\frac{3}{4}$	$F_y =$ 36 ksi	$F_y =$ 50 ksi	$F_y =$ 36 ksi	$F_y =$ 50 ksi
	Lb.	In.²	In.⁴	In.³	In.	In.							
L 4 x3½x ½	23.8	7.00	7.58	3.03	1.04	1.00	1.76	1.89	2.04	—	—	—	—
⅜	18.2	5.34	5.97	2.35	1.06	.955	1.74	1.87	2.01	—	—	—	—
5/16	15.4	4.49	5.10	1.99	1.07	.932	1.73	1.86	2.00	—	—	.997	.935
¼	12.4	3.63	4.19	1.62	1.07	.909	1.72	1.85	1.99	.995	.921	.911	.834
L 4 x3 x ½	22.2	6.50	4.85	2.23	.864	.827	1.82	1.96	2.11	—	—	—	—
⅜	17.0	4.97	3.84	1.73	.879	.782	1.80	1.94	2.08	—	—	—	—
5/16	14.4	4.18	3.29	1.47	.887	.759	1.79	1.93	2.07	—	—	.997	.935
¼	11.6	3.38	2.71	1.20	.896	.736	1.78	1.92	2.06	.995	.921	.911	.834
L 3½x3 x ⅜	15.8	4.59	3.69	1.70	.897	.830	1.53	1.67	1.82	—	—	—	—
5/16	13.2	3.87	3.17	1.44	.905	.808	1.52	1.66	1.80	—	—	—	.986
¼	10.8	3.13	2.61	1.18	.914	.785	1.52	1.65	1.79	—	.982	.965	.897
L 3½x2½x ⅜	14.4	4.22	2.18	1.18	.719	.660	1.60	1.74	1.89	—	—	—	—
5/16	12.2	3.55	1.88	1.01	.727	.637	1.59	1.73	1.88	—	—	—	.986
¼	9.8	2.88	1.55	.824	.735	.614	1.58	1.72	1.86	—	.982	.965	.897
L 3 x2½x ⅜	13.2	3.84	2.08	1.16	.736	.706	1.33	1.47	1.62	—	—	—	—
¼	9.0	2.63	1.49	.808	.753	.661	1.31	1.45	1.60	—	—	—	.961
3/16	6.77	1.99	1.15	.620	.761	.638	1.30	1.44	1.58	.995	.921	.911	.834
L 3 x2 x ⅜	11.8	3.47	1.09	.743	.559	.539	1.40	1.55	1.70	—	—	—	—
5/16	10.0	2.93	.941	.634	.567	.516	1.39	1.53	1.68	—	—	—	—
¼	8.2	2.38	.784	.520	.574	.493	1.38	1.52	1.67	—	—	—	.961
3/16	6.1	1.80	.613	.401	.583	.470	1.37	1.51	1.66	.995	.921	.911	.834
L 2½x2 x ⅜	10.6	3.09	1.03	.725	.577	.581	1.13	1.28	1.43	—	—	—	—
5/16	9.0	2.62	.893	.620	.584	.559	1.12	1.26	1.42	—	—	—	—
¼	7.2	2.13	.745	.509	.592	.537	1.11	1.25	1.40	—	—	—	—
3/16	5.5	1.62	.583	.392	.600	.514	1.10	1.24	1.39	—	—	.982	.919

* Where no value of Q_s is shown, the angles comply with Specification Sect. 1.9.1.2 and may be considered fully effective.

For $F_y = 36$ ksi: $C'_c = 126.1/\sqrt{Q_s}$

For $F_y = 50$ ksi: $C'_c = 107.0/\sqrt{Q_s}$

AMERICAN INSTITUTE OF STEEL CONSTRUCTION

Table S-7c Dimensions and physical properties of double angles *(continued)*

Round Pipe and Square Tube

Pipe is not manufactured from ASTM A36 steel. Pipe is manufactured from ASTM A53 Grade B steel or from ASTM A501 steel. For either steel, the value of F_y may be taken as 36 ksi and the value of F_u may be taken as 58 ksi.

Structural tubing is manufactured from ASTM A500 Grade B steel. For square tube, the value of F_y is 46 ksi and the value of F_u is 58 ksi. These are the values used in computing the allowable column loads of Table S-13 for square tubes.

For use as beams, all pipe satisfies the requirements of ASD Table B5.1 for compact sections and all structural tube satisfies requirements for noncompact sections.

Square tube is manufactured with rounded corners. The outside radius may be as little as two times the wall thickness or as much as three times the wall thickness. Welding or otherwise fabricating structural tube must allow for some variation in the rounded corners.

PIPE
Dimensions and properties

	Dimensions			Weight per Foot Lbs. Plain Ends	Properties			
Nominal Diameter In.	Outside Diameter In.	Inside Diameter In.	Wall Thickness In.		A In.^{2}	I In.^{4}	S In.^{3}	r In.
Standard Weight								
$1/2$	.840	.622	.109	.85	.250	.017	.041	.261
$3/4$	1.050	.824	.113	1.13	.333	.037	.071	.334
1	1.315	1.049	.133	1.68	.494	.087	.133	.421
$1^1/4$	1.660	1.380	.140	2.27	.669	.195	.235	.540
$1^1/2$	1.900	1.610	.145	2.72	.799	.310	.326	.623
2	2.375	2.067	.154	3.65	1.07	.666	.561	.787
$2^1/2$	2.875	2.469	.203	5.79	1.70	1.53	1.06	.947
3	3.500	3.068	.216	7.58	2.23	3.02	1.72	1.16
$3^1/2$	4.000	3.548	.226	9.11	2.68	4.79	2.39	1.34
4	4.500	4.026	.237	10.79	3.17	7.23	3.21	1.51
5	5.563	5.047	.258	14.62	4.30	15.2	5.45	1.88
6	6.625	6.065	.280	18.97	5.58	28.1	8.50	2.25
8	8.625	7.981	.322	28.55	8.40	72.5	16.8	2.94
10	10.750	10.020	.365	40.48	11.9	161	29.9	3.67
12	12.750	12.000	.375	49.56	14.6	279	43.8	4.38
Extra Strong								
$1/2$	.840	.546	.147	1.09	.320	.020	.048	.250
$3/4$	1.050	.742	.154	1.47	.433	.045	.085	.321
1	1.315	.957	.179	2.17	.639	.106	.161	.407
$1^1/4$	1.660	1.278	.191	3.00	.881	.242	.291	.524
$1^1/2$	1.900	1.500	.200	3.63	1.07	.391	.412	.605
2	2.375	1.939	.218	5.02	1.48	.868	.731	.766
$2^1/2$	2.875	2.323	.276	7.66	2.25	1.92	1.34	.924
3	3.500	2.900	.300	10.25	3.02	3.89	2.23	1.14
$3^1/2$	4.000	3.364	.318	12.50	3.68	6.28	3.14	1.31
4	4.500	3.826	.337	14.98	4.41	9.61	4.27	1.48
5	5.563	4.813	.375	20.78	6.11	20.7	7.43	1.84
6	6.625	5.761	.432	28.57	8.40	40.5	12.2	2.19
8	8.625	7.625	.500	43.39	12.8	106	24.5	2.88
10	10.750	9.750	.500	54.74	16.1	212	39.4	3.63
12	12.750	11.750	.500	65.42	19.2	362	56.7	4.33
Double-Extra Strong								
2	2.375	1.503	.436	9.03	2.66	1.31	1.10	.703
$2^1/2$	2.875	1.771	.552	13.69	4.03	2.87	2.00	.844
3	3.500	2.300	.600	18.58	5.47	5.99	3.42	1.05
4	4.500	3.152	.674	27.54	8.10	15.3	6.79	1.37
5	5.563	4.063	.750	38.55	11.3	33.6	12.1	1.72
6	6.625	4.897	.864	53.16	15.6	66.3	20.0	2.06
8	8.625	6.875	.875	72.42	21.3	162	37.6	2.76

The listed sections are available in conformance with ASTM Specification A53 Grade B or A501. Other sections are made to these specifications. Consult with pipe manufacturers or distributors for availability.

AMERICAN INSTITUTE OF STEEL CONSTRUCTION

Table S-8 Dimensions and physical properties of round pipe

STRUCTURAL TUBING
Square
Dimensions and properties

DIMENSIONS			PROPERTIES**			
Nominal* Size	Wall Thickness		Weight per Foot	Area	I	S	r
In.	In.		Lb.	In.²	In.⁴	In.³	In.
16 x 16	.5000	1/2	103.30	30.4	1200	150	6.29
	.3750	3/8	78.52	23.1	931	116	6.35
	.3125	5/16	65.87	19.4	789	98.6	6.38
14 x 14	.5000	1/2	89.68	26.4	791	113	5.48
	.3750	3/8	68.31	20.1	615	87.9	5.54
	.3125	5/16	57.36	16.9	522	74.6	5.57
12 x 12	.5000	1/2	76.07	22.4	485	80.9	4.66
	.3750	3/8	58.10	17.1	380	63.4	4.72
	.3125	5/16	48.86	14.4	324	54.0	4.75
	.2500	1/4	39.43	11.6	265	44.1	4.78
10 x 10	.6250	5/8	76.33	22.4	321	64.2	3.78
	.5000	1/2	62.46	18.4	271	54.2	3.84
	.3750	3/8	47.90	14.1	214	42.9	3.90
	.3125	5/16	40.35	11.9	183	36.7	3.93
	.2500	1/4	32.63	9.59	151	30.1	3.96
8 x 8	.6250	5/8	59.32	17.4	153	38.3	2.96
	.5000	1/2	48.85	14.4	131	32.9	3.03
	.3750	3/8	37.69	11.1	106	26.4	3.09
	.3125	5/16	31.84	9.36	90.9	22.7	3.12
	.2500	1/4	25.82	7.59	75.1	18.8	3.15
	.1875	3/16	19.63	5.77	58.2	14.6	3.18
7 x 7	.5000	1/2	42.05	12.4	84.6	24.2	2.62
	.3750	3/8	32.58	9.58	68.7	19.6	2.68
	.3125	5/16	27.59	8.11	59.5	17.0	2.71
	.2500	1/4	22.42	6.59	49.4	14.1	2.74
	.1875	3/16	17.08	5.02	38.5	11.0	2.77

* Outside dimensions across flat sides.
** Properties are based upon a nominal outside corner radius equal to two times the wall thickness.

AMERICAN INSTITUTE OF STEEL CONSTRUCTION

Table S-9 Dimensions and physical properties of square tube

STRUCTÚRAL TUBING
Square
Dimensions and properties

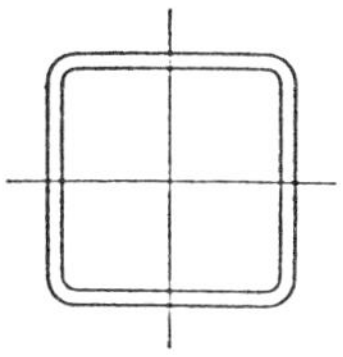

DIMENSIONS			PROPERTIES**			
Nominal* Size	Wall Thickness		Weight per Foot	Area	I	S	r
In.	In.		Lb.	In.²	In.⁴	In.³	In.
6 x 6	.5000	½	35.24	10.4	50.5	16.8	2.21
	.3750	⅜	27.48	8.08	41.6	13.9	2.27
	.3125	⁵⁄₁₆	23.34	6.86	36.3	12.1	2.30
	.2500	¼	19.02	5.59	30.3	10.1	2.33
	.1875	³⁄₁₆	14.53	4.27	23.8	7.93	2.36
5 x 5	.5000	½	28.43	8.36	27.0	10.8	1.80
	.3750	⅜	22.37	6.58	22.8	9.11	1.86
	.3125	⁵⁄₁₆	19.08	5.61	20.1	8.02	1.89
	.2500	¼	15.62	4.59	16.9	6.78	1.92
	.1875	³⁄₁₆	11.97	3.52	13.4	5.36	1.95
4 x 4	.5000	½	21.63	6.36	12.3	6.13	1.39
	.3750	⅜	17.27	5.08	10.7	5.35	1.45
	.3125	⁵⁄₁₆	14.83	4.36	9.58	4.79	1.48
	.2500	¼	12.21	3.59	8.22	4.11	1.51
	.1875	³⁄₁₆	9.42	2.77	6.59	3.30	1.54
3.5 x 3.5	.3125	⁵⁄₁₆	12.70	3.73	6.09	3.48	1.28
	.2500	¼	10.51	3.09	5.29	3.02	1.31
	.1875	³⁄₁₆	8.15	2.39	4.29	2.45	1.34
3 x 3	.3125	⁵⁄₁₆	10.58	3.11	3.58	2.39	1.07
	.2500	¼	8.81	2.59	3.16	2.10	1.10
	.1875	³⁄₁₆	6.87	2.02	2.60	1.73	1.13
2.5 x 2.5	.2500	¼	7.11	2.09	1.69	1.35	.899
	.1875	³⁄₁₆	5.59	1.64	1.42	1.14	.930
2 x 2	.2500	¼	5.41	1.59	.766	.766	.694
	.1875	³⁄₁₆	4.32	1.27	.668	.668	.726

* Outside dimensions across flat sides.
** Properties are based upon a nominal outside corner radius equal to two times the wall thickness.

AMERICAN INSTITUTE OF STEEL CONSTRUCTION

Table S-9 Dimensions and physical properties of square tube *(continued)*

Column Design Tables

The following tables give the allowable concentric column load on various steel sections for various unsupported lengths. Tables include *W* and *S* shapes, *WT* and *ST* shapes, pipes and square tubes. Except for square tubes, all tables are computed for a yield stress of 36 ksi.

Allowable stresses on axially loaded compact and noncompact sections are prescribed by ASD Chapter E. Allowable stresses on slender element sections are prescribed by ASD Appendix B. Of the sections listed in the following tables, only the tee shapes contain slender element sections.

All the tables generally follow the same format. The sections are listed in order of descending weight with unsupported lengths increasing to the right. In all cases, the unsupported length *KL* is the length between points of zero moment, *not* the total unsupported length of the member.

When the value of *KL/r* is greater than 200, no entry is made in the table.

The following examples illustrate the use of the tables.

Given: Concentric column load of 134 kips. Column is hinged at both ends. Unsupported length is 16'-0".

To Find: Suitable *W* shape in A36 steel to support the load.

Solution: The unsupported length *KL* between points of zero moment is 16'-0".
Tables S-10a and S-10b contain the allowable loads on the *W* shapes. Enter Table S-10b at the bottom of the listing for a distance *KL* of 16 ft. Proceed upward to find that a section W8 × 35 is the first (and lightest) section that will sustain the given load of 134 kips; it will support 141 kips on the *yy* axis and 185 kips on the *xx* axis.

Given: Concentric column load of 99 kips. Column is fixed at the top and hinged at the bottom. Total actual length top to bottom is 30 feet. If necessary, the weak axis could be supported each 10 ft.

To Find: Suitable *S* shape in A36 steel to support the load.

Solution: The unsupported length *KL* between points of zero moment is 0.7 × 30 = 21 ft.
Table S-10b contains the allowable loads on the S shapes. Enter Table S-10b at the bottom of the listing. Proceed upward to find that there are no S sections that will support a load of 99 kips on its weak axis over an unsupported length of 30 feet.

If the weak axis is supported each 10 feet, the capacity of the column sections is found by proceeding upward on the 21 ft listing (interpolated) for the strong *xx* axis and upward on the 10 ft listing for the weak *yy* axis. The first (and lightest) *S* section that will sustain 99 kips on both axis is the S12 × 35 section; it will support 102 kips on the *yy* axis and 188 kips on the *xx* axis.

Given: Concentric column load of 112 kips. Column is fixed both at top and bottom. Total actual length top to bottom is 28 feet.

To Find: Suitable *WT* shape in A36 steel to support the load.

Solution: The unsupported length between points of zero moment is $0.5 \times 28 = 14$ ft.

Tables 11a and 11b contain the allowable loads on *WT* shapes. Enter Table S-11b at the bottom of the listing, for a distance *KL* of 14 feet. Proceed upward to find the first (and lightest) section that will sustain 112 kips on both the *xx* and *yy* axis. This section is found to be a *WT* 7×26.5; it will support 113 kips on the *yy* axis and 112 kips on the *xx* axis.

Given: Concentric column load of 61 kips. Column is fixed at the bottom, free at the top (free standing pole). Total actual length top to bottom is 15 feet.

To Find: Suitable pipe section or square tube to support the load, whichever is lighter.

Solution: The unsupported length between points of zero moment is $2.0 \times 15 = 30$ feet.

Table S-12 contains the allowable loads on structural steel pipe. Enter Table S-12 at the bottom of the listing, for a distance *KL* of 30 feet. Proceed upward to find the first (and lightest) section that will sustain 61 kips. The section is found to be an 8 inch diameter schedule 40 pipe; its weight is 28.55 lbs/ft.

Table S-13 contains the allowable loads on structural steel tubes. Enter Table S-13 at the bottom of the listing, for a distance *KL* of 30 feet. Proceed upward to find the first (and lightest) section that will sustain the 61 kips. The section is found to be a section 8 in. $\times$ 8 in. square; its weight is 25.82 lbs/ft.

A square tube 8 in. $\times$ 8 in. weighing 25.82 lbs/ft is found to be the lightest section to support the given load of 61 kips.

ALLOWABLE COLUMN LOAD IN KIPS
ON WIDE FLANGE SECTIONS AND
AMERICAN STANDARD I SECTIONS

ASTM A36 Steel

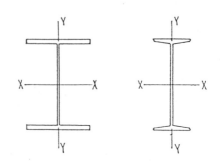

DISTANCE KL BETWEEN POINTS OF ZERO MOMENT

SECTION	4 XX AXIS	4 YY AXIS	6 XX AXIS	6 YY AXIS	8 XX AXIS	8 YY AXIS	10 XX AXIS	10 YY AXIS	12 XX AXIS	12 YY AXIS	14 XX AXIS	14 YY AXIS	16 XX AXIS	16 YY AXIS
W12X136	847	833	838	815	829	795	818	772	807	747	795	721	783	693
W14X132	825	815	818	801	810	786	801	768	792	750	782	730	772	708
W14X120	751	742	744	729	737	714	729	699	720	682	711	663	702	644
W12X120	749	737	741	721	733	702	723	682	713	660	703	636	691	611
W10X112	696	682	686	663	677	642	666	619	654	593	641	565	628	535
W14X109	681	672	674	661	668	647	661	633	653	618	645	601	636	583
W12X106	662	651	655	637	647	620	639	602	630	583	621	561	611	539
W10X100	621	609	613	592	604	573	594	551	584	528	572	503	560	476
W14X99	619	611	613	600	607	589	600	575	593	561	586	546	578	529
W12X96	598	588	592	575	585	560	577	544	569	526	561	506	551	486
W14X90	563	557	558	547	553	536	547	524	540	511	533	497	526	482
W10X88	547	536	540	521	532	504	523	485	513	464	503	442	492	417
W12X87	543	534	537	522	531	508	524	493	516	477	508	459	500	440
W14X82	512	497	507	482	502	465	497	446	491	425	484	402	478	377
W12X79	492	484	487	473	481	460	474	446	468	431	460	415	453	398
W10X77	477	468	471	454	464	439	456	422	447	404	438	384	429	362
W14X74	463	450	459	436	454	421	449	403	444	384	438	363	432	341
W12X72	447	440	442	430	437	418	431	406	425	392	418	377	411	361
W10X68	422	414	416	402	410	388	403	373	395	357	387	339	379	320
W14X68	425	413	421	400	417	385	412	369	407	351	402	332	396	311
W8X67	414	403	407	387	399	370	390	350	380	328	370	304	359	279
W12X65	405	398	400	389	396	378	390	367	384	354	378	341	372	326
W14X61	380	369	377	358	373	345	369	330	364	314	359	297	354	278
W10X60	372	364	366	353	361	341	354	328	347	313	340	297	332	280
W12X58	360	351	356	341	352	329	347	315	342	301	337	285	331	268
W8X58	359	349	353	336	345	320	337	303	329	283	320	263	310	240
W10X54	333	327	329	317	324	306	318	294	312	281	305	267	298	251
W14X53	331	316	328	302	325	286	321	268	317	248	313	226	308	202
W12X53	331	322	327	312	323	301	318	288	314	275	309	260	303	244
W12X50	312	299	308	286	304	271	300	254	295	236	291	216	285	195
W10X49	304	298	300	289	295	279	290	268	284	256	278	242	271	228
W14X48	300	286	297	273	293	258	290	242	286	224	282	204	278	182

DISTANCE KL BETWEEN POINTS OF ZERO MOMENT

Table S-10a Allowable column loads on W and S sections

ALLOWABLE COLUMN LOAD IN KIPS
ON WIDE FLANGE SECTIONS AND
AMERICAN STANDARD I SECTIONS

ASTM A36 Steel

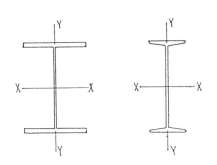

	DISTANCE KL BETWEEN POINTS OF ZERO MOMENT												
	18		20		22		24		26		28		30	
SECTION	XX AXIS	YY AXIS	XX AXIS	YY AXIS	XX AXIS	YY AXIS	XX AXIS	YY AXIS	XX AXIS	YY AXIS	XX AXIS	YY AXIS	XX AXIS	YY AXIS
W12X136	770	662	756	630	742	597	726	561	711	524	695	485	678	444
W14X132	761	686	750	662	738	637	726	610	713	583	700	554	686	524
W14X120	692	623	682	601	671	578	660	554	648	528	636	502	623	475
W12X120	680	584	667	555	654	525	641	493	627	460	612	425	597	388
W10X112	614	503	599	469	584	433	568	395	551	355	533	313	515	272
W14X109	627	564	618	544	608	523	597	501	587	478	576	454	564	429
W12X106	600	514	589	489	577	462	565	433	553	404	540	372	526	340
W10X100	547	446	534	416	520	383	505	348	490	312	474	273	457	238
W14X99	570	512	561	494	552	475	542	454	533	433	522	411	512	388
W12X96	542	464	532	440	521	416	510	390	499	362	487	334	474	304
W14X90	518	466	510	449	502	432	493	413	484	394	475	374	465	353
W10X88	481	392	469	364	456	335	443	304	429	271	415	237	400	206
W12X87	491	420	482	398	472	376	462	352	451	327	440	301	429	273
W14X82	471	351	463	323	455	293	447	261	439	227	430	196	421	171
W12X79	444	379	436	360	427	339	418	317	408	294	398	270	387	245
W10X77	418	339	408	315	397	289	385	261	373	232	360	202	347	176
W14X74	426	317	419	292	412	265	404	236	397	206	389	177	381	154
W12X72	404	344	396	326	388	308	379	288	370	267	361	245	351	222
W10X68	369	299	360	278	350	255	339	230	328	204	317	177	305	155
W14X68	390	289	384	266	377	241	371	214	364	186	356	160	349	139
W8X67	347	251	335	221	322	190	308	159	294	136	279	117	264	102
W12X65	365	311	358	294	350	277	343	259	335	240	326	220	317	199
W14X61	349	258	343	237	337	214	331	190	325	165	318	142	312	124
W10X60	324	262	316	243	307	222	297	201	287	177	277	154	266	134
W12X58	325	249	319	230	312	209	305	187	298	164	290	142	283	123
W8X58	299	216	288	190	277	162	265	136	252	116	239	100	225	87
W10X54	291	235	283	217	275	199	266	179	257	158	248	137	238	119
W14X53	303	177	298	149	293	123	288	104	282	88	276	76	270	66
W12X53	298	227	292	209	286	189	279	169	272	147	265	127	258	111
W12X50	280	171	274	146	268	121	262	102	256	87	249	75	242	65
W10X49	265	213	258	197	250	180	242	161	234	142	225	123	217	107
W14X48	274	159	269	133	265	110	260	93	255	79	249	68	244	59
	18		20		22		24		26		28		30	

DISTANCE KL BETWEEN POINTS OF ZERO MOMENT

Table S-10a Allowable column loads on W and S sections *(continued)*

ALLOWABLE COLUMN LOAD IN KIPS
ON WIDE FLANGE SECTIONS AND
AMERICAN STANDARD I SECTIONS

ASTM A36 Steel

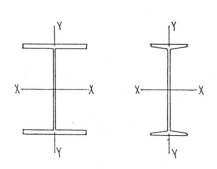

DISTANCE KL BETWEEN POINTS OF ZERO MOMENT

SECTION	4 XX AXIS	4 YY AXIS	6 XX AXIS	6 YY AXIS	8 XX AXIS	8 YY AXIS	10 XX AXIS	10 YY AXIS	12 XX AXIS	12 YY AXIS	14 XX AXIS	14 YY AXIS	16 XX AXIS	16 YY AXIS
W8X48	296	288	290	276	284	263	278	249	271	233	263	215	255	196
W12X45	280	268	276	256	273	243	269	228	265	211	261	193	256	173
W10X45	281	271	277	260	272	247	267	232	262	216	256	199	250	180
W14X43	268	255	265	244	262	230	259	215	256	199	252	181	249	161
W12X40	250	239	247	229	244	217	241	203	237	188	233	172	229	154
W8X40	245	238	241	229	236	218	230	205	224	192	217	177	210	160
W10X39	243	234	239	224	235	213	231	200	226	186	221	170	216	154
W8X35	216	210	212	201	207	191	202	180	197	168	191	155	185	141
W10X33	205	197	202	189	198	179	195	167	191	155	186	142	182	127
W8X31	191	186	188	178	184	170	179	160	174	149	169	137	163	124
W8X28	173	165	170	155	166	144	162	132	157	118	152	103	147	87
W6X25	152	146	148	136	143	126	138	114	133	100	126	85	120	69
W8X24	148	141	145	133	142	124	139	113	135	101	130	88	126	74
W8X21	129	119	127	109	124	97	121	84	118	68	114	52	110	40
W6X20	122	116	118	109	114	100	110	90	106	79	101	67	95	54
W8X18	110	101	108	92	106	82	103	70	100	56	97	42	94	32
W6X16	98	87	95	76	92	62	89	46	85	32	80	23	76	18
W8X15	93	80	91	67	89	52	86	35	84	25	81	18	78	
W6X15	92	87	89	82	86	75	82	67	79	58	75	49	70	38
W6X12	73	64	71	55	69	44	66	31	63	22	59	16	56	
S12X50	311	274	306	242	302	203	297	158	291	112	286	83	279	63
S12X40.8	254	225	251	200	247	170	243	135	239	97	235	71	230	55
S12X35	218	190	215	165	212	136	209	102	205	71	201	52	197	40
S10X35	216	186	213	158	209	125	204	87	199	60	194	44	188	
S12X31.8	198	173	195	152	193	126	190	96	187	67	183	49	180	38
S10X25.4	157	137	155	118	152	96	149	70	146	49	142	36	139	
S8X23	141	118	138	96	135	69	131	45	126	31	122		117	
S7X20	122	99	119	78	115	51	111	33	106	23	101		96	
S8X18.4	113	95	111	79	108	59	105	39	102	27	99		95	
S6X17.25	104	83	101	61	96	37	92	24	87		81		75	
S7X15.3	94	77	91	62	89	43	86	27	82	19	79		75	
S6X12.5	76	61	73	47	71	30	68	19	64		61		57	

DISTANCE KL BETWEEN POINTS OF ZERO MOMENT

Table S-10b Allowable column loads on W and S sections

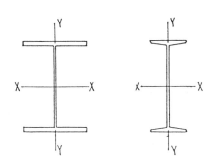

ALLOWABLE COLUMN LOAD IN KIPS
ON WIDE FLANGE SECTIONS AND
AMERICAN STANDARD I SECTIONS

ASTM A36 Steel

DISTANCE KL BETWEEN POINTS OF ZERO MOMENT

SECTION	18 XX AXIS	18 YY AXIS	20 XX AXIS	20 YY AXIS	22 XX AXIS	22 YY AXIS	24 XX AXIS	24 YY AXIS	26 XX AXIS	26 YY AXIS	28 XX AXIS	28 YY AXIS	30 XX AXIS	30 YY AXIS
W8X48	246	176	237	154	227	131	217	110	206	94	195	81	183	70
W12X45	251	152	246	129	241	106	235	89	229	76	223	66	217	57
W10X45	244	160	237	138	230	115	223	97	215	82	207	71	199	62
W14X43	245	140	241	117	236	96	232	81	227	69	222	60	217	52
W12X40	224	135	220	114	215	94	210	79	205	67	199	58	194	51
W8X40	203	143	195	124	186	104	178	88	168	75	159	64	149	56
W10X39	210	136	205	116	198	97	192	81	185	69	178	60	171	52
W8X35	178	125	171	109	164	91	156	76	148	65	139	56	130	49
W10X33	177	112	172	95	166	78	161	66	155	56	149	48	143	42
W8X31	157	110	151	95	144	80	137	67	130	57	122	49	114	43
W8X28	142	69	136	56	130	46	123	39	117	33	110		102	
W6X25	113	54	105	44	97	36	89	31	80		71		62	
W8X24	121	59	116	48	111	39	105	33	100	28	93		87	
W8X21	106	31	102	25	98		93		88		83		77	
W6X20	89	42	83	34	77	28	70	24	63		55		48	
W8X18	90	25	86	21	83		78		74		70		65	
W6X16	71		66		61		55		49		42		37	
W8X15	75		72		68		64		60		56		52	
W6X15	66	30	61	24	56	20	50	17	44		38		33	
W6X12	52		48		43		39		34		29		25	
S12X50	273		266		259		251		244		236		227	
S12X40.8	225		220		214		209		203		197		190	
S12X35	193		188		183		179		173		168		162	
S10X35	182		176		169		162		155		148		140	
S12X31.8	176		172		168		163		159		154		149	
S10X25.4	135		131		126		122		117		112		107	
S8X23	111		106		100		94		87		80		73	
S7X20	90		84		78		71		64		56		49	
S8X18.4	91		87		82		78		73		68		63	
S6X17.25	69		62		55		47		40		35		30	
S7X15.3	71		67		63		58		53		48		42	
S6X12.5	53		49		44		39		34		29		25	
	18		20		22		24		26		28		30	

DISTANCE KL BETWEEN POINTS OF ZERO MOMENT

Table S-10b Allowable column loads on W and S sections *(continued)*

ALLOWABLE COLUMN LOAD IN KIPS
ON TEES CUT FROM WIDE FLANGE
AND AMERICAN STANDARD I SECTIONS

ASTM A36 Steel

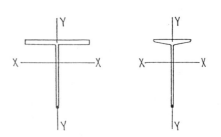

DISTANCE KL BETWEEN POINTS OF ZERO MOMENT

SECTION	4 XX AXIS	4 YY AXIS	6 XX AXIS	6 YY AXIS	8 XX AXIS	8 YY AXIS	10 XX AXIS	10 YY AXIS	12 XX AXIS	12 YY AXIS	14 XX AXIS	14 YY AXIS	16 XX AXIS	16 YY AXIS
WT6X68	399	418	375	409	348	398	317	387	283	375	245	361	203	347
WT7X66	390	408	370	401	346	393	320	384	291	375	259	365	224	354
WT7X60	355	372	337	365	315	358	290	350	263	342	234	333	201	323
WT6X60	350	367	329	359	305	350	277	340	246	329	212	317	175	304
WT5X56	321	342	296	333	266	322	233	310	195	297	152	283	116	268
WT7X54.5	321	336	303	330	283	324	261	317	235	309	208	300	177	292
WT6X53	310	326	290	318	268	310	242	301	214	291	182	281	148	269
WT5X50	285	304	262	296	235	286	203	276	168	264	129	251	99	238
WT7X49.5	292	307	276	301	258	295	237	289	214	282	188	274	160	266
WT6X48	279	294	262	288	241	280	217	272	191	263	162	253	130	243
WT7X45	264	277	250	272	233	267	214	261	193	254	169	247	144	240
WT5X44	250	267	229	260	204	251	176	242	145	231	110	220	84	208
WT6X43.5	253	267	237	261	218	254	197	247	173	238	146	229	117	220
WT7X41	243	248	231	240	218	231	203	222	187	211	169	200	150	188
WT6X39.5	229	242	215	236	197	230	177	223	155	216	131	208	104	199
WT5X38.5	218	234	199	227	177	220	151	211	122	202	92	192	70	181
WT7X37	220	225	210	218	197	210	184	202	168	192	152	182	134	170
WT6X36	210	221	196	216	180	210	161	204	141	197	119	189	94	181
WT5X34	192	207	175	201	155	194	132	186	106	178	79	169	60	160
WT7X34	202	206	192	200	181	192	168	184	154	176	139	166	122	156
WT4X33.5	184	201	163	193	138	185	109	175	78	164	57	152	44	139
WT6X32.5	188	199	176	194	161	189	145	183	126	177	106	170	84	163
WT7X30.5	181	185	172	179	162	173	150	165	138	157	124	149	109	139
WT5X30	169	182	154	177	136	171	115	164	92	157	68	149	52	141
WT6X29	169	176	158	171	145	165	131	158	115	151	97	143	78	134
WT4X29	159	175	141	168	118	160	92	151	65	142	48	131	37	120
WT5X27	152	164	137	159	121	153	102	147	80	141	59	133	45	126
WT7X26.5	158	158	151	151	143	143	133	134	123	124	112	113	99	101
WT6X26.5	154	161	144	156	133	150	120	144	105	137	89	130	72	122
WT6X25	146	149	138	143	128	135	117	127	104	118	91	108	75	97
WT5X24.5	138	149	125	145	110	140	92	134	72	128	53	121	41	114
WT7X24	143	143	137	137	129	130	120	121	111	112	101	102	89	91

DISTANCE KL BETWEEN POINTS OF ZERO MOMENT

Table S-11a Allowable column loads on WT and ST sections

ALLOWABLE COLUMN LOAD IN KIPS
ON TEES CUT FROM WIDE FLANGE
AND AMERICAN STANDARD I SECTIONS

ASTM A36 Steel

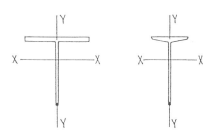

DISTANCE KL BETWEEN POINTS OF ZERO MOMENT

SECTION	18 XX AXIS	18 YY AXIS	20 XX AXIS	20 YY AXIS	22 XX AXIS	22 YY AXIS	24 XX AXIS	24 YY AXIS	26 XX AXIS	26 YY AXIS	28 XX AXIS	28 YY AXIS	30 XX AXIS	30 YY AXIS
WT6X68	162	332	131	316	108	299	91	281	78	263		243		223
WT7X66	186	343	151	331	124	318	105	305	89	291	77	277		262
WT7X60	166	312	134	301	111	290	93	278	79	265	68	252		238
WT6X60	139	291	112	277	93	262	78	246	67	229		212		193
WT5X56	92	252	75	235	62	217		198		178		157		137
WT7X54.5	145	282	117	272	97	262	81	251	69	239	60	227		215
WT6X53	117	257	95	244	78	231	66	217		202		186		170
WT5X50	78	223	63	208		191		174		156		137		119
WT7X49.5	130	257	106	248	87	238	73	228	62	217		206		195
WT6X48	103	232	83	220	69	208	58	195		181		167		152
WT7X45	116	232	94	224	78	215	65	206	56	196		186		176
WT5X44	67	195	54	181		167		151		135		118		103
WT6X43.5	92	210	75	199	62	188	52	176		163		150		137
WT7X41	129	175	106	161	88	146	74	130	63	113	54	98	47	85
WT6X39.5	82	190	67	180	55	169	46	159		147		135		123
WT5X38.5	56	170	45	157		144		131		116		101		88
WT7X37	114	159	94	146	77	132	65	118	55	103	48	89	42	77
WT6X36	74	173	60	164	50	155	42	145		134		123		111
WT5X34	48	150	39	139		127		115		102		89		77
WT7X34	104	144	85	133	70	120	59	107	50	93	43	80	38	70
WT4X33.5		125		111		95		80		68		58		51
WT6X32.5	66	155	53	147	44	138	37	129		120		110		99
WT7X30.5	92	129	75	119	62	107	52	95	45	83	38	71	33	62
WT5X30	41	131	33	122		111		100		89		77		67
WT6X29	61	125	50	115	41	105	35	94		82		71		62
WT4X29		108		95		81		68		58		50		43
WT5X27	36	118		109		99		90		79		69		60
WT7X26.5	86	89	72	75	59	62	50	52	42	44	37	38	32	33
WT6X26.5	57	113	46	104	38	94	32	84		73		63		55
WT6X25	60	85	49	73	40	60	34	51	29	43		37		32
WT5X24.5	32	107		99		90		81		71		62		54
WT7X24	77	80	64	67	53	55	45	46	38	40	33	34	28	30

DISTANCE KL BETWEEN POINTS OF ZERO MOMENT

Table S-11a Allowable column loads on WT and ST sections *(continued)*

ALLOWABLE COLUMN LOAD IN KIPS
ON TEES CUT FROM WIDE FLANGE
AND AMERICAN STANDARD I SECTIONS

ASTM A36 Steel

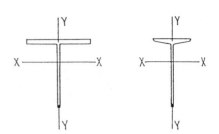

DISTANCE KL BETWEEN POINTS OF ZERO MOMENT

SECTION	4 XX AXIS	4 YY AXIS	6 XX AXIS	6 YY AXIS	8 XX AXIS	8 YY AXIS	10 XX AXIS	10 YY AXIS	12 XX AXIS	12 YY AXIS	14 XX AXIS	14 YY AXIS	16 XX AXIS	16 YY AXIS
WT4X24	130	144	114	138	94	132	71	124	49	116	36	108	28	98
WT6X22.5	132	134	124	128	115	122	104	114	93	106	80	97	67	87
WT5X22.5	128	135	117	129	104	123	89	116	72	108	54	99	41	90
WT7X21.5	121	121	116	116	110	110	103	103	95	96	86	87	77	78
WT6X20	117	120	110	114	102	108	93	101	82	94	71	86	59	77
WT4X20	108	120	95	115	78	109	59	103	41	96	30	89	23	80
WT5X19.5	111	117	101	112	90	106	77	100	62	93	47	85	36	77
WT4X17.5	94	105	82	100	67	96	50	90	35	84	25	77	19	70
WT5X16.5	94	98	86	94	77	89	66	84	54	78	41	71	31	64
WT4X15.5	84	93	73	89	60	85	44	80	31	74	23	68	17	62
WT4X14	76	82	67	78	56	72	43	66	30	59	22	52	17	43
WT3X12.5	64	73	52	68	37	63	24	57	16	50		43		34
WT4X12	66	71	57	67	48	62	36	56	25	51	19	44	14	37
WT4X10.5	58	60	52	55	45	49	37	42	28	34	20	26	16	20
WT3X10	51	58	41	54	29	50	18	45	13	40		34		27
WT4X9	50	51	45	46	39	41	32	35	25	28	18	21	14	16
WT3X8	42	44	35	38	26	31	18	23	12	16	9	12		9
WT4X7.5	43	40	39	34	34	26	29	18	23	12	17	9	13	
WT3X7.5	38	44	31	41	23	37	15	33	10	29		24		19
WT3X6	32	32	27	28	20	22	14	16	10	11	7	8		
ST6X25	149	137	142	121	134	102	125	79	115	56	104	41	92	32
ST6X20.4	121	113	115	100	108	85	100	67	92	49	82	36	72	27
ST6X17.5	104	95	99	83	93	68	87	51	80	36	72	26	64	20
ST5X17.5	102	93	96	79	89	62	81	43	72	30	62	22	51	
ST6X15.9	94	87	90	76	84	63	78	48	71	34	64	25	56	19
ST5X12.7	74	68	69	59	63	48	56	35	49	24	41	18	32	
ST4X11.5	65	59	59	48	52	35	45	22	36	16	27		20	
ST3.5X10	55	50	49	39	42	26	33	16	24	11	18		14	
ST4X9.2	51	48	46	39	40	29	33	19	25	13	19		14	
ST3X8.63	46	41	39	31	31	19	22	12	15		11			
ST3.5X7.65	42	39	36	31	30	21	23	14	16	10	12		9	
ST3X6.25	32	30	27	23	20	15	13	9	9					

DISTANCE KL BETWEEN POINTS OF ZERO MOMENT

Table S-11b Allowable column loads on WT and ST sections

ALLOWABLE COLUMN LOAD IN KIPS
ON TEES CUT FROM WIDE FLANGE
AND AMERICAN STANDARD I SECTIONS

ASTM A36 Steel

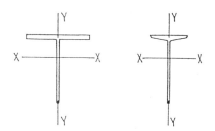

DISTANCE KL BETWEEN POINTS OF ZERO MOMENT

SECTION	18 XX AXIS	18 YY AXIS	20 XX AXIS	20 YY AXIS	22 XX AXIS	22 YY AXIS	24 XX AXIS	24 YY AXIS	26 XX AXIS	26 YY AXIS	28 XX AXIS	28 YY AXIS	30 XX AXIS	30 YY AXIS
WT4X24		88		77		65		55		47		40		35
WT6X22.5	53	76	43	64	35	53	30	45	25	38		33		29
WT5X22.5	33	80	26	69		57		48		41		35		31
WT7X21.5	67	69	57	58	47	48	39	41	33	35	29	30	25	26
WT6X20	46	67	38	57	31	47	26	39	22	34		29		25
WT4X20		72		62		52		44		37		32		28
WT5X19.5	28	68	23	58		48		40		34		30		26
WT4X17.5		62		54		45		38		32		28		24
WT5X16.5	25	56	20	47		39		33		28		24		21
WT4X15.5		55		48		40		33		29		25		21
WT4X14		35		28		23		19		17				
WT3X12.5		27		22		18		15						
WT4X12		29		24		20		17		14				
WT4X10.5	12	16		13										
WT3X10		21		17		14		12						
WT4X9	11	13		10										
WT3X8														
WT4X7.5	11		9											
WT3X7.5		15		12		10		8						
WT3X6														
ST6X25	79		65		54		45		39		33		29	
ST6X20.4	61		49		41		34		29		25			
ST6X17.5	54		45		37		31		26		23		20	
ST5X17.5	40		32		27		23		19					
ST6X15.9	47		38		32		27		23		20			
ST5X12.7	25		20		17		14							
ST4X11.5	16		13											
ST3.5X10														
ST4X9.2	11													
ST3X8.65														
ST3.5X7.65														
ST3X6.25														

| | 18 | 20 | 22 | 24 | 26 | 28 | 30 |

DISTANCE KL BETWEEN POINTS OF ZERO MOMENT

Table S-11b Allowable column loads on WT and ST sections *(continued)*

ALLOWABLE COLUMN LOAD IN KIPS
ON STRUCTURAL PIPE SECTIONS

ASTM A501 or ASTM A53 Steel
$F_Y = 36$ ksi

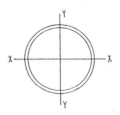

SCHEDULE	DIA IN	WEIGHT LB/FT	DISTANCE KL IN FEET BETWEEN POINTS OF ZERO MOMENT													
			4	6	8	10	12	14	16	18	20	22	24	26	28	30
120	8	72.42	442	430	417	403	387	369	351	331	310	288	265	240	214	187
80	12	65.42	405	399	393	386	378	370	362	353	343	333	322	311	300	288
80	10	54.74	338	332	325	318	309	301	291	281	271	260	248	236	224	211
120	6	53.16	318	305	291	275	256	237	216	193	168	142	119	102	88	76
40	12	49.56	308	304	299	294	288	282	276	269	262	254	246	238	229	221
80	8	43.39	266	260	252	244	235	225	215	204	192	179	166	152	138	122
40	10	40.48	250	245	240	235	229	223	216	209	201	193	185	176	167	157
120	5	38.55	227	215	201	186	169	150	129	107	87	72	60	51	44	
80	6	28.57	172	166	159	151	142	132	122	110	99	86	73	62	53	46
40	8	28.55	175	171	166	161	155	149	142	135	127	120	111	102	93	83
120	4	27.54	159	147	133	117	100	80	62	49	39	33				
80	5	20.78	123	118	111	103	95	86	76	65	54	44	37	32	27	24
40	6	18.97	114	110	106	101	95	89	82	75	68	59	51	43	37	33
120	3	18.58	102	91	77	61	43	32	24							
80	4	14.98	87	81	75	67	59	49	39	31	25	21	17			
40	5	14.62	87	83	79	73	68	61	55	47	39	33	27	23	20	18
120	2.5	13.69	71	59	45	30	21	15								
80	3.5	12.50	72	66	59	52	43	33	26	20	16					
40	4	10.79	63	59	54	49	43	36	29	23	19	15	13			
80	3	10.25	57	52	45	37	28	21	16	13						
40	3.5	9.11	52	48	44	38	32	25	19	15	12	10				
120	2	9.03	44	34	21	14										
80	2.5	7.66	41	35	28	20	14	10								
40	3	7.58	43	38	34	28	22	16	12	10						
			4	6	8	10	12	14	16	18	20	22	24	26	28	30
			DISTANCE KL IN FEET BETWEEN POINTS OF ZERO MOMENT													

Note: Schedule 40 pipe is termed "standard" pipe by AISC
      Schedule 80 pipe is termed "extra strong" pipe by AISC
      Schedule 120 pipe is termed "double extra strong" pipe by AISC

Table S-12 Allowable column loads on structural pipe

ALLOWABLE COLUMN LOAD IN KIPS
ON SQUARE STRUCTURAL TUBING

ASTM A500 Steel, Gr.B
Fy = 46ksi

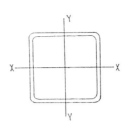

SIZE	WEIGHT LB/FT	DISTANCE KL IN FEET BETWEEN POINTS OF ZERO MOMENT													
		4	6	8	10	12	14	16	18	20	22	24	26	28	30
8 X 8	59.32	460	447	432	416	398	378	358	335	311	286	260	231	202	176
8 X 8	54.17	421	409	396	381	365	348	329	309	288	265	241	216	189	165
8 X 8	48.85	381	371	359	346	332	316	299	281	262	242	221	199	175	152
7 X 7	46.51	360	347	333	318	300	282	261	240	217	192	165	141	122	106
7 X 7	42.05	326	315	302	288	273	256	238	219	198	177	153	131	113	98
6 X 6	38.86	296	283	268	251	232	212	190	166	140	116	98	83	72	62
8 X 8	37.69	294	286	278	268	257	245	233	219	205	190	174	158	140	122
7 X 7	32.58	252	244	234	224	212	200	187	172	157	141	123	106	91	79
6 X 6	35.24	270	258	245	230	213	195	175	154	131	109	91	78	67	59
8 X 8	31.84	248	242	234	226	217	207	197	186	174	162	149	135	120	105
5 X 5	28.43	213	200	186	169	151	131	109	87	70	58	49	42	36	31
7 X 7	27.59	213	207	199	190	181	170	159	147	134	121	106	91	79	69
6 X 6	27.48	210	201	191	180	168	154	140	124	107	89	75	64	55	48
8 X 8	25.82	201	196	190	184	177	169	160	152	142	132	122	111	99	87
6 X 6	23.34	179	171	163	154	143	132	120	107	93	78	65	56	48	42
5 X 5	22.37	168	159	148	136	122	107	90	73	59	49	41	35	30	26
7 X 7	22.42	174	168	162	155	147	139	130	120	110	100	88	76	65	57
4 X 4	21.63	157	143	127	108	87	65	50	39	32	26				
5 X 5	19.08	144	136	127	116	105	93	79	64	52	43	36	31	27	23
6 X 6	19.02	146	140	133	126	117	108	99	88	77	65	55	47	40	35
4 X 4	17.27	126	116	103	89	74	57	43	34	28	23	19			
5 X 5	15.62	118	111	104	96	87	77	66	54	44	36	30	26	22	19
4 X 4	14.83	108	100	90	78	65	51	39	31	25	20	17			
4 X 4	12.21	89	83	74	65	55	43	33	26	21	18	15			
5 X 5	11.97	90	86	80	74	67	60	52	43	35	29	24	21	18	15
3 X 3	10.58	73	63	51	37	26	19	14							
4 X 4	9.42	69	64	58	51	43	35	27	21	17	14	12			
3 X 3	8.81	61	53	44	32	23	17	13	10						
3 X 3	6.87	48	42	35	27	19	14	10	8						
2 X 2	6.32	37	25	14	9										
2 X 2	5.41	32	22	12	8										
2 X 2	4.32	26	18	11	7	5									
		4	6	8	10	12	14	16	18	20	22	24	26	28	30

DISTANCE KL IN FEET BETWEEN POINTS OF ZERO MOMENT

Table S-13 Allowable column loads on square table

Beam
Design Tables

The following tables give the moment capacity of various steel sections for various unsupported lengths along the compression flange. Tables include *W* shapes, *S* shapes and channels. All tables are computed for ASTM A36, $F_y = 36$ ksi.

Allowable stresses on beams are prescribed by ASD Specifications Chapter F. Where the unsupported length of the compression flange is less than L_c, as prescribed by ASD EQ (F1-2), the allowable compressive stress is 24 ksi. Where the unsupported length of the compression flange is greater than L_c but less than L_u as prescribed by ASD Eq (F1-2), the allowable compressive stress is 22 ksi. For unsupported lengths greater than L_w the allowable stress decreases continuously and sharply as the unsupported length increases.

In the tables, an asterisk marks the lengths where L_c and L_u occur. For example, for section W21 × 62 in Table S-14a, the first of the two asterisks indicate the L_c is between 8 and 9 feet; the second indicates that L_u is between 11 and 12 feet. The only exception to this format occurs for the section W6 × 15 in Table S-14c. Section W6 × 15 is not a compact section; the only asterisk shown is that for its value of L_u between 12 and 13 feet.

As usual, the sections are listed in order of descending weights with unsupported length of the compression flange increasing to the right. To use the tables, go to the lightest beams at the end of the table, then enter the table at the appropriate value of unsupported length. Proceed upward until the first section is reached that will sustain the given moment. That section is the lightest section available that will support the given moment at that unsupported length.

Some examples will illustrate the use of the tables.

200

Given: Design moment of 164 kip•ft on a wide flange beam. The compression flange is supported at 4 ft. intervals by open web floor joists.

To Find: Suitable W sections in A36 steel to support the load.

Solution: Tables 14a, b and c contain the allowable moments on *W* shapes for various values of unsupported length. Enter Table S-14c at the bottom of the page at an unsupported length of 4 feet. Proceed upward through Table 14c to 14b until the first beam is found that will sustain a moment of 164 kip•ft. The first (and lightest) beam is found to be a W18 × 50.

Given: Design moment of 99 kip•ft on a beam that spans 20 feet. The compression flange is supported laterally only at the end supports.

To Find: Suitable *S* section in A36 steel support the load.

Solution: Table S-15 contains the allowable moments on *S* sections for various values of unsupported lengths of the compression flange. Enter Table S-15 at the bottom of the page at an unsupported length of 20 feet. Proceed upward to find that the first (and lightest) *S* section that will sustain a moment of 99 kip•ft is a S18 × 70.

Given: Design moment of 25 kip•ft on an American standard channel section. The load is placed at the shear center. Lateral supports of the compression flange are placed at 5 ft. along the span.

To Find: Suitable channel section in A36 steel to support the load.

Solution: Table S-16 contains the allowable moments on channel sections loaded through the shear center. Enter Table S-16 at the bottom of the page at an unsupported length of 5 feet. Proceed upward to find that the first (and lightest) channel section that will sustain 25 kip•ft is a C10 × 20.

ALLOWABLE MOMENT IN KIP-FEET FOR LATERALLY UNSUPPORTED WIDE FLANGE BEAMS

ASTM A36 Steel

UNSUPPORTED LENGTH OF THE COMPRESSION FLANGE IN FEET
FIRST ASTERISK INDICATES L_c. SECOND ASTERISK INDICATES L_u

SECTION	1	2	3	4	5	6	7	8	9	10	11	12	13	14	15	16	17
W12X136	372	372	372	372	372	372	372	372	372	372	372	372	372*	341	341	341	341
W14X132	418	418	418	418	418	418	418	418	418	418	418	418	418	418	418*	383	383
W14X120	380	380	380	380	380	380	380	380	380	380	380	380	380	380	380*	348	348
W12X120	326	326	326	326	326	326	326	326	326	326	326	326	326*	298	298	298	298
W10X112	252	252	252	252	252	252	252	252	252	252*	231	231	231	231	231	231	231
W14X109	346	346	346	346	346	346	346	346	346	346	346	346	346	346	346*	317	317
W12X106	290	290	290	290	290	290	290	290	290	290	290	290*	265	265	265	265	265
W16X100	350	350	350	350	350	350	350	350	350	350	350*	320	320	320	320	320	320
W10X100	224	224	224	224	224	224	224	224	224	224*	205	205	205	205	205	205	205
W14X99	314	314	314	314	314	314	314	314	314	314	314	314	314	314	314*	287	287
W18X97	376	376	376	376	376	376	376	376	376	376	376*	344	344	344	344	344	344
W12X96	262	262	262	262	262	262	262	262	262	262	262	262*	240	240	240	240	240
W21X93	384	384	384	384	384	384	384	384*	352	352	352	352	352	352	352	352*	340
W14X90	286	286	286	286	286	286	286	286	286	286	286	286	286	286	286*	262	262
W16X89	310	310	310	310	310	310	310	310	310	310*	284	284	284	284	284	284	284
W10X88	197	197	197	197	197	197	197	197	197	197*	180	180	180	180	180	180	180
W12X87	236	236	236	236	236	236	236	236	236	236	236	236*	216	216	216	216	216
W18X86	332	332	332	332	332	332	332	332	332	332	332*	304	304	304	304	304	304
W24X84	392	392	392	392	392	392	392	392	392*	359	359	359	359*	336	313	296	284
W21X83	342	342	342	342	342	342	342	342*	313	313	313	313	313	313	313*	289	272
W14X82	246	246	246	246	246	246	246	246	246	246*	225	225	225	225	225	225	225
W12X79	214	214	214	214	214	214	214	214	214	214	214	214*	196	196	196	196	196
W16X77	268	268	268	268	268	268	268	268	268	268*	245	245	245	245	245	245	245
W10X77	171	171	171	171	171	171	171	171	171	171*	157	157	157	157	157	157	157
W24X76	352	352	352	352	352	352	352	352	352*	322	322*	312	294	285	275	264	253
W18X76	292	292	292	292	292	292	292	292	292	292	292*	267	267	267	267	267	267
W14X74	224	224	224	224	224	224	224	224	224	224*	205	205	205	205	205	205	205
W21X73	302	302	302	302	302	302	302	302*	276	276	276	276	276*	259	242	227	213
W12X72	194	194	194	194	194	194	194	194	194	194	194	194*	178	178	178	178	178
W18X71	254	254	254	254	254	254	254	254*	232	232	232	232	232	232	232*	221	208
W24X68	308	308	308	308	308	308	308	308	308*	282*	270	263	256	247	239	229	219
W10X68	151	151	151	151	151	151	151	151	151	151*	138	138	138	138	138	138	138
W21X68	280	280	280	280	280	280	280	280*	256	256	256	256*	240	223	208	198	188
W14X68	206	206	206	206	206	206	206	206	206	206*	188	188	188	188	188	188	188
W16X67	234	234	234	234	234	234	234	234	234	234*	214	214	214	214	214	214	214
W8X67	120	120	120	120	120	120	120	120*	110	110	110	110	110	110	110	110	110
W18X65	234	234	234	234	234	234	234	234*	214	214	214	214	214	214*	201	189	177
W12X65	175	175	175	175	175	175	175	175	175	175	175	175*	161	161	161	161	161
W24X62	262	262	262	262	262	262	262*	240*	225	216	206	196	185	172	159	145	130
W21X62	254	254	254	254	254	254	254	254*	232	232	232*	212	204	196	188	179	169
	1	2	3	4	5	6	7	8	9	10	11	12	13	14	15	16	17

UNSUPPORTED LENGTH OF THE COMPRESSION FLANGE IN FEET

Table S-14a Allowable moment on laterally unsupported W shapes

ALLOWABLE MOMENT IN KIP-FEET FOR LATERALLY UNSUPPORTED WIDE FLANGE BEAMS

ASTM A36 Steel

UNSUPPORTED LENGTH OF THE COMPRESSION FLANGE IN FEET
FIRST ASTERISK INDICATES L_o. SECOND ASTERISK INDICATES L_u

SECTION	18	19	20	21	22	23	24	25	26	27	28	29	30	31	32	33	34
W12X136	341	341	341	341	341	341	341	341	341	341	341	341	341	341	341	341	341
W14X132	383	383	383	383	383	383	383	383	383	383	383	383	383	383	383	383	383
W14X120	348	348	348	348	348	348	348	348	348	348	348	348	348	348	348	348	348
W12X120	298	298	298	298	298	298	298	298	298	298	298	298	298	298	298	298	298
W10X112	231	231	231	231	231	231	231	231	231	231	231	231	231	231	231	231	231
W14X109	317	317	317	317	317	317	317	317	317	317	317	317	317	317	317	317	317
W12X106	265	265	265	265	265	265	265	265	265	265	265	265	265	265	265	265	265
W16X100	320	320	320	320	320	320	320	320	320	320	320*	304	294	284	275	267	259
W10X100	205	205	205	205	205	205	205	205	205	205	205	205	205	205	205	205	205
W14X99	287	287	287	287	287	287	287	287	287	287	287	287	287	287	287	287	287
W18X97	344	344	344	344	344	344	344*	326	314	302	291	281	272	263	255	247	240
W12X96	240	240	240	240	240	240	240	240	240	240	240	240	240	240	240	240	240
W21X93	321	305	289	275	263	251	241	231	222	214	206	199	193	186	181	175	170
W14X90	262	262	262	262	262	262	262	262	262	262	262	262	262	262	262	262	262*
W16X89	284	284	284	284	284	284	284	284*	268	259	249	241	233	225	218	211	205
W10X88	180	180	180	180	180	180	180	180	180	180	180	180	180	180	180	180	180
W12X87	216	216	216	216	216	216	216	216	216	216	216	216	216	216	216	216	216
W18X86	304	304	304	304*	291	279	267	256	247	237	229	221	214	207	200	194	188
W24X84	271	257	242	227	213	204	196	188	181	174	168	162	156	151	147	142	138
W21X83	257	244	231	220	210	201	193	185	178	171	165	159	154	149	144	140	136
W14X82	225	225	225	225	225	225	225	225	225	225	225*	213	206	200	193	187	182
W12X79	196	196	196	196	196	196	196	196	196	196	196	196	196	196	196	196*	188
W16X77	245	245	245	245*	240	229	220	211	203	195	188	182	176	170	165	160	155
W10X77	157	157	157	157	157	157	157	157	157	157	157	157	157	157	157	157	157
W24X76	241	228	215	201	186	171	157	149	144	138	133	129	124	120	117	113	110
W18X76	267	267*	250	238	227	217	208	200	192	185	179	172	167	161	156	151	147
W14X74	205	205	205	205	205	205	205	205*	200	192	185	179	173	167	162	157	153
W21X73	202	191	181	173	165	158	151	145	139	134	129	125	121	117	113	110	106
W12X72	178	178	178	178	178	178	178	178	178	178	178	178	178*	172	167	161	157
W18X71	196	186	177	168	161	154	147	141	136	131	126	122	118	114	110	107	104
W24X68	208	197	185	172	159	146	134	123	114	106	101	97	94	91	88	85	83
W10X68	138	138	138	138	138	138	138	138	138	138	138	138	138	138	138	138	138*
W21X68	177	165	156	148	142	135	130	125	120	115	111	107	104	100	97	94	91
W14X68	188	188	188	188	188	188*	184	176	169	163	157	152	147	142	138	133	129
W16X67	214	214*	203	193	184	176	169	162	156	150	145	140	135	131	126	123	119
W8X67	110	110	110	110	110	110	110	110	110	110	110	110	110	110	110	‡10	110
W18X65	168	159	151	144	137	131	126	120	116	112	108	104	100	97	94	91	88
W12X65	161	161	161	161	161	161	161	161	161	161*	156	151	146	141	137	132	129
W24X62	116	104	95	90	86	83	79	76	73	70	68	65	63	61	59	57	56
W21X62	159	148	136	124	116	111	106	102	98	94	91	88	85	82	79	77	75
	18	19	20	21	22	23	24	25	26	27	28	29	30	31	32	33	34

UNSUPPORTED LENGTH OF THE COMPRESSION FLANGE IN FEET

Table S-14a Allowable moment on laterally unsupported W shapes *(continued)*

ALLOWABLE MOMENT IN KIP-FEET FOR LATERALLY UNSUPPORTED WIDE FLANGE BEAMS

ASTM A36 Steel

UNSUPPORTED LENGTH OF THE COMPRESSION FLANGE IN FEET
FIRST ASTERISK INDICATES L_c. SECOND ASTERISK INDICATES L_u

SECTION	1	2	3	4	5	6	7	8	9	10	11	12	13	14	15	16	17
W14X61	184	184	184	184	184	184	184	184	184	184*	169	169	169	169	169	169	169
W18X60	216	216	216	216	216	216	216*	198	198	198	198	198	198*	185	172	161	152
W10X60	133	133	133	133	133	133	133	133	133	133*	122	122	122	122	122	122	122
W12X58	156	156	156	156	156	156	156	156	156	156*	143	143	143	143	143	143	143
W8X58	104	104	104	104	104	104	104	104*	95	95	95	95	95	95	95	95	95
W21X57	222	222	222	222	222	222*	203	203	203*	187	171	161	151	139	127	116	110
W16X57	184	184	184	184	184	184	184*	169	169	169	169	169	169	169*	158	148	140
W24X55	228	228	228	228	228	228*	209*	201	194	186	178	168	158	147	135	122	109
W18X55	196	196	196	196	196	196	196*	180	180	180	180	180*	165	153	143	134	126
W10X54	120	120	120	120	120	120	120	120	120	120*	110	110	110	110	110	110	110
W14X53	155	155	155	155	155	155	155	155*	142	142	142	142	142	142	142	142	142*
W12X53	141	141	141	141	141	141	141	141	141	141*	129	129	129	129	129	129	129
W21X50	189	189	189	189	189	189*	173*	165	158	151	143	134	125	115	104	92	82
W18X50	177	177	177	177	177	177	177*	162	162	162*	159	146	137	130	123	116	108
W16X50	162	162	162	162	162	162	162*	148	148	148	148	148*	142	132	123	115	108
W12X50	129	129	129	129	129	129	129	129*	118	118	118	118	118	118	118	118	118
W10X49	109	109	109	109	109	109	109	109	109	109*	100	100	100	100	100	100	100
W14X48	140	140	140	140	140	140	140	140*	128	128	128	128	128	128	128	128*	119
W8X48	86	86	86	86	86	86	86	86*	79	79	79	79	79	79	79	79	79
W18X46	157	157	157	157	157	157*	144	144	144*	133	121	111	102	95	88	83	78
W16X45	145	145	145	145	145	145	145*	133	133	133	133*	124	114	106	99	93	87
W12X45	116	116	116	116	116	116	116	116*	106	106	106	106	106	106	106	106	106*
W10X45	98	98	98	98	98	98	98	98*	90	90	90	90	90	90	90	90	90
W21X44	163	163	163	163	163	163*	146	141	135	129	122	114	106	97	87	77	68
W14X43	125	125	125	125	125	125	125	125*	114	114	114	114	114	114*	108	101	95
W18X40	136	136	136	136	136	136*	125	125*	112	106	100	93	85	77	69	62	59
W16X40	129	129	129	129	129	129	129*	118	118	118*	108	100	95	90	84	78	72
W12X40	103	103	103	103	103	103	103	103*	95	95	95	95	95	95	95*	93	87
W8X40	71	71	71	71	71	71	71	71*	65	65	65	65	65	65	65	65	65
W10X39	84	84	84	84	84	84	84	84*	77	77	77	77	77	77	77	77	77
W14X38	109	109	109	109	109	109	109*	100	100	100	100*	93	86	80	75	70	66
W16X36	113	113	113	113	113	113	113*	103*	99	95	91	87	82	77	72	67	61
W18X35	115	115	115	115	115	115*	102	98	93	88	83	77	70	63	55	49	43
W12X35	91	91	91	91	91	91*	83	83	83	83	83	83*	79	74	69	64	61
W8X35	62	62	62	62	62	62	62	62*	57	57	57	57	57	57	57	57	57
W14X34	97	97	97	97	97	97	97*	89	89	89*	80	74	70	65	61	56	52
W10X33	70	70	70	70	70	70	70	70*	64	64	64	64	64	64	64	64*	61
W16X31	94	94	94	94	94*	86	86*	78	74	69	64	58	52	45	40	37	35
W8X31	55	55	55	55	55	55	55	55*	50	50	50	50	50	50	50	50	50
W14X30	84	84	84	84	84	84	84*	77*	72	69	66	63	60	56	52	47	43
	1	2	3	4	5	6	7	8	9	10	11	12	13	14	15	16	17

UNSUPPORTED LENGTH OF THE COMPRESSION FLANGE IN FEET

Table S-14b Allowable moment on laterally unsupported W shapes

ALLOWABLE MOMENT IN KIP-FEET FOR LATERALLY UNSUPPORTED WIDE FLANGE BEAMS

ASTM A36 Steel

UNSUPPORTED LENGTH OF THE COMPRESSION FLANGE IN FEET
FIRST ASTERISK INDICATES L_c. SECOND ASTERISK INDICATES L_u

SECTION	18	19	20	21	22	23	24	25	26	27	28	29	30	31	32	33	34
W14X61	169	169	169	169*	162	155	148	142	137	132	127	122	118	115	111	108	104
W18X60	143	136	129	123	117	112	107	103	99	95	92	89	86	83	80	78	76
W10X60	122	122	122	122	122	122	122	122	122	122	122	122	122	122*	116	112	109
W12X58	143	143	143	143	143	143	143*	136	131	126	122	117	113	110	106	103	100
W8X58	95	95	95	95	95	95	95	95	95	95	95	95	95	95	95	95	95
W21X57	103	98	93	89	85	81	77	74	71	69	66	64	62	60	58	56	55
W16X57	132	125	119	113	108	103	99	95	91	88	85	82	79	76	74	72	70
W24X55	97	87	79	71	65	61	59	57	54	52	50	49	47	45	44	43	41
W18X55	119	112	107	102	97	93	89	85	82	79	76	73	71	69	67	65	63
W10X54	110	110	110	110	110	110	110	110	110	110	110*	105	101	98	95	92	89
W14X53	137	130	123	117	112	107	103	99	95	91	88	85	82	79	77	75	72
W12X53	129	129	129	129	129*	121	116	112	107	103	100	96	93	90	87	84	82
W21X50	73	69	66	62	60	57	55	52	50	48	47	45	44	42	41	40	38
W18X50	100	92	87	83	79	76	73	70	67	65	62	60	58	56	54	53	51
W16X50	102	97	92	88	84	80	77	73	71	68	66	63	61	59	57	56	54
W12X50	118	118*	114	108	103	99	95	91	87	84	81	78	76	73	71	69	67
W10X49	100	100	100	100	100	100	100	100*	98	94	91	88	85	82	79	77	75
W14X48	112	106	101	96	92	88	84	81	78	75	72	69	67	65	63	61	59
W8X48	79	79	79	79	79	79	79	79	79	79	79	79	79*	76	73	71	69
W18X46	74	70	66	63	60	57	55	53	51	49	47	45	44	43	41	40	39
W16X45	82	78	74	71	67	64	62	59	57	55	53	51	49	48	46	45	43
W12X45	103	97	92	88	84	80	77	74	71	68	66	64	61	59	58	56	54
W10X45	90	90	90	90	90*	87	83	80	77	74	71	69	67	64	62	61	59
W21X44	61	54	49	45	43	41	40	38	37	35	34	33	32	31	30	29	28
W14X43	90	85	81	77	73	70	67	64	62	60	57	55	54	52	50	49	47
W18X40	55	52	50	47	45	43	41	40	38	37	35	34	33	32	31	30	29
W16X40	66	62	59	56	54	51	49	47	45	44	42	41	39	38	37	36	34
W12X40	82	78	74	71	67	64	62	59	57	55	53	51	49	48	46	45	43
W8X40	65	65	65	65	65	65	65	65*	62	60	57	55	54	52	50	49	47
W10X39	77	77*	74	71	68	65	62	59	57	55	53	51	49	48	46	45	44
W14X38	62	59	56	53	51	48	46	45	43	41	40	38	37	36	35	34	33
W16X36	54	49	44	42	40	38	37	35	34	33	31	30	29	28	27	27	26
W18X35	38	36	34	32	31	30	28	27	26	25	24	23	23	22	21	20	20
W12X35	57	54	51	49	47	45	43	41	39	38	37	35	34	33	32	31	30
W8X35	57	57	57	57	57*	55	52	50	48	47	45	43	42	41	39	38	37
W14X34	49	46	44	42	40	38	37	35	34	32	31	30	29	28	27	26	26
W10X33	57	54	51	49	47	45	43	41	39	38	37	35	34	33	32	31	30
W16X31	33	31	30	28	27	26	25	24	23	22	21	20	20	19	18	18	17
W8X31	50	50	50*	47	45	43	41	39	38	36	35	34	33	32	31	30	29
W14X30	38	34	32	31	29	28	27	26	25	24	23	22	21	21	20	19	19
	18	19	20	21	22	23	24	25	26	27	28	29	30	31	32	33	34

UNSUPPORTED LENGTH OF THE COMPRESSION FLANGE IN FEET

Table S-14b Allowable moment on laterally unsupported W shapes *(continued)*

ALLOWABLE MOMENT IN KIP-FEET FOR LATERALLY UNSUPPORTED WIDE FLANGE BEAMS

ASTM A36 Steel

UNSUPPORTED LENGTH OF THE COMPRESSION FLANGE IN FEET
FIRST ASTERISK INDICATES L_c. SECOND ASTERISK INDICATES L_u

SECTION	1	2	3	4	5	6	7	8	9	10	11	12	13	14	15	16	17
W12X30	77	77	77	77	77	77*	70	70	70	70*	67	62	57	53	49	46	43
W10X30	64	64	64	64	64	64*	59	59	59	59	59	59	59*	54	50	47	44
W8X28	48	48	48	48	48	48*	44	44	44	44	44	44	44	44	44	44	44*
W16X26	76	76	76	76	76*	64	66	63	59	55	51	46	41	35	31	27	24
W14X26	70	70	70	70	70*	64	64*	56	52	48	44	39	34	31	29	27	26
W12X26	66	66	66	66	66	66*	61	61	61*	56	52	50	47	44	40	37	33
W10X26	55	55	55	55	55	55*	51	51	51	51	51*	47	43	40	38	35	33
W6X25	33	33	33	33	33	33*	30	30	30	30	30	30	30	30	30	30	30
W8X24	41	41	41	41	41	41*	38	38	38	38	38	38	38	38	38*	35	33
W14X22	58	58	58	58	58*	51	48	45	42	39	35	30	26	22	19	18	17
W12X22	50	50	50	50*	46	46*	42	36	32	29	26	24	22	21	19	18	17
W10X22	46	46	46	46	46	46*	42	42	42*	39	35	32	30	28	26	24	23
W8X21	36	36	36	36	36*	33	33	33	33	33	33*	32	29	27	25	24	22
W6X20	26	26	26	26	26	26*	24	24	24	24	24	24	24	24	24	24*	23
W12X19	42	42	42	42*	39*	34	31	28	25	20	18	17	15	14	13	12	12
W10X19	37	37	37	37*	34	34	34*	30	26	24	22	20	18	17	16	15	14
W5X19	20	20	20	20	20*	18	18	18	18	18	18	18	18	18	18	18	18
W8X18	30	30	30	30	30*	27	27	27	27*	26	24	22	20	19	17	16	15
W10X17	32	32	32	32*	29	29*	25	22	19	17	16	14	13	12	11	11	10
W12X16	34	34	34	34*	29	27	24	22	18	15	12	10	9	8	8	7	7
W6X16	20	20	20	20*	18	18	18	18	18	18	18	18*	16	15	14	13	12
W5X16	17	17	17	17	17*	15	15	15	15	15	15	15	15	15	15	15*	14
W10X15	27	27	27	27*	25*	22	20	18	16	13	11	10	9	8	8	7	7
W8X15	23	23	23	23*	21	21	21*	19	17	15	13	12	11	10	10	9	9
W6X15	17	17	17	17	17	17	17	17	17	17	17	17*	16	15	14	13	12
W12X14	29	29	29*	27*	25	23	21	19	16	13	10	9	7	6	6	5	5
W8X13	19	19	19	19*	18*	17	15	13	11	10	9	8	8	7	7	6	6
W4X13	10	10	10	10*	10	10	10	10	10	10	10	10	10	10	10*	9	9
W10X12	21	21	21*	19*	18	17	15	14	12	9	8	6	5	5	5	4	4
W6X12	14	14	14	14*	13	13	13	13*	12	11	10	9	8	8	7	7	6
W8X10	15	15	15	15*	13	12	11	10	9	7	6	5	5	4	4	4	3
W6X9	11	11	11	11*	10	10*	9	8	7	6	6	5	5	4	4	4	3
	1	2	3	4	5	6	7	8	9	10	11	12	13	14	15	16	17

UNSUPPORTED LENGTH OF THE COMPRESSION FLANGE IN FEET

Table S-14c Allowable moment on laterally unsupported W shapes

ALLOWABLE MOMENT IN KIP-FEET FOR LATERALLY UNSUPPORTED WIDE FLANGE BEAMS

ASTM A36 Steel

UNSUPPORTED LENGTH OF THE COMPRESSION FLANGE IN FEET
FIRST ASTERISK INDICATES L_c. SECOND ASTERISK INDICATES L_u

SECTION	18	19	20	21	22	23	24	25	26	27	28	29	30	31	32	33	34
W12X30	41	39	37	35	33	32	31	29	28	27	26	25	24	24	23	22	21
W10X30	42	40	38	36	34	33	31	30	29	28	27	26	25	24	23	23	22
W8X28	42	40	38	36	34	33	31	30	29	28	27	26	25	24	23	23	22
W16X26	21	20	19	18	17	16	16	15	14	14	13	13	12	12	12	11	11
W14X26	24	23	22	21	20	19	18	17	17	16	15	15	14	14	13	13	13
W12X26	31	29	28	26	25	24	23	22	21	20	20	19	18	18	17	17	16
W10X26	31	30	28	27	25	24	23	22	21	21	20	19	19	18	17	17	16
W6X25	30	30	30*	28	27	26	25	24	23	22	21	20	20	19	18	18	17
W8X24	31	30	28	27	25	24	23	22	21	21	20	19	19	18	17	17	16
W14X22	16	15	14	14	13	12	12	11	11	10	10	10	9	9	9	8	8
W12X22	16	15	14	14	13	12	12	11	11	10	10	10	9	9	9	8	8
W10X22	21	20	19	18	17	17	16	15	15	14	14	13	13	12	12	11	11
W8X21	21	20	19	18	17	16	16	15	14	14	13	13	12	12	12	11	11
W6X20	21	20	19	18	17	17	16	15	15	14	14	13	13	12	12	11	11
W12X19	11	10	10	9	9	8	8	8	7	7	7	7	6	6	6	6	6
W10X19	13	12	12	11	11	10	10	9	9	8	8	8	8	7	7	7	7
W5X19	18	18*	17	16	16	15	14	14	13	13	12	12	11	11	11	10	10
W8X18	14	14	13	12	12	11	11	10	10	9	9	9	8	8	8	8	7
W10X17	9	9	8	8	8	7	7	7	6	6	6	6	5	5	5	5	5
W12X16	6	6	6	5	5	5	5	5	4	4	4	4	4	4	3	3	3
W6X16	12	11	11	10	10	9	9	8	8	8	7	7	7	7	6	6	6
W5X16	14	13	12	12	11	11	10	10	9	9	9	8	8	8	7	7	7
W10X15	6	6	6	5	5	5	5	4	4	4	4	4	4	3	3	3	3
W8X15	8	8	7	7	6	6	6	6	5	5	5	5	5	4	4	4	4
W6X15	11	11	10	10	9	9	8	8	8	7	7	7	7	6	6	6	6
W12X14	5	4	4	4	4	4	3	3	3	3	3	3	3	3	2	2	2
W8X13	5	5	5	5	4	4	4	4	3	3	3	3	3	3	3	3	3
W4X13	8	8	7	7	6	6	6	6	5	5	5	5	5	4	4	4	4
W10X12	4	4	3	3	3	3	3	3	2	2	2	2	2	2	2	2	2
W6X12	6	5	5	5	5	4	4	4	4	4	4	3	3	3	3	3	3
W8X10	3	3	3	3	3	2	2	2	2	2	2	2	2	2	2	2	1
W6X9	3	3	3	3	3	2	2	2	2	2	2	2	2	2	2	2	1
	18	19	20	21	22	23	24	25	26	27	28	29	30	31	32	33	34

UNSUPPORTED LENGTH OF THE COMPRESSION FLANGE IN FEET

Table S-14c Allowable moment on laterally unsupported W shapes *(continued)*

ALLOWABLE MOMENT IN KIP-FEET FOR
LATERALLY UNSUPPORTED AMERICAN STANDARD I SHAPES

ASTM A36 Steel

UNSUPPORTED LENGTH OF THE COMPRESSION FLANGE IN FEET
FIRST ASTERISK INDICATES L_c.   SECOND ASTERISK INDICATES L_u.

SECTION	1	2	3	4	5	6	7	8	9	10	11	12	13	14	15	16	17
S20X75	256	256	256	256	256	256*	234	234	234	234	234*	225	208	193	180	169	159
S20X66	238	238	238	238	238	238*	218	218	218	218	218*	205	189	176	164	154	143
S18X70	206	206	206	206	206	206*	188	188	188	188	188*	171	158	147	137	128	121
S18X54.7	178	178	178	178	178	178*	163	163	163	163*	156	143	132	122	114	107	100
S15X50	129	129	129	129	129*	118	118	118	118	118*	114	105	97	90	84	78	74
S15X42.9	119	119	119	119	119*	109	109	109	109	109*	102	94	87	80	75	70	66
S12X50	101	101	101	101	101*	93	93	93	93	93	93	93	93*	90	84	79	74
S12X40.8	90	90	90	90	90*	83	83	83	83	83	83	83	83*	77	72	68	64
S12X35	76	76	76	76	76*	70	70	70	70	70*	66	61	56	52	48	45	43
S12X31.8	72	72	72	72	72*	66	66	66	66	66*	62	57	52	49	45	42	40
S10X35	58	58	58	58	58*	53	53	53	53	53	53*	49	45	42	39	37	34
S10X25.4	49	49	49	49*	45	45	45	45	45	45*	42	39	36	33	31	29	27
S8X23	32	32	32	32*	29	29	29	29	29	29*	27	24	23	21	19	18	17
S8X18.4	28	28	28	28*	26	26	26	26	26*	25	23	21	19	18	17	15	15
S7X20	24	24	24	24*	22	22	22	22	22	22*	19	18	16	15	14	13	12
S7X15.3	21	21	21*	19	19	19	19	19	19*	17	16	14	13	12	11	11	10
S6X17.25	17	17	17*	16	16	16	16	16	16*	15	14	12	11	11	10	9	9
S6X12.5	14	14	14*	13	13	13	13	13	13*	12	11	10	9	8	8	7	7
S5X14.75	12	12	12*	11	11	11	11	11	11*	10	9	9	8	7	7	6	6
S5X10	9	9	9*	9	9	9	9	9	9*	8	7	6	6	5	5	5	4
S4X9.5	6	6*	6	6	6	6	6	6	6*	5	5	4	4	4	3	3	3
S4X7.7	6	6*	5	5	5	5	5	5	5*	4	4	4	3	3	3	3	2
S3X7.5	3	3*	3	3	3	3	3	3	3	3*	3	2	2	2	2	2	2
S3X5.7	3	3*	3	3	3	3	3	3	3*	2	2	2	2	2	1	1	1
	1	2	3	4	5	6	7	8	9	10	11	12	13	14	15	16	17

UNSUPPORTED LENGTH OF THE COMPRESSION FLANGE IN FEET

Table S-15 Allowable moment on laterally unsupported S shapes

ALLOWABLE MOMENT IN KIP-FEET FOR
LATERALLY UNSUPPORTED AMERICAN STANDARD I SHAPES

ASTM A36 Steel

UNSUPPORTED LENGTH OF THE COMPRESSION FLANGE IN FEET
FIRST ASTERISK INDICATES L_c.   SECOND ASTERISK INDICATES L_u.

SECTION	18	19	20	21	22	23	24	25	26	27	28	29	30	31	32	33	34
S20X75	150	142	135	128	123	117	112	108	104	100	96	93	90	87	84	82	79
S20X66	136	129	123	117	112	107	102	98	94	91	88	85	82	79	77	74	72
S18X70	114	108	102	98	93	89	85	82	79	76	73	71	68	66	64	62	
S18X54.7	95	90	85	81	78	74	71	68	66	63	61	59	57	55			
S15X50	70	66	63	60	57	54	52	50	48	46	45	43	42	40	39		
S15X42.9	62	59	56	53	51	49	47	45	43	41	40	39	37	36			
S12X50	70	67	63	60	57	55	53	50	48	47	45	43	42	41	39	38	37
S12X40.8	60	57	54	51	49	47	45	43	41	40	38	37	36	35	34	33	32
S12X35	40	38	36	34	33	31	30	29	28	27	26	25	24	23			
S12X31.8	38	36	34	32	31	29	28	27	26	25	24	23	22	22			
S10X35	33	31	29	28	27	25	24	23	22	22	21	20	19	19	18	18	
S10X25.4	26	24	23	22	21	20	19	18	18	17	16	16	15	15			
S8X23	16	15	14	14	13	13	12	11	11	11	10	10	9				
S8X18.4	14	13	12	12	11	11	10	10	9	9	9	8					
S7X20	12	11	10	10	9	9	9	8	8	8	7	7	7				
S7X15.3	9	9	8	8	8	7	7	7	6	6	6						
S6X17.25	8	8	7	7	7	6	6	6	5	5	5	5					
S6X12.5	6	6	6	5	5	5	5	4	4	4							
S5X14.75	6	5	5	5	4	4	4	4	4	4	3	3					
S5X10	4	4	4	3	3	3	3	3	3	2							
S4X9.5	3	3	2	2	2	2	2	2	2	2	2						
S4X7.7	2	2	2	2	2	2	2	1	1	1							
S3X7.5	1	1	1	1	1	1	1	1	1	1	1	1	1				
S3X5.7	1	1	1	1	1	1	1	1	1	1	1						
	18	19	20	21	22	23	24	25	26	27	28	29	30	31	32	33	34

UNSUPPORTED LENGTH OF THE COMPRESSION FLANGE IN FEET

Table S-15 Allowable moment on laterally unsupported S shapes *(continued)*

ALLOWABLE MOMENT IN KIP-FEET FOR LATERALLY UNSUPPORTED
AMERICAN STANDARD CHANNELS LOADED THROUGH THEIR SHEAR CENTERS

ASTM A36 Steel

UNSUPPORTED LENGTH OF THE COMPRESSION FLANGE IN FEET
FIRST ASTERISK INDICATES L_c. SECOND ASTERISK INDICATES L_u.

SECTION	1	2	3	4	5	6	7	8	9	10	11	12	13	14	15	16	17
C12X30	54	54	54*	49	49	49*	42	37	33	29	27	24	22	21	19	18	17
C12X25	48	48	48*	44	44	44*	42	36	31	28	25	23	21	19	18	17	15	15
C12X20.7	43	43	43*	39	39*	36	31	27	24	22	20	18	16	15	14	13	12
C10X30	41	41	41*	37	37	37*	32	28	25	22	20	19	17	16	15	14	13
C10X25	36	36	36*	33	33*	31	27	23	21	19	17	15	14	13	12	11	11
C10X20	31	31*	28	28	28*	26	22	19	17	15	14	13	12	11	10	9	9
C10X15.3	27	27*	24	24	24*	21	18	15	14	12	11	10	9	9	8	7	7
C9X20	27	27*	24	24	24*	22	19	17	15	13	12	11	10	9	9	8	8
C9X15	22	22*	20	20	20*	17	15	13	11	10	9	8	8	7	7	6	6
C9X13.4	21	21*	19	19	19*	16	14	12	10	9	8	8	7	7	6	6	5
C8X18.75	22	22*	20	20	20*	18	16	14	12	11	10	9	8	8	7	7	6
C8X13.75	18	18*	16	16	16*	14	12	10	9	8	7	7	6	6	5	5	5
C8X11.5	16	16*	14	14	14*	12	10	9	8	7	6	6	5	5	4	4	4
C7X14.75	15	15*	14	14	14*	12	11	9	8	7	7	6	5	5	5	4	4
C7X12.25	13	13*	12	12	12*	11	9	8	7	6	6	5	5	4	4	4	3
C7X9.8	12	12*	11	11	11*	9	7	6	6	5	5	4	4	3	3	3	3
C6X13	11	11*	10	10	10*	9	8	7	6	5	5	4	4	4	3	3	3
C6X10.5	10	10*	9	9	9*	8	7	6	5	4	4	4	3	3	3	3	2
C6X8.2	8	8*	8	8	8*	6	5	5	4	4	3	3	3	2	2	2	2
C5X9	7*	6	6	6	6*	5	5	4	3	3	3	2	2	2	2	2	2
C5X6.7	6*	5	5	5	5*	4	4	3	3	2	2	2	2	2	1	1	1
C4X7.25	4*	4	4	4	4*	4	3	3	2	2	2	2	1	1	1	1	1
C4X5.4	3*	3	3	3	3*	3	2	2	2	1	1	1	1	1	1	1	1
C3X6	2*	2	2	2	2	2*	2	2	1	1	1	1	1	1	1	1	0
C3X5	2*	2	2	2	2	2*	2	1	1	1	1	1	1	1	0	0	0
C3X4.1	2*	2	2	2	2*	1	1	1	1	1	1	0	0	0	0	0	0
	1	2	3	4	5	6	7	8	9	10	11	12	13	14	15	16	17

UNSUPPORTED LENGTH OF THE COMPRESSION FLANGE IN FEET

Table S-16 Allowable moment on laterally unsupported channels

Beam Design Tables
For WT Sections

The design tables for tee sections are considered separately from other beams. In many of the tee sections, the slender stem causes the section to be classified as a section having slender elements. The ASD Specification is vague about using sections having slender elements as beams, though the ASD Appendix B does include provisions for using such sections as columns.

In giving the allowable stresses in sections having no slender elements, ASD (F1.3) includes all rolled sections having an axis of symmetry in, and loaded in the plane of the web. The Specification also requires that the compression flange be solid and approximately rectangular in cross section, and that it be larger in cross sectional area than the tension flange. The Specification does not require that the tension flange be rectangular.

Under these limitations, the only tees that may be used as beams are those that do not have slender elements. Further, a tee may be used as a beam only where its flange is in compression and the stem is in tension. And further, tees may not be used as a beam in any span having an inflection point, that is, a point in the span where the stress changes from compression on the top fibers of the beam to compression on the bottom fibers.

The following table, Table S-17, adheres to these limitations. As usual, the table lists the sections in descending order of weight. For a detailed list of the limitations placed on the allowable flexure stresses, see Table 27-2 of the text.

An example will illustrate the use of Table S-17.

Given: Design moment of 22 kip•ft to be carried by a tee section cut from a wide flange section. The compression flange is supported at regular intervals of 10 feet.

To Find: Suitable WT section in A36 steel to support the load

Solution: Tables S-17a and b contain the allowable moments on WT shapes for various values of unsupported lengths. Enter Table S-17b at the bottom of the page at an unsupported length of 10 ft and proceed upward through Table S-17b into Table S-17a. The first section (and the lightest section) that will sustain a moment of 22 kip•ft is found to be a WT10.5 × 41.5.

ALLOWABLE MOMENT IN KIP-FEET FOR LATERALLY UNSUPPORTED TEES CUT FROM WIDE FLANGE BEAMS

ASTM A36 Steel

DISTANCE IN FEET BETWEEN REGULARLY SPACED
LATERAL SUPPORTS ALONG THE COMPRESSION FLANGE

SECTION	1	2	3	4	5	6	7	8	9	10	11	12	13	14	15	16	17
WT6X68	16	16	16	16	16	16	16	15	15	15	15	15	14	14	14	14	13
WT7X66	17	17	16	16	16	16	16	16	16	15	15	15	15	15	15	14	14
WT7X60	15	15	15	15	15	14	14	14	14	14	14	13	13	13	13	13	13
WT6X60	14	14	14	14	14	14	13	13	13	13	13	13	12	12	12	12	12
WT5X56	11	11	11	11	10	10	10	10	10	10	10	9	9	9	9	9	8
WT7X54.5	13	13	13	13	13	13	12	12	12	12	12	12	12	12	11	11	11
WT6X53	12	12	12	12	11	11	11	11	11	11	11	11	10	10	10	10	10
WT8X50	20	20	19	19	19	19	19	18	18	18	17	17	17	16	16	16	15
WT5X50	9	9	9	9	9	9	9	9	9	8	8	8	8	8	8	7	7
WT7X49.5	12	12	12	12	11	11	11	11	11	11	11	11	11	10	10	10	10
WT9X48.5	22	22	22	22	21	21	21	21	20	20	20	19	19	19	18	18	18
WT6X48	10	10	10	10	10	10	10	10	10	10	9	9	9	9	9	9	8
WT10.5X46.5	31	31	31	30	30	29	29	28	27	27	26	25	25	24	23	22	21
WT7X45	11	10	10	10	10	10	10	10	10	10	10	10	9	9	9	9	9
WT8X44.5	18	17	17	17	17	17	16	16	16	16	15	15	15	14	14	14	13
WT5X44	8	8	8	8	8	8	7	7	7	7	7	7	7	7	6	6	6
WT6X43.5	10	9	9	9	9	9	9	9	9	9	9	8	8	8	8	8	8
WT9X43	20	19	19	19	19	19	18	18	18	18	17	17	17	16	16	16	15
WT10.5X41.5	27	27	27	26	26	26	25	25	24	23	23	22	22	21	20	19	19
WT7X41	12	12	12	12	12	12	11	11	11	11	11	10	10	10	10	10	9
WT6X39.5	8	8	8	8	8	8	8	8	8	8	8	7	7	7	7	7	7
WT8X38.5	15	15	15	14	14	14	14	14	13	13	13	13	12	12	12	12	11
WT5X38.5	7	7	7	7	6	6	6	6	6	6	6	6	6	5	5	5	5
WT7X37	11	11	10	10	10	10	10	10	10	9	9	9	9	9	8	8	8
WT6X36	8	8	7	7	7	7	7	7	7	7	7	7	7	6	6	6	6
WT9X35.5	19	19	19	19	18	18	17	17	17	16	16	15	15	14	13	13	12
WT5X34	6	6	6	6	5	5	5	5	5	5	5	5	5	5	5	4	4
WT7X34	10	10	9	9	9	9	9	9	9	9	8	8	8	8	8	7	7
WT8X33.5	13	13	12	12	12	12	12	12	11	11	11	11	11	10	10	10	10
WT4X33.5	5	5	5	5	5	5	4	4	4	4	4	4	4	4	4	3	3
WT9X32.5	17	17	17	17	16	16	16	15	15	15	14	14	13	13	12	12	11
WT6X32.5	7	7	7	7	7	6	6	6	6	6	6	6	6	6	6	6	5
WT7X30.5	9	8	8	8	8	8	8	8	8	8	7	7	7	7	7	7	6
WT5X30	5	5	5	5	5	5	5	4	4	4	4	4	4	4	4	4	4
WT6X29	6	6	6	6	6	6	6	6	6	5	5	5	5	5	5	5	5
WT4X29	4	4	4	4	4	4	4	4	4	3	3	3	3	3	3	3	3
	1	2	3	4	5	6	7	8	9	10	11	12	13	14	15	16	17

DISTANCE IN FEET BETWEEN REGULARLY SPACED
LATERAL SUPPORTS ALONG THE COMPRESSION FLANGE

Table S-17a Allowable moment on laterally unsupported WT shapes

ALLOWABLE MOMENT IN KIP-FEET FOR LATERALLY UNSUPPORTED TEES CUT FROM WIDE FLANGE BEAMS

ASTM A36 Steel

DISTANCE IN FEET BETWEEN REGULARLY SPACED
LATERAL SUPPORTS ALONG THE COMPRESSION FLANGE

SECTION	18	19	20	21	22	23	24	25	26	27	28	29	30	31	32	33	34
WT6X68	13	13	13	12	12	12	12	11	11	11	10	10	10	9	9	9	8
WT7X66	14	14	14	13	13	13	13	13	12	12	12	12	11	11	11	11	10
WT7X60	13	12	12	12	12	12	11	11	11	11	11	10	10	10	10	9	9
WT6X60	11	11	11	11	10	10	10	10	9	9	9	9	8	8	8	7	7
WT5X56	8	8	8	7	7	7	7	6	6	6	6	5	5	5	4	4	4
WT7X54.5	11	11	11	10	10	10	10	10	10	9	9	9	9	9	8	8	8
WT6X53	9	9	9	9	9	8	8	8	8	8	7	7	7	7	6	6	6
WT8X50	15	14	14	14	13	13	12	12	11	11	10	10	9	9	8	8	7
WT5X50	7	7	7	6	6	6	6	5	5	5	5	4	4	4	4	3	3
WT7X49.5	10	10	10	9	9	9	9	9	9	8	8	8	8	8	8	7	7
WT9X48.5	17	17	16	16	15	15	14	14	14	13	12	12	11	11	10	10	9
WT6X48	8	8	8	8	8	7	7	7	7	7	6	6	6	6	5	5	5
WT10.5X46.5	21	20	19	18	17	16	15	13	12	12	11	10	9	9	8	8	
WT7X45	9	9	9	8	8	8	8	8	8	8	7	7	7	7	7	6	6
WT8X44.5	13	13	12	12	12	11	11	10	10	9	9	8	8	8	7	7	6
WT5X44	6	6	6	5	5	5	5	5	4	4	4	4	3	3	3	3	3
WT6X43.5	8	7	7	7	7	7	7	6	6	6	6	6	5	5	5	5	5
WT9X43	15	15	14	14	13	13	13	12	12	11	11	10	10	9	9	8	8
WT10.5X41.5	18	17	16	15	14	14	13	12	11	10	9	8	8	7	7		
WT7X41	9	9	8	8	8	8	7	7	7	6	6	6	5	5	5	4	4
WT6X39.5	7	7	6	6	6	6	6	6	5	5	5	5	5	5	4	4	4
WT8X38.5	11	11	10	10	10	9	9	9	8	8	7	7	7	6	6	5	5
WT5X38.5	5	5	5	4	4	4	4	4	4	3	3	3	3	3	2	2	2
WT7X37	8	8	7	7	7	7	6	6	6	5	5	5	5	4	4	4	3
WT6X36	6	6	6	6	5	5	5	5	5	5	5	4	4	4	4	4	4
WT9X35.5	12	11	10	10	9	8	7	7	6	6	5	5	5				
WT5X34	4	4	4	4	4	3	3	3	3	3	3	3	2	2	2	2	2
WT7X34	7	7	7	6	6	6	6	5	5	5	5	4	4	4	3	3	3
WT8X33.5	9	9	9	8	8	8	8	7	7	7	6	6	6	5	5	5	4
WT4X33.5	3	3	3	3	2	2	2	2	2	2							
WT9X32.5	10	10	9	8	8	7	7	6	5	5	5						
WT6X32.5	5	5	5	5	5	5	5	4	4	4	4	4	4	4	3	3	3
WT7X30.5	6	6	6	6	5	5	5	5	5	4	4	4	4	3	3	3	3
WT5X30	4	3	3	3	3	3	3	3	3	2	2	2	2	2	2	2	
WT6X29	4	4	4	4	4	4	4	3	3	3	3	3	2	2	2	2	2
WT4X29	3	2	2	2	2	2	2	2									
	18	19	20	21	22	23	24	25	26	27	28	29	30	31	32	33	34

DISTANCE IN FEET BETWEEN REGULARLY SPACED
LATERAL SUPPORTS ALONG THE COMPRESSION FLANGE

Table S-17a Allowable moment on laterally unsupported WT shapes *(continued)*

ALLOWABLE MOMENT IN KIP-FEET FOR LATERALLY UNSUPPORTED TEES CUT FROM WIDE FLANGE BEAMS

ASTM A36 Steel

DISTANCE IN FEET BETWEEN REGULARLY SPACED
LATERAL SUPPORTS ALONG THE COMPRESSION FLANGE

SECTION	1	2	3	4	5	6	7	8	9	10	11	12	13	14	15	16	17
WT8X28.5	13	13	13	13	12	12	12	12	11	11	10	10	10	9	9	8	8
WT12X27.5	15	15	15	15	14	14	14	14	13	13	13	12	12	12	11	11	10
WT5X27	4	4	4	4	4	4	4	4	4	4	4	4	3	3	3	3	3
WT7X26.5	8	8	8	8	8	8	8	7	7	7	7	7	6	6	6	6	5
WT6X26.5	6	6	6	6	6	5	5	5	5	5	5	5	5	5	5	4	4
WT6X25	6	6	6	6	6	6	6	6	5	5	5	5	5	5	4	4	4
WT5X24.5	4	4	4	4	4	4	3	3	3	3	3	3	3	3	3	3	3
WT7X24	7	7	7	7	7	7	7	7	6	6	6	6	6	6	5	5	5
WT4X24	3	3	3	3	3	3	3	3	3	2	2	2	2	2	2	2	2
WT6X22.5	6	5	5	5	5	5	5	5	5	5	4	4	4	4	4	4	4
WT5X22.5	4	4	4	4	4	4	4	3	3	3	3	3	3	3	3	3	2
WT6X20	5	5	5	5	4	4	4	4	4	4	4	4	4	3	3	3	3
WT4X20	3	2	2	2	2	2	2	2	2	2	2	2	2	2	2	2	2
WT5X19.5	3	3	3	3	3	3	3	3	3	3	3	3	2	2	2	2	2
WT6X17.5	5	5	5	5	5	5	5	4	4	4	4	4	4	3	3	3	3
WT4X17.5	2	2	2	2	2	2	2	2	2	2	2	2	1	1	1	1	1
WT5X16.5	3	3	3	3	3	3	3	3	2	2	2	2	2	2	2	2	2
WT4X15.5	2	2	2	2	2	2	2	2	1	1	1	1	1	1	1	1	1
WT5X15	3	3	3	3	3	3	3	3	3	3	2	2	2	2	2	2	1
WT4X14	2	2	2	2	2	2	2	1	1	1	1	1	1	1	1	1	1
WT5X13	3	3	3	3	3	3	2	2	2	2	2	2	2	2	1	1	1
WT3X12.5	1	1	1	1	1	1	1	1	1	1	1	1	1				
WT4X12	1	1	1	1	1	1	1	1	1	1	1	1	1		1	1	1
WT4X10.5	2	2	2	1	1	1	1	1	1	1	1	1	1	1	1		
WT3X10	1	1	1	1	1	1	1	1									
WT5X9.5	3	2	2	2	2	2	2	2	1	1	1	1	1				
WT4X9	1	1	1	1	1	1	1	1	1	1	1	1	1				
WT5X8.5	2	2	2	2	2	2	2	1	1	1	1	1	1				
WT3X8	1	1	1	1	1												
WT4X7.5	1	1	1	1	1	1	1	1	1	1							
WT3X7.5	1	1															
WT4X6.5	1	1	1	1	1	1	1	1	1								

| 1 | 2 | 3 | 4 | 5 | 6 | 7 | 8 | 9 | 10 | 11 | 12 | 13 | 14 | 15 | 16 | 17 |

DISTANCE IN FEET BETWEEN REGULARLY SPACED
LATERAL SUPPORTS ALONG THE COMPRESSION FLANGE

Table S-17b Allowable moment on laterally unsupported WT shapes

ALLOWABLE MOMENT IN KIP-FEET FOR LATERALLY UNSUPPORTED TEES CUT FROM WIDE FLANGE BEAMS

ASTM A36 Steel

DISTANCE IN FEET BETWEEN REGULARLY SPACED
LATERAL SUPPORTS ALONG THE COMPRESSION FLANGE

SECTION	18	19	20	21	22	23	24	25	26	27	28	29	30	31	32	33	34
WT8X28.5	7	7	6	6	5	5	4	4	4								
WT12X27.5	10	10	9	9	8	8	7	7	6	6							
WT5X27	3	3	3	3	3	2	2	2	2	2	2	2	2				
WT7X26.5	5	5	5	4	4	4	3	3	3	3							
WT6X26.5	4	4	4	4	4	3	3	3	3	3	3	2	2	2	2	2	2
WT6X25	4	4	3	3	3	3	3	2	2	2	2	2					
WT5X24.5	3	3	2	2	2	2	2	2	2	2	2	2	1	1	1	1	1
WT7X24	5	4	4	4	4	3	3	3	3	2	2	2	2	2			
WT4X24	2	2	2	1	1	1	1	1	1	1	1	1	1				
WT6X22.5	3	3	3	3	3	2	2	2	2	2							
WT5X22.5	2	2	2	2	2	2											
WT6X20	3	3	3	2	2	2	2	2									
WT4X20	1	1	1	1	1	1	1	1	1	1	1						
WT5X19.5	2	2	2	2	1	1	1	1	1	1	1	1	1	1			
WT6X17.5	2	2	2	2	2												
WT4X17.5	1	1	1	1	1	1	1	1									
WT5X16.5	2	2	1	1	1	1	1	1	1	1	1	1					
WT4X15.5	1	1	1	1	1	1	1										
WT5X15	1	1	1	1	1	1											
WT4X14	1	1															
WT5X13	1	1	1	1													
WT3X12.5																	
WT4X12																	
WT4X10.5																	
WT3X10																	
WT5X9.5																	
WT4X9																	
WT5X8.5																	
WT3X8																	
WT4X7.5																	
WT3X7.5																	
WT4X6.5																	

	18	19	20	21	22	23	24	25	26	27	28	29	30	31	32	33	34

DISTANCE IN FEET BETWEEN REGULARLY SPACED
LATERAL SUPPORTS ALONG THE COMPRESSION FLANGE

Table S-17b Allowable moment on laterally unsupported WT shapes *(continued)*

Miscellaneous Tables

The following tables provide data for an assortment of applications.

Table S-18 lists the section modulus and bracing lengths for each of the wide flange sections listed in earlier tables. To use the table, compute the required minimum value of the section modulus S_{xx} using the allowable stress F_b. (Remember that the allowable stress F_b is 24 ksi if lateral supports are provided at a distance not farther apart than L_c but not greater than L_u.)

1. Enter Table S-18 in the lower right corner, proceed upward along the S_{xx} column until the first numerical value of S_{xx} is reached which is equal to or larger than the computed minimum value. The beam having this section modulus is the lightest section that will sustain the given moment at the assumed allowable stress.

2. Check the listed value of L_c or L_u for this beam to see that the actual lateral support length is not longer than these maxima. If the actual lateral support length is longer than the listed value, continue upward along the S_{xx} column until a section is found which has *both* a value of S_{xx} larger than the computed minimum value *and* a value of L_c (or L_u) larger than the actual support length. The beam thus found will sustain the given moment at the assumed allowable stress.

3. For convenience, the moment capacity M_c is also listed in Table S-18 for continuously supported beams, for which F_b is 24 ksi. If it is known that lateral supports on the compression flange will be provided not farther apart than L_c, the table may be entered in the M_c column rather than the S_{xx} column.

4. For estimating the size of a section, the load constant $wL^2 = WL$ for a simple span may be computed, where w is the sum of the dead and live uniform loads. Table S-18 may then be entered in the WL column rather than the S_{xx} column. The use of the WL column is particularly useful when a steel member is used to support a timber roof structure above; such conditions commonly arise at long spans such as wide door openings. The dead and live load of the timber structure tributary to the steel beam is estimated (usually 40 to 50 psf) and a steel member is selected for that load. Bracing limits on the compression flange will, of course, remain an essential requirement of the final design.

The remaining miscellaneous tables S-19 through S-29 contain miscellaneous design data; the applications are given in the tables as appropriate.

SECTION MODULUS, ALLOWABLE MOMENTS AND LOAD CONSTANTS*
FOR LATERALLY SUPPORTED WIDE FLANGE BEAMS,
ARRANGED IN ORDER OF DESCENDING WEIGHTS

ASTM A36 STEEL

Computed values for moment M_C and load constant WL are valid only where lateral supports exist along the compression flange, spaced not farther apart than L_C, for which case the allowable flexural stress is 24000 psi.

SECTION	S_{xx} in^3	L_C ft.	L_u ft.	M_C k·ft	WL k·ft	SECTION	S_{xx} in^3	L_C ft.	L_u ft.	M_C k·ft	WL k·ft	Section	S_{xx} in^3	L_C ft.	L_u ft.	M_C k·ft	WL k·ft
W12X136	186.00	13.1	53.2	372	2976	W14X61	92.20	10.6	21.5	184	1475	W12X30	38.60	6.9	10.8	77	618
W14X132	209.00	15.5	47.7	418	3344	W18X60	108.00	8.0	13.3	216	1728	W10X30	32.40	6.1	13.1	64	518
W14X120	190.00	15.5	44.1	380	3040	W10X60	66.70	10.6	31.1	133	1067	W8X28	24.30	6.9	17.5	48	389
W12X120	163.00	13.0	48.2	326	2608	W12X58	78.00	10.6	24.4	156	1248	W16X26	38.40	5.6	6.0	76	614
W10X112	126.00	11.0	53.2	252	2016	W8X58	52.00	8.7	35.3	104	832	W14X26	35.30	5.3	7.0	70	565
W14X109	173.00	15.4	40.6	346	2768	W21X57	111.00	6.9	9.4	222	1776	W12X26	33.40	6.9	9.4	66	534
W12X106	145.00	12.9	43.3	290	2320	W16X57	92.20	7.5	14.3	184	1475	W10X26	27.90	6.1	11.4	55	446
W16X100	175.00	11.0	28.1	350	2800	W24X55	114.00	6.9	7.5	228	1824	W6X25	16.70	6.4	20.0	33	267
W10X100	112.00	10.9	48.2	224	1792	W18X55	98.30	7.9	12.1	196	1573	W8X24	20.90	6.9	15.2	41	334
W14X99	157.00	15.4	37.0	314	2512	W10X54	60.00	10.6	28.2	120	960	W14X22	29.00	5.3	5.6	58	464
W18X97	188.00	11.8	24.1	376	3008	W14X53	77.80	8.5	17.7	155	1245	W12X22	25.40	4.3	6.4	50	406
W12X96	131.00	12.8	39.9	262	2096	W12X53	70.60	10.6	22.0	141	1130	W10X22	23.20	6.1	9.4	46	371
W21X93	192.00	8.9	16.8	384	3072	W21X50	94.50	6.9	7.8	189	1512	W8X21	18.20	5.6	11.8	36	291
W14X90	143.00	15.3	34.0	286	2288	W18X50	88.90	7.9	11.0	177	1422	W6X20	13.40	6.4	16.4	26	214
W16X89	155.00	10.9	25.0	310	2480	W16X50	81.00	7.5	12.7	162	1296	W12X19	21.30	4.2	5.3	42	341
W10X88	98.50	10.8	43.3	197	1576	W12X50	64.70	8.5	19.6	129	1035	W10X19	18.80	4.2	7.2	37	301
W12X87	118.00	12.8	36.2	236	1888	W10X49	54.60	10.6	20.6	109	874	W5X19	10.20	5.3	19.5	20	163
W18X86	166.00	11.7	21.5	332	2656	W14X48	70.30	8.5	16.0	140	1125	W8X18	15.20	5.5	9.9	30	243
W24X84	196.00	9.5	13.3	392	3136	W8X48	43.30	8.6	30.3	86	693	W10X17	16.20	4.2	6.1	32	259
W21X83	171.00	8.8	15.1	342	2736	W18X46	78.80	6.4	9.4	157	1261	W12X16	17.10	4.1	4.3	34	274
W14X82	123.00	10.7	28.1	246	1968	W16X45	72.70	7.4	11.4	145	1163	W6X16	10.20	4.3	12.0	20	163
W12X79	107.00	12.8	33.3	214	1712	W12X45	58.10	8.5	17.7	116	930	W5X16	8.51	5.3	16.7	17	136
W16X77	134.00	10.9	21.9	268	2144	W10X45	49.10	8.5	22.8	98	786	W10X15	13.80	4.2	5.0	27	221
W10X77	85.90	10.8	38.6	171	1374	W21X44	81.60	6.6	7.0	163	1306	W8X15	11.80	4.2	7.2	23	189
W24X76	176.00	9.5	11.8	352	2816	W14X43	62.70	8.4	14.4	125	1003	W6X15					
W18X76	146.00	11.6	19.1	292	2336	W18X40	68.40	6.3	8.2	136	1094	W12X14	14.90	3.5	4.2	29	238
W14X74	112.00	10.6	25.9	224	1792	W16X40	64.70	7.4	10.2	129	1035	W8X13	9.91	4.2	5.9	19	159
W21X73	151.00	8.8	13.4	302	2416	W12X40	51.90	8.4	16.0	103	830	W4X13	5.46	4.3	15.6	10	87
W12X72	97.40	12.7	30.5	194	1558	W8X40	35.50	8.5	25.3	71	568	W10X12	10.90	3.9	4.3	21	174
W18X71	127.00	8.1	15.5	254	2032	W10X39	42.10	8.4	19.8	84	674	W6X12	7.31	4.2	8.6	14	117
W24X68	154.00	9.5	10.2	308	2464	W14X38	54.60	7.1	11.5	109	874	W8X10	7.81	4.2	4.7	15	125
W10X68	75.70	10.7	34.8	151	1211	W16X36	56.50	7.4	8.8	113	904	W6X9	5.56	4.2	6.7	11	89
W21X68	140.00	8.7	12.4	280	2240	W18X35	57.60	6.3	6.7	115	922						
W14X68	103.00	10.6	23.9	206	1648	W12X35	45.60	6.9	12.6	91	730						
W16X67	117.00	10.8	19.3	234	1872	W8X35	31.20	8.5	22.6	62	499						
W8X67	60.40	8.7	39.9	120	966	W14X34	48.60	7.1	10.2	97	778						
W18X65	117.00	8.0	14.4	234	1872	W10X33	35.00	8.4	16.5	70	560						
W12X65	87.90	12.7	27.7	175	1406	W16X31	47.20	5.8	7.1	94	755						
W24X62	131.00	7.4	8.1	262	2096	W8X31	27.50	8.4	20.1	55	440						
W21X62	127.00	8.7	11.2	254	2032	W14X30	42.00	7.1	8.7	84	672						

*Load constant WL is the total dead and live uniform load wL multiplied by the length L. For A36 steel, WL = 16·S_{xx}, W in kips, L in ft., S_{xx} in in^3.

Table S-18 Section modulus table for laterally supported beams

Type of Connection and Type of Member	Configuration Conditions	Value of U
Bolted Connections		
W, M, or S shapes having $b_f > {}^2/_3 d$	Min. of 3 fasteners in direction of load	0.90
Plates and all other W, M, or S shapes	Min. of 3 fasteners in direction of load	0.85
All members and all shapes	Two fasteners in direction of load	0.75
Welded Connections		
W, M, S, WT, and ST shapes	Load transferred by fillets transverse to the direction of load	1.00
From any shape of width w to a plate	Load transferred by longitudinal fillets of length l at both edges, with min. total $l \geq w$	
	1. When $l > 2w$	1.00
	2. When $2w > l > 1.5w$	0.87
	3. when $1.5w > l \geq w$	0.75

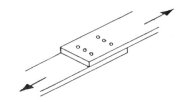

Two fasteners in
direction of load

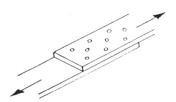

Three fasteners in
direction of load

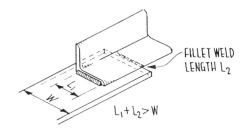

Fillet welds transverse
to direction of load

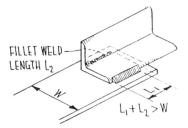

Longitudinal
Fillet welds

Table S-19 Area reduction coefficients for tension members

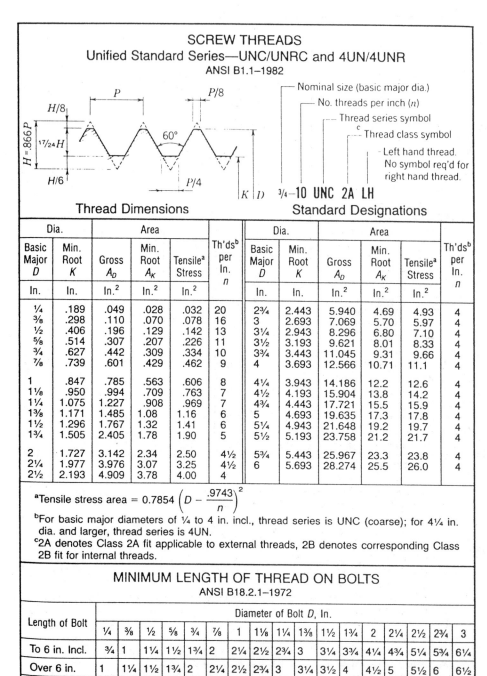

SCREW THREADS
Unified Standard Series—UNC/UNRC and 4UN/4UNR
ANSI B1.1–1982

Thread Dimensions

Standard Designations

¾–10 UNC 2A LH

Dia.		Area			Th'ds[b] per In. n	Dia.		Area			Th'ds[b] per In. n
Basic Major D	Min. Root K	Gross A_D	Min. Root A_K	Tensile[a] Stress		Basic Major D	Min. Root K	Gross A_D	Min. Root A_K	Tensile[a] Stress	
In.	In.	In.²	In.²	In.²		In.	In.	In.²	In.²	In.²	
¼	.189	.049	.028	.032	20	2¾	2.443	5.940	4.69	4.93	4
⅜	.298	.110	.070	.078	16	3	2.693	7.069	5.70	5.97	4
½	.406	.196	.129	.142	13	3¼	2.943	8.296	6.80	7.10	4
⅝	.514	.307	.207	.226	11	3½	3.193	9.621	8.01	8.33	4
¾	.627	.442	.309	.334	10	3¾	3.443	11.045	9.31	9.66	4
⅞	.739	.601	.429	.462	9	4	3.693	12.566	10.71	11.1	4
1	.847	.785	.563	.606	8	4¼	3.943	14.186	12.2	12.6	4
1⅛	.950	.994	.709	.763	7	4½	4.193	15.904	13.8	14.2	4
1¼	1.075	1.227	.908	.969	7	4¾	4.443	17.721	15.5	15.9	4
1⅜	1.171	1.485	1.08	1.16	6	5	4.693	19.635	17.3	17.8	4
1½	1.296	1.767	1.32	1.41	6	5¼	4.943	21.648	19.2	19.7	4
1¾	1.505	2.405	1.78	1.90	5	5½	5.193	23.758	21.2	21.7	4
2	1.727	3.142	2.34	2.50	4½	5¾	5.443	25.967	23.3	23.8	4
2¼	1.977	3.976	3.07	3.25	4½	6	5.693	28.274	25.5	26.0	4
2½	2.193	4.909	3.78	4.00	4						

[a]Tensile stress area = $0.7854 \left(D - \dfrac{.9743}{n} \right)^2$

[b]For basic major diameters of ¼ to 4 in. incl., thread series is UNC (coarse); for 4¼ in. dia. and larger, thread series is 4UN.

[c]2A denotes Class 2A fit applicable to external threads, 2B denotes corresponding Class 2B fit for internal threads.

MINIMUM LENGTH OF THREAD ON BOLTS
ANSI B18.2.1–1972

Length of Bolt	Diameter of Bolt D, In.																
	¼	⅜	½	⅝	¾	⅞	1	1⅛	1¼	1⅜	1½	1¾	2	2¼	2½	2¾	3
To 6 in. Incl.	¾	1	1¼	1½	1¾	2	2¼	2½	2¾	3	3¼	3¾	4¼	4¾	5¼	5¾	6¼
Over 6 in.	1	1¼	1½	1¾	2	2¼	2½	2¾	3	3¼	3½	4	4½	5	5½	6	6½

Thread length for bolts up to 6 in. long is 2D + ¼. For bolts over 6-in. long, thread length is 2D + ½. These proportions may be used to compute thread length for diameters not shown in the table. Bolts which are too short for listed or computed thread lengths are threaded as close to the head as possible.

For thread lengths for high-strength bolts, refer to *Allowable Stress Design Specification for Structural Joints Using ASTM A325 or A490 Bolts.*

AMERICAN INSTITUTE OF STEEL CONSTRUCTION

Table S-20 Thread sizes and root areas

Description of Fasteners	Allow-able Tension[g] (F_t)	Allowable Shear[g] (F_v)					Bearing-type Connec-tions[i]
		Slip-critical Connections[e,i]					
		Standard size Holes	Oversized and Short-slotted Holes	Long-slotted holes			
				Transverse[j] Load	Parallel[j] Load		
A502, Gr. 1, hot-driven rivets	23.0[a]						17.5[f]
A502, Gr. 2 and 3, hot-driven rivets	29.0[a]						22.0[f]
A307 bolts	20.0[a]						10.0[b,f]
Threaded parts meeting the requirements of Sects. A3.1 and A3.4 and A449 bolts meeting the requirements of Sect. A3.4, when threads are not excluded from shear planes	$0.33F_u$[a,c,h]						$0.17F_u$[h]
Threaded parts meeting the requirements of Sects. A3.1 and A3.4, and A449 bolts meeting the requirements of Sect. A3.4, when threads are excluded from shear planes	$0.33F_u$[a,h]						$0.22F_u$[h]
A325 bolts, when threads are not excluded from shear planes	44.0[d]	17.0	15.0	12.0	10.0		21.0[f]
A325 bolts, when threads are excluded from shear planes	44.0[d]	17.0	15.0	12.0	10.0		30.0[f]
A490 bolts, when threads are not excluded from shear planes	54.0[d]	21.0	18.0	15.0	13.0		28.0[f]
A490 bolts, when threads are excluded from shear planes	54.0[d]	21.0	18.0	15.0	13.0		40.0[f]

[a]Static loading only.

[b]Threads permitted in shear planes.

[c]The tensile capacity of the threaded portion of an upset rod, based upon the cross-sectional area at its major thread diameter A_b shall be larger than the nominal body area of the rod before upsetting times $0.60F_y$.

[d]For A325 and A490 bolts subject to tensile fatigue loading, see Appendix K4.3.

[e]Class A (slip coefficient 0.33). Clean mill scale and blast-cleaned surfaces with Class A coatings. When specified by the designer, the allowable shear stress, F_v, for slip-critical connections having special faying surface conditions may be increased to the applicable value given in the RCSC Specification.

[f]When bearing-type connections used to splice tension members have a fastener pattern whose length, measured parallel to the line of force, exceeds 50 in., tabulated values shall be reduced by 20%.

[g]See Sect. A5.2

[h]See Table 2, Numerical Values Section for values for specific ASTM steel specifications.

[i]For limitations on use of oversized and slotted holes, see Sect. J3.2.

[j]Direction of load application relative to long axis of slot.

AMERICAN INSTITUTE OF STEEL CONSTRUCTION

Table S-21 Allowable stresses on bolts and threaded members

	Leg size, inches								
Gage	8	7	6	5	4	$3^1/_2$	3	$2^1/_2$	2
g	$4^1/_2$	4	$3^1/_2$	3	$2^1/_2$	2	$1^3/_4$	$1^3/_8$	$1^1/_8$
g_1	3	$2^1/_2$	$2^1/_4$	2					
g_2	3	3	$2^1/_2$	$1^3/_4$					
Maximum bolt size, in., to satisfy bearing stress limitations	$1^1/_8$	$1^1/_8$	$^7/_8$	$^5/_8$					

Table S-22 Gage distances for angles

WIRE AND SHEET METAL GAGES
Equivalent thickness in decimals of an inch

Gage No.	U.S. Standard Gage for Uncoated Hot & Cold Rolled Sheets[b]	Galvanized Sheet Gage for Hot-Dipped Zinc Coated Sheets[b]	USA Steel Wire Gage	Gage No.	U.S. Standard Gage for Uncoated Hot & Cold Rolled Sheets[b]	Galvanized Sheet Gage for Hot-Dipped Zinc Coated Sheets[b]	USA Steel Wire Gage
7/0	—	—	.490	13	.0897	.0934	.092[a]
6/0	—	—	.462[a]	14	.0747	.0785	.080
5/0	—	—	.430[a]	15	.0673	.0710	.072
4/0	—	—	.394[a]	16	.0598	.0635	.062[a]
3/0	—	—	.362[a]	17	.0538	.0575	.054
2/0	—	—	.331	18	.0478	.0516	.048[a]
1/0	—	—	.306	19	.0418	.0456	.041
1	—	—	.283	20	.0359	.0396	.035[a]
2	—	—	.262[a]	21	.0329	.0366	—
3	.2391	—	.244[a]	22	.0299	.0336	—
4	.2242	—	.225[a]	23	.0269	.0306	—
5	.2092	—	.207	24	.0239	.0276	—
6	.1943	—	.192	25	.0209	.0247	—
7	.1793	—	.177	26	.0179	.0217	—
8	.1644	.1681	.162	27	.0164	.0202	—
9	.1495	.1532	.148[a]	28	.0149	.0187	—
10	.1345	.1382	.135	29	—	.0172	—
11	.1196	.1233	.120[a]	30	—	.0157	—
12	.1046	.1084	.106[a]				

[a] Rounded value. The steel wire gage has been taken from ASTM A510 "General Requirements for Wire Rods and Coarse Round Wire, Carbon Steel". Sizes originally quoted to 4 decimal equivalent places have been rounded to 3 decimal places in accordance with rounding procedures of ASTM "Recommended Practice" E29.

[b] The equivalent thicknesses are for information only. The product is commonly specified to decimal thickness, not to gage number.

Table S-23 Wire and sheet metal gages

Summary of symbols used in the following requirements:

d = Bolt diameter, in.
t = Thickness of the connected plate, in.
s = Spacing (pitch) of bolts in the direction of load, in.
g = Spacing (gage) of bolts transverse to the direction of load, in.
L_e = Edge distance or end distance from the center of a standard hole to the edge or end of the connected plate, in.
F_u = Ultimate tensile strength of the connected plate, ksi
F_p = Allowable bearing stress between bolt and plate, ksi
P = Load per bolt, kips

Requirements for bolt layout and corresponding stresses:

1. Standard hole diameter = $d + \frac{1}{16}$ in.

2. Effective bearing area = projected area = $d \times t$

3. Allowable bearing stress on the projected area of bolts in shear when end distance is $\geq 1\frac{1}{2}d$ and both g and s are $\geq 3d$ and when there are two or more bolts in the line of force,

$$F_p \leq 1.2 \times F_u$$

or, for the same conditions but with wrinkling in the plate permitted,

$$F_p \leq 1.5 \times F_u$$

4. Where any or all of the conditions of Requirement 3 are not met,

$$F_p \leq L_e F_u / 2d, \text{ but not greater than the value}$$
$$\text{of } F_p \text{ as given in Requirement 3.}$$

5. Minimum center-to-center spacing of bolts:

gage g may not be less than $2\frac{2}{3}d$ with $3d$ preferred

pitch s may not be less than $2\frac{2}{3}d$ with $3d$ preferred
nor less than $3d$ if F_p from Requirement 3 is used
nor less than $2P/F_u t + \frac{d}{2}$ in any case

6. Minimum edge or end distance from the center of a standard hole to the edge or end of a connected plate is given in the following table:

Bolt dia. in.	At sheared edges in.	At rolled, gas-cut or sawn edges in.
$\frac{1}{2}$	$\frac{7}{8}$	$\frac{3}{4}$
$\frac{5}{8}$	$1\frac{1}{8}$	$\frac{7}{8}$
$\frac{3}{4}$	$1\frac{1}{4}$	1
$\frac{7}{8}$	$1\frac{1}{2}$	$1\frac{1}{8}$
1	$1\frac{3}{4}$	$1\frac{1}{4}$
$1\frac{1}{8}$	2	$1\frac{1}{2}$
$1\frac{1}{4}$	$2\frac{1}{4}$	$1\frac{5}{8}$

In addition, the end distance in the direction of load from the center of a standard hole to the end of the plate may not be less than $1\frac{1}{2}d$ if F_p from Requirement 3 is used nor less than $2P/F_u t$ in any case.

7. Maximum distance from the center of a standard hole to the nearest edge or end of a connected plate may not exceed $12t$ nor 6 in.

Table S-24 Layout and stress requirements for bearing connections

MINIMUM THICKNESS OF CONNECTED MEMBERS IN INCHES[1]
TO DEVELOP BOLTS BY BEARING

ASTM A36 STEEL

Listed values of plate thickness are the minimum thickness required to avoid "large bolt/thin-plate" problems in a connection design. Due to other factors, plates will generally be much thicker than these minimums.

Maximum Bearing Stress F_p	Type of Connection	No. of Shear Planes	Bolt diameter in inches					
			$5/8$	$3/4$	$7/8$	1	$1\,1/8$	$1\,1/4$
$1.2F_u$	A325N[2]	Single Shear	$3/16$	$3/16$	$1/4$	$1/4$	$5/16$	$5/16$
	N	Double Shear	$5/16$	$3/8$	$7/16$	$1/2$	$9/16$	$5/8$
	X[3]	Single Shear	$1/4$	$5/16$	$5/16$	$3/8$	$3/8$	$7/16$
	X	Double Shear	$7/16$	$1/2$	$5/8$	$11/16$	$13/16$	$7/8$
	A490N	Single Shear	$1/4$	$1/4$	$5/16$	$3/8$	$3/8$	$7/16$
	N	Double Shear	$7/16$	$1/2$	$9/16$	$11/16$	$3/4$	$13/16$
	X	Single Shear	$5/16$	$3/8$	$7/16$	$1/2$	$9/16$	$5/8$
	X	Double Shear	$9/16$	$11/16$	$13/16$	$15/16$	1	$1\,1/8$
$1.5F_u$	A325N	Single Shear	$1/8$	$3/16$	$3/16$	$3/16$	$1/4$	$1/4$
	N	Double Shear	$1/4$	$5/16$	$3/8$	$7/16$	$7/16$	$1/2$
	X	Single Shear	$3/16$	$1/4$	$1/4$	$5/16$	$5/16$	$3/8$
	X	Double Shear	$3/8$	$7/16$	$1/2$	$9/16$	$5/8$	$11/16$
	A490N	Single Shear	$3/16$	$1/4$	$1/4$	$5/16$	$5/16$	$3/8$
	N	Double Shear	$3/8$	$7/16$	$1/2$	$9/16$	$5/8$	$11/16$
	X	Single Shear	$1/4$	$5/16$	$3/8$	$3/8$	$7/16$	$1/2$
	X	Double Shear	$1/2$	$9/16$	$11/16$	$3/4$	$13/16$	$15/16$

[1] Wrinkling of plate occurs when bearing stress F_p exceeds $1.2F_u$.
[2] N means that threads are not excluded from the shear planes.
[3] X means that threads are excluded from the shear planes.

Table S-25 Minimum plate thickness to develop bolts in bearing

TYPICAL BOLT VALUES IN KIPS[1] FOR BEARING TYPE CONNECTIONS

Type of Connection	F_v ksi	No. of Shear Planes	Bolt diameter in inches					
			$5/8$	$3/4$	$7/8$	1	$1^1/8$	$1^1/4$
A307N[2]	10.0	Single Shear	3.1	4.4	6.0	7.9	9.9	12.3
N	10.0	Double Shear	6.1	8.8	12.0	15.7	19.9	24.5
X[3]	10.0	Single Shear	3.1	4.4	6.0	7.9	9.9	12.3
X	10.0	Double Shear	6.1	8.8	12.0	15.7	19.9	24.5
A325N	21.0	Single Shear	6.4	9.3	12.6	16.5	20.9	25.8
N	21.0	Double Shear	12.9	18.6	25.3	33.0	41.7	51.5
X	30.0	Single Shear	9.2	13.3	18.0	23.6	29.8	36.8
X	30.0	Double Shear	18.4	26.5	36.1	47.1	59.6	73.6
A490N	28.0	Single Shear	8.6	12.4	16.8	22.0	27.8	34.4
N	28.0	Double Shear	17.2	24.7	33.7	44.0	55.7	68.7
X	40.0	Single Shear	12.3	17.7	24.1	31.4	39.8	49.1
X	40.0	Double Shear	24.5	35.3	48.1	62.8	79.5	98.2
Gross Area of Bolts, in^2			0.3068	0.4418	0.6013	0.7854	0.9940	1.2270

[1]Values are computed for hole diameter $1/16$ in. larger than bolt diameter.
[2]N means that threads are not excluded from the shear planes.
[3]X means that threads are excluded from the shear planes.

Table S-26 Typical allowable bolt values for bearing connections

REQUIRED TENSION FORCE AND ALLOWABLE SHEAR LOAD PER BOLT
IN A SLIP-CRITICAL CONNECTION

ASTM Bolt	Type of Force	Bolt diameter in inches					
		$5/8$	$3/4$	$7/8$	1	$1^1/8$	$1^1/4$
A325	Required Tension Force T, kips	19	28	39	51	56	71
	Shear Load, kips, single shear	5.2	7.5	10.2	13.4	16.9	20.9
	double shear	10.4	15.0	20.4	26.7	33.8	41.7
A490	Required Tension Force T, kips	24	35	49	64	80	102
	Shear Load, kips, single shear	6.4	9.3	12.6	16.5	20.9	25.8
	double shear	12.9	18.6	25.3	33.0	41.7	51.5

Table S-27 Required tension force and allowable shear loads in slip-critical connections

Type of Weld and Stress[a]	Allowable Stress	Required Weld Strength Level[b,c]
Complete-penetration Groove Welds		
Tension normal to effective area	Same as base metal	"Matching" weld metal shall be used.
Compression normal to effective area	Same as base metal	Weld metal with a strength level equal to or less than "matching" weld metal is permitted.
Tension or compression parallel to axis of weld	Same as base metal	
Shear on effective area	0.30 × nominal tensile strength of weld metal (ksi)	
Partial-penetration Groove Welds[d]		
Compression normal to effective area	Same as base metal	Weld metal with a strength level equal to or less than "matching" weld metal is permitted.
Tension or compression parallel to axis of weld[e]	Same as base metal	
Shear parallel to axis of weld	0.30 × nominal tensile strength of weld metal (ksi)	
Tension normal to effective area	0.30 × nominal tensile strength of weld metal (ksi), except tensile stress on base metal shall not exceed 0.60 × yield stress of base metal	
Fillet Welds		
Shear on effective area	0.30 × nominal tensile strength of weld metal (ksi)	Weld metal with a strength level equal to or less than "matching" weld metal is permitted.
Tension or compression Parallel to axis of weld[e]	Same as base metal	
Plug and Slot Welds		
Shear parallel to faying surfaces (on effective area)	0.30 × nominal tensile strength of weld metal (ksi)	Weld metal with a strength level equal to or less than "matching" weld metal is permitted.

[a]For definition of effective area, see Sect. J2.
[b]For "matching" weld metal, see Table 4.1.1, AWS D1.1.
[c]Weld metal one strength level stronger than "matching" weld metal will be permitted.
[d]See Sect. J2.1b for a limitation on use of partial-penetration groove welded joints.
[e]Fillet welds and partial-penetration groove welds joining the component elements of built-up members, such as flange-to-web connections, may be designed without regard to the tensile or compressive stress in these elements parallel to the axis of the welds.
[f]The design of connected material is governed by Chapters D through G. Also see Commentary Sect. J2.4.

AMERICAN INSTITUTE OF STEEL CONSTRUCTION

Table S-28 Allowable stresses on welds

Thickness of thicker part of material joined, inches	Minimum effective throat thickness, inches
To $1/4$ inclusive	$1/8$
Over $1/4$ to $1/2$	$3/16$
Over $1/2$ to $3/4$	$1/4$
Over $3/4$ to $1^1/2$	$5/16$
Over $1^1/2$ to $2^1/4$	$3/8$
Over $2^1/4$ to 6	$1/2$
Over 6	$5/8$

Table S-29 Minimum effective throat thickness of partial-penetration groove welds

Thickness of thicker part of material joined, inches	Minimum size of fillet weld, inches
To $1/4$ inclusive	$1/8$
Over $1/4$ to $1/2$	$3/16$
Over $1/2$ to $3/4$	$1/4$
Over $3/4$	$5/16$

Table S-30 Minimum size of fillet welds

Electrode	Size of Fillet Leg , in.									
	$3/16$	$1/4$	$5/16$	$3/8$	$7/16$	$1/2$	$9/16$	$5/8$	$11/16$	$3/4$
E60xx	2.4	3.2	4.0	4.8	5.6	6.4	7.2	8.0	8.7	9.5
E70xx	2.8	3.7	4.6	5.6	6.5	7.4	8.4	9.3	10.2	11.1
E80xx	3.3	4.2	5.3	6.4	7.4	8.5	9.5	10.6	11.7	12.7

Table S-31 Capacity of SMAW fillet welds

SMAW is the abbreviation for Shielded Metal Arc Welding, the type of welding commonly made manually using flux-coated electrodes.

SAW is the abbreviation for Submerged Arc Welding, the type of welding commonly made by automatic-feed welding machine using a bare metal electrode with the arc submerged under a granular fusible material.

Allowable shear stress F_v = 0.30 x nominal strength of the electrode

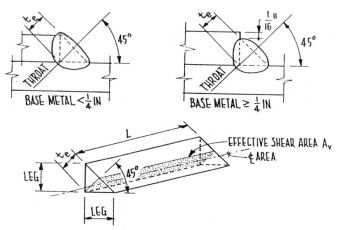

Effective Areas, Sizes and Limitations for Fillet Welds

1. Effective shear area A_v = effective throat t_e x effective length L_{eff}
   L_{eff} is the overall length of the fillet, including returns.
   Minimum L_{eff} ≥ 4 x leg size
   For intermittent welds, minimum L_{eff} ≥ 4 x leg size, but ≥ $1^1/_2$ in.

   Effective throat thickness t_e for Shielded Metal Arc Welds (SMAW)
   t_e = 0.707 x leg size, but ≤ $^1/_4$ x L_{eff}

   Effective throat thickness t_e for Submerged Arc Welds (SAW)
   For legs ≤ $^3/_8$ in., t_e = leg size, but ≤ $^1/_4$ x L_{eff}
   For legs > $^3/_8$ in., t_e = 0.707 x leg size + 0.11 in., but ≤ $^1/_4$ x L_{eff}

2. Minimum leg size of fillet welds

Thickness of thicker part of material joined, in.	Minimum leg size of fillet weld, in.
To $^1/_4$ inclusive	$^1/_8$
Over $^1/_4$ to $^1/_2$	$^3/_{16}$
Over $^1/_2$ to $^3/_4$	$^1/_4$
Over $^3/_4$	$^5/_{16}$

3. Maximum leg size of fillet welds
   For material thickness < $^1/_4$ in., leg size ≤ mat'l thickness
   For material thickness ≥ $^1/_4$ in., leg size ≤ mat'l thickness − $^1/_{16}$ in.

4. Minimum lap length in lapped joints
   Minimum lap length ≥ 5 x thickness of thinner part

Table S-32 Layout and stress requirements for fillet welded connections

APPENDIX **T**

Timber Design Tables

The data and tables in this section of the appendix are included in support of the design procedures for lumber and timber given in this textbook. The initial entry is a listing of the standard symbols used in Chapters 29 through 33 for the development of the design equations and theory.

The following tables T-1 through T-28 provide design data for use in designing structural members in lumber and timber.

The major parts of the design criteria and tables are taken from (or are based on) data provided in the *National Design Specification for Wood Construction,* American Forest and Paper Association, Wash., D.C., revised 1991 edition. Additional design tables for joists and rafters are taken from the *Uniform Building Code,* International Conference of Building Officials, Whittier, California. And finally, the plywood design values are taken from publications of the American Plywood Association, Tacoma, Washington, 98411. All data herein is reproduced with the permission of the publishers.

Though the NDS gives design criteria for some 40 species of wood, the criteria listed in this appendix is limited for the sake of brevity to three species:

- Douglas fir-larch
- Southern pine
- Spruce-pine-fir

For a complete listing of other species, the reader is referred to the complete NDS criteria.

STANDARD SYMBOLS USED IN STRUCTURAL LUMBER AND TIMBER

A = Area of cross section, in.^{2}

A_m = In fastener group analysis, gross cross sectional area of main wood member before boring and grooving, in.^{2}

A_s = In fastener group analysis, sum of gross cross sectional areas of wood side members before boring or drilling, in.^{2}

A_s = Area of steel member, in.^{2}

C_D = Duration-of-load factor

C_F = Size factor

C_H = Shear stress factor

C_L = Lateral stability of beams factor

C_M = Wet service factor

C_P = Lateral stability of columns factor

C_f = Form factor

C_{fu} = Flat use factor

C_g = Group action factor

C_t = Temperature factor

D = Generic symbol for diameter, in.

E = Modulus of elasticity, psi

F_b = Design value in bending, psi

F_c = Design value in compression parallel to grain, psi

$F_c\perp$ = Design value in compression perpendicular to grain, psi

F_g = Design value for end grain in bearing, psi

F_t = Design value in tension parallel to grain, psi

F_v = Design value in horizontal shear, psi

G = Shear modulus of elasticity, psi

I = Moment of inertia, in.^{2}

K_e = Effective buckling length factor

L = Span, ft.

L_e = Effective length for shear

M = Moment capacity, lb-in.

N = Fastener capacity at angle in direction of grain, lbs

P = Axial load; fastener value for load parallel to grain, lbs

Q = Fastener value for load perpendicular to grain, lbs

Q = Statical moment of area, in.^{3}

R_v = Vertical reaction, lbs

T = Tensile force, lbs

V = Vertical shear force, lbs

W = Total uniform load, lbs

a = Generic symbol for the dimension of a member, in.

b = Generic symbol for the width of a rectangular member, in.

c = Distance from neutral axis to outer surface of beam, in.

d = Generic symbol for the depth of bending of a rectangular member; least dimension of a compression member in the direction of buckling, in.

e = Eccentricity of load, in.

f_b = Computed bending stress, psi

f_c = Computed stress parallel to grain, psi

$f_c\perp$ = Computed stress perpendicular to grain, psi

f_{cr} = Column critical stress as buckling impends, psi

f_t = Computed stress in tension parallel to grain, psi

f_v = Computed stress in horizontal shear, psi

l = Generic symbol for the length of a member, in.

l_e = Effective length of beam; effective unsupported column length, in.

l_u = Unsupported beam length, in.

t = Generic symbol for thickness

w = Uniform load, lbs/ft

x = Variable distance, in.; horizontal location, ft.

y = Vertical location, ft.

Δ = Deflection

Nominal size b x h in.	Dressed size b x h in.	Moment of Inertia I_{xx}	Area of Section A	Section Modulus S_{xx}	Radius of Gyration r_{xx}
2 x 4	1½ x 3½	5.4	5.3	3.1	1.01
2 x 6	1½ x 5½	20.8	8.3	7.6	1.59
2 x 8	1½ x 7¼	47.6	10.9	13.1	2.09
2 x 10	1½ x 9¼	98.9	13.9	21.4	2.67
2 x 12	1½ x 11¼	178.0	16.9	31.6	3.25
2 x 14	1½ x 13¼	290.8	19.9	43.9	3.82
4 x 1	3½ x ¾	0.1	2.6	0.3	0.22
4 x 2	3½ x 1½	1.0	5.3	1.3	0.43
4 x 4	3½ x 3½	12.5	12.3	7.1	1.01
4 x 6	3½ x 5½	48.5	19.3	17.6	1.59
4 x 8	3½ x 7¼	111.1	25.4	30.7	2.09
4 x 10	3½ x 9¼	230.8	32.4	49.9	2.67
4 x 12	3½ x 11¼	415.3	39.4	73.8	3.25
4 x 14	3½ x 13¼	678.5	46.4	102.4	3.82
4 x 16	3½ x 15¼	1034.4	53.4	135.7	4.40
6 x 1	5½ x ¾	0.2	4.1	0.5	0.22
6 x 2	5½ x 1½	1.5	8.3	2.1	0.43
6 x 4	5½ x 3½	19.7	19.3	11.2	1.01
6 x 6	5½ x 5½	76.3	30.3	27.7	1.59
6 x 8	5½ x 7½	193.4	41.3	51.6	2.17
6 x 10	5½ x 9½	393.0	52.3	82.7	2.74
6 x 12	5½ x 11½	697.1	63.3	121.2	3.32
6 x 14	5½ x 13½	1127.7	74.3	167.1	3.90
6 x 16	5½ x 15½	1706.8	85.3	220.2	4.47
6 x 18	5½ x 17½	2456.4	96.3	280.7	5.05
8 x 1	7¼ x ¾	0.3	5.4	0.7	0.22
8 x 2	7¼ x 1½	2.0	10.9	2.7	0.43
8 x 4	7¼ x 3½	25.9	25.4	14.8	1.01
8 x 6	7½ x 5½	104.0	41.3	37.8	1.59
8 x 8	7½ x 7½	263.7	56.3	70.3	2.17
8 x 10	7½ x 9½	535.9	71.3	112.8	2.74
8 x 12	7½ x 11½	950.5	86.3	165.3	3.32
8 x 14	7½ x 13½	1537.7	101.3	227.8	3.90
8 x 16	7½ x 15½	2327.4	116.3	300.3	4.47
8 x 18	7½ x 17½	3349.6	131.3	382.8	5.05

Nominal size b x h in.	Dressed size b x h in.	Moment of Inertia I_{xx}	Area of Section A	Section Modulus S_{xx}	Radius of Gyration r_{xx}
10 x 2	9¼ x 1½	2.6	13.9	3.5	0.43
10 x 4	9¼ x 3½	33.0	32.4	18.9	1.01
10 x 6	9½ x 5½	131.7	52.3	47.9	1.59
10 x 8	9½ x 7½	334.0	71.3	89.1	2.17
10 x 10	9½ x 9½	678.8	90.3	142.9	2.74
10 x 12	9½ x 11½	1204.0	109.3	209.4	3.32
10 x 14	9½ x 13½	1947.8	128.3	288.6	3.90
10 x 16	9½ x 15½	2948.1	147.3	380.4	4.47
10 x 18	9½ x 17½	4242.8	166.3	484.9	5.05
12 x 2	11¼ x 1½	3.2	16.9	4.2	0.43
12 x 4	11¼ x 3½	40.2	39.4	23.0	1.01
12 x 6	11½ x 5½	159.4	63.3	58.0	1.59
12 x 8	11½ x 7½	404.3	86.3	107.8	2.17
12 x 10	11½ x 9½	821.7	109.3	173.0	2.74
12 x 12	11½ x 11½	1457.5	132.3	253.5	3.32
12 x 14	11½ x 13½	2357.9	155.3	349.3	3.90
12 x 16	11½ x 15½	3568.7	178.3	460.5	4.47
12 x 18	11½ x 17½	5136.1	201.3	587.0	5.05
14 x 2	13¼ x 1½	3.7	19.9	5.0	0.43
14 x 4	13¼ x 3½	47.3	46.4	27.1	1.01
14 x 6	13½ x 5½	187.2	74.3	68.1	1.59
14 x 8	13½ x 7½	474.6	101.3	126.6	2.17
14 x 10	13½ x 9½	964.5	128.3	203.1	2.74
14 x 12	13½ x 11½	1711.0	155.3	297.6	3.32
14 x 14	13½ x 13½	2767.9	182.3	410.1	3.90
14 x 16	13½ x 15½	4189.4	209.3	540.6	4.47
14 x 18	13½ x 17½	6029.3	236.3	689.1	5.05
16 x 4	15¼ x 3½	54.5	53.4	31.1	1.01
16 x 6	15½ x 5½	214.9	85.3	78.1	1.59
16 x 8	15½ x 7½	544.9	116.3	145.3	2.17
16 x 10	15½ x 9½	1107.4	147.3	233.1	2.74
16 x 12	15½ x 11½	1964.5	178.3	341.6	3.32
16 x 14	15½ x 13½	3178.0	209.3	470.8	3.90
16 x 16	15½ x 15½	4810.0	240.3	620.6	4.47
16 x 18	15½ x 17½	6922.5	271.3	791.1	5.05
18 x 6	17½ x 5½	242.6	96.3	88.2	1.59
18 x 8	17½ x 7½	615.2	131.3	164.1	2.17
18 x 10	17½ x 9½	1250.3	166.3	263.2	2.74
18 x 12	17½ x 11½	2217.9	201.3	385.7	3.32
18 x 14	17½ x 13½	3588.0	236.3	531.6	3.90
18 x 16	17½ x 15½	5430.6	271.3	700.7	4.47
18 x 18	17½ x 17½	7815.8	306.3	893.2	5.05

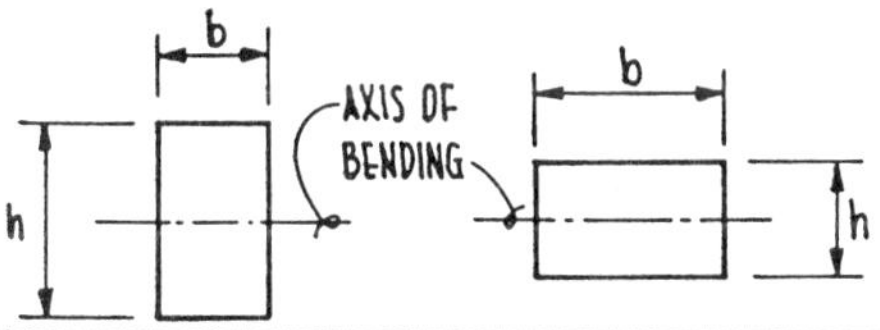
b
AXIS OF BENDING
b
h
h

Table T-1 Section properties of dressed lumber and timber

Generic Name	Size Ranges Thickness	Width	Major Usage	Abbrev- iation
Boards	1 in.	2 in. and wider	Light framing	LF
Decking	1 in. to 4 in.	4 in. and wider	Decking and siding	D&M, T&G[a]
Lumber	2 in. to 4 in.	4 in. and wider	Light framing Joists and Planks	LF J&P
Timber	5 in. and thicker	Not more than thickness+2 in.	Posts and Timbers	P&T
		More than thickness+2 in.	Beams and stringers	B&S

[a] Dressed and Matched, or, Tongue and Groove

Table T-2 Common size and use classification

DOUGLAS FIR-LARCH;   WWPA GRADING RULES;   LUMBER AND LIGHT FRAMING

ALLOWABLE STRESSES IN PSI IN LUMBER LESS THAN 5 INCHES THICK[a]
WIDTHS UP TO 12 INCHES

DRY SERVICE CONDITIONS;   MOISTURE CONTENT 19% OR LESS

COMMERCIAL GRADE	4" wide 2 x 4 3 x 4 4 x 4	6" wide 2 x 6 3 x 6 4 x 6	8 inches wide 2 x 8 3 x 8	4 x 8	10 inches wide 2 x 10 3 x 10	4 x 10	12 inches wide 2 x 12 3 x 12	4 x 12
ALLOWABLE STRESS IN BENDING IN A SINGLE MEMBER, F_b, PSI								
Select structural	2175	1885	1740	1885	1600	1740	1450	1600
No. 1	1500	1300	1200	1300	1100	1200	1000	1100
No. 2	1310	1140	1050	1140	960	1050	875	960
No. 3	750	650	600	650	550	600	500	550
Stud	740	675	—	—	—	—	—	—
Construction	1000	—	—	—	—	—	—	—
Standard	550	—	—	—	—	—	—	—
ALLOWABLE STRESS IN BENDING IN REPETITIVE MEMBERS, F_b, PSI								
Select structural	2500	2170	2000	2170	1830	2000	1670	1830
No. 1	1725	1500	1380	1500	1265	1380	1150	1265
No. 2	1510	1310	1210	1310	1110	1210	1010	1110
No. 3	860	750	690	750	630	690	575	630
Stud	860	775	—	—	—	—	—	—
Construction	1150	—	—	—	—	—	—	—
Standard	630	—	—	—	—	—	—	—
ALLOWABLE STRESS IN TENSION PARALLEL TO GRAIN, F_t, PSI								
Select structural	1500	1300	1200	1200	1100	1100	1000	1000
No. 1	1010	880	810	810	740	740	675	675
No. 2	860	750	690	690	630	630	575	575
No. 2	490	420	390	390	360	360	325	325
Stud	495	450	—	—	—	—	—	—
Construction	650	—	—	—	—	—	—	—
Standard	375	—	—	—	—	—	—	—
ALLOWABLE STRESS IN SHEAR PARALLEL TO GRAIN, F_v, PSI								
Select structural	95	95	95	95	95	95	95	95
No. 1	95	95	95	95	95	95	95	95
No. 2	95	95	95	95	95	95	95	95
No. 3	95	95	95	95	95	95	95	95
Stud	95	95	—	—	—	—	—	—
Construction	95	—	—	—	—	—	—	—
Standard	95	—	—	—	—	—	—	—
ALLOWABLE STRESS IN COMPRESSION PARALLEL TO GRAIN, F_c, PSI								
Select structural	1955	1870	1785	1700	1700	1700	1700	1700
No. 1	1670	1595	1520	1520	1450	1450	1450	1450
No. 2	1495	1430	1365	1365	1300	1300	1300	1300
No. 3	860	825	790	790	750	750	750	750
Stud	870	825	—	—	—	—	—	—
Construction	1600	—	—	—	—	—	—	—
Standard	1350	—	—	—	—	—	—	—
ALLOWABLE STRESS IN COMPRESSION PERPENDICULAR TO GRAIN, $F_{c\perp}$, PSI								
Select structural	625	625	625	625	625	625	625	625
No. 1	625	625	625	625	625	625	625	625
No. 2	625	625	625	625	625	625	625	625
No. 3	625	625	625	625	625	625	625	625
Stud	625	625	—	—	—	—	—	—
Construction	625	—	—	—	—	—	—	—
Standard	625	—	—	—	—	—	—	—

Table T-3a Allowable stresses in lumber and timber

DOUGLAS FIR-LARCH; WWPA GRADING RULES

ALLOWABLE STRESSES IN PSI IN TIMBERS 5 INCHES THICK OR MORE[a]

DRY SERVICE CONDITIONS; MOISTURE CONTENT 19% OR LESS

	BEAMS AND STRINGERS; WIDTHS GREATER THAN THICKNESS + 2 INCHES				
	Bending Single or repetitive	Tension Parallel to grain	Shear Parallel to grain	Compression	
				Parallel to grain	Perpendicular to grain
	F_b	F_t	F_v	F_c	$F_{c\perp}$
Dense select structural	1850	1100	85	1300	730
Select structural	1600	950	85	1100	625
Dense No. 1	1550	775	85	1100	730
No. 1	1350	675	85	925	625
Dense No. 2	1000	500	85	700	730
No. 2	875	425	85	600	625
	POSTS AND TIMBERS; WIDTHS NOT GREATER THAN THICKNESS + 2 INCHES				
Dense select structural	1750	1150	85	1350	730
Select structural	1500	1000	85	1150	625
Dense No. 1	1400	950	85	1200	730
No. 1	1200	825	85	1000	625
Dense No. 2	800	550	85	550	730
No. 2	700	475	85	475	625

MODULUS OF ELASTICITY FOR TABULATED COMMERCIAL GRADES

LUMBER GRADE	Modulus of Elasticity, psi	TIMBER GRADE	Modulus of Elasticity, psi
Select structural	1,900,000	Dense select structural	1,700,000
No. 1	1,700,000	Select structural	1,600,000
No. 2	1,600,000		
No. 3	1,400,000	Dense No. 1	1,700,000
		No. 1	1,600,000
Stud	1,400,000	Dense No. 2	1,400,000
Construction	1,500,000	No. 2	1,300,000
Standard	1,400,000		

[a] Source: National Design Specifications for Wood Construction

For lumber sizes 2" to 4" thick and up to 12" wide, the tabulated stresses include the size factor C_F.

For all lumber thicknesses, when the width of the member is greater than 12" the tabulated stresses for lumber 12" wide shall apply, except that for bending, for tension and for compression parallel to grain, the tabulated stresses shall be adjusted by the size factor $C_F = 0.90$; this adjustment need not be applied to bending stresses in members 4" wide.

For all timber thicknesses, when the depth d exceeds 12", the tabulated bending stresses shall be adjusted by the size factor $C_F = (12/d)^{1/9}$.

For wet service conditions where the moisture content is greater than 19% for extended periods, the tabulated stresses shall be adjusted by the following wet service factors C_M.

	Values of C_M for the indicated stresses					
	F_b	F_t	F_v	F_c	$F_{c\perp}$	E
Lumber 2" to 4" thick	0.85[b]	1.00	0.97	0.80[c]	0.67	0.90
Timber 5" thick or more	1.00	1.00	1.00	1.00	0.67	1.00

[b] When $(F_b)(C_F) \leq 1150$ psi, $C_M = 1.00$

[c] When $(F_b)(C_F) \leq 750$ psi, $C_M = 1.00$

Table T-3a Allowable stresses in lumber and timber *(continued)*

SOUTHERN PINE;   SPIB GRADING RULES;   LUMBER AND LIGHT FRAMING·

ALLOWABLE STRESSES IN PSI IN LUMBER LESS THAN 5 INCHES THICK[a]
WIDTHS UP TO 12 INCHES

DRY SERVICE CONDITIONS;   MOISTURE CONTENT 19% OR LESS

COMMERCIAL GRADE	4" wide 2 x 4 3 x 4 4 x 4	6" wide 2 x 6 3 x 6 4 x 6	8 inches wide		10 inches wide		12 inches wide	
			2 x 8 3 x 8	4 x 8	2 x 10 3 x 10	4 x 10	2 x 12 3 x 12	4 x 12
ALLOWABLE STRESS IN BENDING IN A SINGLE MEMBER, F_b, PSI								
Select structural	2850	2550	2300	2530	2050	2255	1900	2090
No. 1	1850	1650	1500	1650	1300	1450	1250	1375
No. 2	1500	1250	1200	1320	1050	1155	975	1070
No. 3	850	750	700	770	600	660	575	630
Stud	875	775	—	—	—	—	—	—
Construction	1100	—	—	—	—	—	—	—
Standard	625	—	—	—	—	—	—	—
ALLOWABLE STRESS IN BENDING IN REPETITIVE MEMBERS, F_b, PSI								
Select structural	3280	2930	2645	2910	2360	2590	2185	2400
No. 1	2130	1900	1725	1900	1495	1640	1440	1580
No. 2	1725	1440	1380	1520	1210	1330	1120	1230
No. 3	980	860	805	890	690	760	660	730
Stud	1010	890	—	—	—	—	—	—
Construction	1265	—	—	—	—	—	—	—
Standard	720	—	—	—	—	—	—	—
ALLOWABLE STRESS IN TENSION PARALLEL TO GRAIN, F_t, PSI								
Select structural	1600	1400	1300	1300	1100	1100	1050	1050
No. 1	1050	900	825	825	725	725	675	675
No. 2	825	725	650	650	575	575	550	550
No. 2	475	425	400	400	325	325	325	325
Stud	500	425	—	—	—	—	—	—
Construction	625	—	—	—	—	—	—	—
Standard	350	—	—	—	—	—	—	—
ALLOWABLE STRESS IN SHEAR PARALLEL TO GRAIN, F_v, PSI								
Select structural	100	90	90	90	90	90	90	90
No. 1	100	90	90	90	90	90	90	90
No. 2	90	90	90	90	90	90	90	90
No. 3	90	90	90	90	90	90	90	90
Stud	90	90	—	—	—	—	—	—
Construction	100	—	—	—	—	—	—	—
Standard	90	—	—	—	—	—	—	—
ALLOWABLE STRESS IN COMPRESSION PARALLEL TO GRAIN, F_c, PSI								
Select structural	2100	2000	1900	1900	1850	1850	1800	1800
No. 1	1850	1750	1650	1650	1600	1600	1600	1600
No. 2	1650	1600	1550	1550	1500	1500	1450	1450
No. 3	975	925	875	875	850	850	825	825
Stud	925	925	—	—	—	—	—	—
Construction	1800	—	—	—	—	—	—	—
Standard	1500	—	—	—	—	—	—	—
ALLOWABLE STRESS IN COMPRESSION PERPENDICULAR TO GRAIN, $F_{c\perp}$, PSI								
Select structural	565	565	565	565	565	565	565	565
No. 1	565	565	565	565	565	565	565	565
No. 2	565	565	565	565	565	565	565	565
No. 3	475	565	565	565	565	565	565	565
Stud	500	565	—	—	—	—	—	—
Construction	625	—	—	—	—	—	—	—
Standard	350	—	—	—	—	—	—	—

Table T-3b Allowable stresses in lumber and timber

SOUTHERN PINE; SPIB GRADING RULES

ALLOWABLE STRESSES IN PSI IN TIMBERS 5 INCHES THICK OR MORE[a]

DRY OR WET SERVICE CONDITIONS;
MOISTURE CONTENT MAY EXCEED 19% FOR EXTENDED PERIODS

BEAMS AND STRINGERS; WIDTHS GREATER THAN THICKNESS + 2 INCHES

POSTS AND TIMBERS; WIDTHS NOT GREATER THAN THICKNESS + 2 INCHES

	Bending Single or repetitive	Tension Parallel to grain	Shear Parallel to grain	Compression Parallel to grain	Compression Perpendicular to grain
	F_b	F_t	F_v	F_C	$F_{C\perp}$
Dense select structural	1750	1200	110	1100	440
Select structural	1500	1000	110	950	375
No. 1 Dense	1550	1050	110	975	440
No. 1	1350	900	110	825	375
No. 2 Dense	975	650	100	625	440
No. 2	850	550	100	525	375

MODULUS OF ELASTICITY FOR TABULATED COMMERCIAL GRADES

LUMBER GRADE	Modulus of Elasticity, psi	TIMBER GRADE	Modulus of Elasticity, psi
Select structural	1,800,000	Dense select structural	1,600,000
No. 1	1,700,000	Select structural	1,500,000
No. 2	1,600,000		
No. 3	1,400,000	No. 1 Dense	1,600,000
Stud	1,400,000	No. 1	1,500,000
Construction	1,500,000	No. 2 Dense	1,300,000
Standard	1,300,000	No. 2	1,200,000

[a] Source: National Design Specifications for Wood Construction

For lumber sizes 2" to 4" thick and up to 12" wide, the tabulated stresses include the size factor C_F.

For all lumber thicknesses (but not for timber), when the width of the member is greater than 12" the tabulated stresses for lumber 12" wide shall apply, except that for bending, for tension and for compression parallel to grain, the tabulated stresses shall be adjusted by the size factor $C_F = 0.90$.

For all timber thicknesses, when the depth d exceeds 12", the tabulated bending stresses shall be adjusted by the size factor $C_F = (12/d)^{1/9}$.

For all lumber sizes under wet service conditions, where the moisture content is greater than 19% for extended periods, the tabulated stresses shall be adjusted by the following wet service factors C_M.

Values of C_M for the indicated stresses

	F_b	F_t	F_v	F_C	$F_{C\perp}$	E
Lumber 2" to 4" thick	0.85[b]	1.00	0.97	0.80[c]	0.67	0.90
Timber 5" thick or more	1.00	1.00	1.00	1.00	1.00	1.00

[b] When $(F_b)(C_F) \leq 1150$ psi, $C_M = 1.00$
[c] When $(F_b) \leq 750$ psi, $C_M = 1.00$

Table T-3b Allowable stresses in lumber and timber *(continued)*

SPRUCE-PINE-FIR; NLGA GRADING RULES; LUMBER AND LIGHT FRAMING

ALLOWABLE STRESSES IN PSI IN LUMBER LESS THAN 5 INCHES THICK[a]
WIDTHS UP TO 12 INCHES

DRY SERVICE CONDITIONS; MOISTURE CONTENT 19% OR LESS

COMMERCIAL GRADE	4" wide 2 x 4 3 x 4 4 x 4	6" wide 2 x 6 3 x 6 4 x 6	8 inches wide 2 x 8 3 x 8	4 x 8	10 inches wide 2 x 10 3 x 10	4 x 10	12 inches wide 2 x 12 3 x 12	4 x 12
ALLOWABLE STRESS IN BENDING IN A SINGLE MEMBER, F_b, PSI								
Select structural	1875	1625	1500	1625	1375	1500	1250	1375
No. 1	1310	1140	1050	1140	960	1050	875	960
No. 2	1310	1140	1050	1140	960	1050	875	960
No. 3	750	650	600	650	550	600	500	550
Stud	740	675	—	—	—	—	—	—
Construction	975	—	—	—	—	—	—	—
Standard	550	—	—	—	—	—	—	—
ALLOWABLE STRESS IN BENDING IN REPETITIVE MEMBERS, F_b, PSI								
Select structural	2160	1870	1725	1870	1580	1725	1440	1580
No. 1	1510	1310	1210	1310	1110	1210	1010	1110
No. 2	1510	1310	1210	1310	1110	1210	1010	1110
No. 3	860	750	690	750	630	690	575	630
Stud	850	780	—	—	—	—	—	—
Construction	1120	—	—	—	—	—	—	—
Standard	630	—	—	—	—	—	—	—
ALLOWABLE STRESS IN TENSION PARALLEL TO GRAIN, F_t, PSI								
Select structural	1010	880	810	810	740	740	675	675
No. 1	640	550	510	510	470	470	425	425
No. 2	640	550	510	510	470	470	425	425
No. 3	375	325	300	300	275	275	250	250
Stud	360	325	—	—	—	—	—	—
Construction	475	—	—	—	—	—	—	—
Standard	275	—	—	—	—	—	—	—
ALLOWABLE STRESS IN SHEAR PARALLEL TO GRAIN, F_v, PSI								
Select structural	70	70	70	70	70	70	70	70
No. 1	70	70	70	70	70	70	70	70
No. 2	70	70	70	70	70	70	70	70
No. 3	70	70	70	70	70	70	70	70
Stud	70	70	—	—	—	—	—	—
Construction	70	—	—	—	—	—	—	—
Standard	70	—	—	—	—	—	—	—
ALLOWABLE STRESS IN COMPRESSION PARALLEL TO GRAIN, F_c, PSI								
Select structural	1610	1540	1470	1470	1400	1400	1400	1400
No. 1	1265	1210	1155	1155	1100	1100	1100	1100
No. 2	1265	1210	1155	1155	1100	1100	1100	1100
No. 3	720	690	660	660	625	625	625	625
Stud	710	675	—	—	—	—	—	—
Construction	1350	—	—	—	—	—	—	—
Standard	1100	—	—	—	—	—	—	—
ALLOWABLE STRESS IN COMPRESSION PERPENDICULAR TO GRAIN, $F_{c\perp}$, PSI								
Select structural	425	425	425	425	425	425	425	425
No. 1	425	425	425	425	425	425	425	425
No. 2	425	425	425	425	425	425	425	425
No. 3	425	425	425	425	425	425	425	425
Stud	425	425	—	—	—	—	—	—
Construction	425	—	—	—	—	—	—	—
Standard	425	—	—	—	—	—	—	—

Table T-3c Allowable stresses in lumber and timber

SPRUCE-PINE-FIR;   NLGA GRADING RULES

ALLOWABLE STRESSES IN PSI IN TIMBERS 5 INCHES THICK OR MORE[a]

DRY SERVICE CONDITIONS;   MOISTURE CONTENT 19% OR LESS

	Bending Single or repetitive F_b	Tension Parallel to grain F_t	Shear Parallel to grain F_v	Compression Parallel to grain F_c	Compression Perpendicular to grain $F_{c\perp}$
BEAMS AND STRINGERS; WIDTHS GREATER THAN THICKNESS + 2 INCHES					
Select structural	1100	650	65	775	425
No. 1	900	450	65	625	425
No. 2	600	300	65	425	425
POSTS AND TIMBERS; WIDTHS NOT GREATER THAN THICKNESS + 2 INCHES					
Select structural	1050	700	65	800	425
No. 1	850	550	65	700	425
No. 2	500	325	65	500	425

MODULUS OF ELASTICITY FOR TABULATED COMMERCIAL GRADES

LUMBER GRADE	Modulus of Elasticity, psi	TIMBER GRADE	Modulus of Elasticity, psi
Select structural	1,500,000	Select structural	1,300,000
No. 1	1,400,000	No. 1	1,300,000
No. 2	1,400,000	No. 2	1,000,000
No. 3	1,200,000		
Stud	1,200,000		
Construction	1,300,000		
Standard	1,200,000		

[a] Source: National Design Specifications for Wood Construction

For lumber sizes 2" to 4" thick and up to 12" wide, the tabulated stresses include the size factor C_F.

For all lumber thicknesses, when the width of the member is greater than 12" the tabulated stresses for lumber 12" wide shall apply, except that for bending, for tension and for compression parallel to grain, the tabulated stresses shall be adjusted by the size factor $C_F = 0.90$; this adjustment need not be applied to bending stresses in members 4" wide.

For all timber thicknesses, when the depth d exceeds 12", the tabulated bending stresses shall be adjusted by the size factor $C_F = (12/d)^{1/9}$.

For wet service conditions where the moisture content is greater than 19% for extended periods, the tabulated stresses shall be adjusted by the following wet service factors C_M.

		F_b	F_t	F_v	F_c	$F_{c\perp}$	E
Values of C_M for the indicated stresses							
Lumber 2" to 4" thick		0.85[b]	1.00	0.97	0.80[c]	0.67	0.90
Timber 5" thick or more		1.00	1.00	1.00	1.00	0.67	1.00

[b] When $(F_b)(C_F) \leq 1150$ psi, $C_M = 1.00$
[c] When $(F_b)(C_F) \leq 750$ psi, $C_M = 1.00$

Table T-3c Allowable stresses in lumber and timber *(continued)*

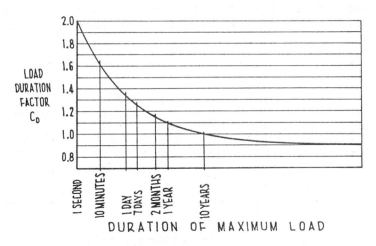

aSource: National Design Specification for Wood Construction

Table T-4 Load duration factor C_Da

Pennyweight	Length (in.)	Wire Diameter (in.)			
		Box Nails	Common Wire Nails	Threaded, Hardened Steel Nails	Common Wire Spikes
6 d	2	0.099	0.113	0.120	
8 d	$2\frac{1}{2}$	0.113	0.131	0.120	
10 d	3	0.128	0.148	0.135	0.192
12 d	$3\frac{1}{4}$	0.128	0.148	0.135	0.192
16 d	$3\frac{1}{2}$	0.135	0.162	0.148	0.207
20 d	4	0.148	0.192	0.177	0.225
30 d	$4\frac{1}{2}$	0.148	0.207	0.177	0.244
40 d	5	0.162	0.225	0.177	0.263
50 d	$5\frac{1}{2}$		0.244	0.177	0.283
60 d	6		0.263	0.177	0.283

aSource: National Design Specification for Wood Construction

Table T-5 Nail and spike sizesa

WITHDRAWAL DESIGN LOADS IN POUNDS PER INCH OF PENETRATION
FOR NAILS AND SPIKES INSTALLED IN SIDE GRAIN OF MAIN MEMBER[a]

Size of Box Nail									
Pennyweight, *d*	6*d*	8*d*	10*d*	12*d*	16*d*	20*d*	30*d*	40*d*		
Diameter, in.	0.099	0.113	0.128	0.128	0.135	0.148	0.148	0.162		
Doug.Fir-Larch	24	28	31	31	33	36	36	40		
Southern Pine	31	35	40	40	42	46	46	50		
Spruce-Pine-Fir	16	18	20	20	21	23	23	26		

Size of Common Nail									
Pennyweight, *d*	6*d*	8*d*	10*d*	12*d*	16*d*	20*d*	30*d*	40*d*	50*d*	60*d*
Diameter, in.	0.113	0.131	0.148	0.148	0.162	0.192	0.207	0.225	0.244	0.263
Doug.Fir-Larch	28	32	36	36	40	47	50	55	60	64
Southern Pine	35	41	46	46	50	59	64	70	76	81
Spruce-Pine-Fir	18	21	23	23	26	30	33	35	38	41

Size of Threaded Hardened Nail									
Pennyweight, *d*	6*d*	8*d*	10*d*	12*d*	16*d*	20*d*	30*d*	40*d*	50*d*	60*d*
Diameter, in.	0.120	0.120	0.135	0.135	0.148	0.177	0.177	0.177	0.177	0.177
Doug.Fir-Larch	32	32	36	36	40	47	47	47	47	47
Southern Pine	41	41	46	46	50	59	59	59	59	59
Spruce-Pine-Fir	21	21	23	23	26	30	30	30	30	30

Size of Common Spike								
Pennyweight, *d*	10*d*	12*d*	16*d*	20*d*	30*d*	40*d*	50*d*	60*d*
Diameter, in.	0.192	0.192	0.207	0.225	0.244	0.263	0.283	0.283
Doug.Fir-Larch	47	47	50	55	60	64	69	69
Southern Pine	59	59	64	70	76	81	88	88
Spruce-Pine-Fir	30	30	33	35	38	41	45	45

[a]Source: National Design Specification for Wood Construction

Design values for withdrawal in toenailed joints are $2/3$ of the tabular values.

Wet service factor C_M does not apply for toenails in withdrawal.

For nails placed in predrilled holes in which the hole diameter is not larger than 75% of the nail diameter, design values may be taken at 100% of the tabular values.

Loading end nailed connections in withdrawal is not recommended. The design values given in this table are not valid for such loading.

Table T-6 Withdrawal design loads for nails and spikes

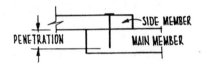

Side member thick-ness in.	Nail size penny-weight	Nail lgth. in.	Required penetration in main member, 12 diameters, in.				DOUGLAS FIR-LARCH				SOUTHERN PINE				SPRUCE-PINE-FIR			
			box nail	cmmn nail	hdnd steel	cmmn spike	box nail	cmmn nail	thrd hdnd steel nail	cmmn spike	box nail	cmmn nail	thrd hdnd steel nail	cmmn spike	box nail	cmmn nail	thrd hdnd steel nail	cmmn spike
	6d	2	1.19	1.36	1.44	–	48	59	71	–	55	67	80	–	38	47	58	–
	8d	2 1/2	1.36	1.57	1.44	–	59	76	71	–	67	85	80	–	47	61	58	–
	10d	3	1.54	1.78	1.62	2.30	73	90	88	124	82	101	98	137	59	73	72	103
	12d	3 1/4	1.78	1.78	1.62	2.30	73	90	88	124	82	101	98	137	59	73	72	103
	16d	3 1/2	1.62	1.94	1.78	2.48	79	105	98	134	89	117	110	148	65	87	81	112
1/2	20d	4	1.78	2.30	2.12	2.70	90	124	133	147	101	137	147	162	73	103	111	123
	30d	4 1/2	1.78	2.48	2.12	2.93	90	134	133	151	101	148	147	166	73	112	111	127
	40d	5	1.94	2.70	2.12	3.16	105	147	133	171	117	162	147	188	87	123	111	144
	50d	5 1/2	–	2.93	2.12	3.40	–	151	133	183	–	166	147	201	–	127	111	155
	60d	6	–	3.16	2.12	3.40	–	171	133	183	–	188	147	201	–	144	111	155
	6d	2	1.19	1.36	1.44	–	55	–	–	–	61	–	–	–	47	–	–	–
	8d	2 1/2	1.36	1.57	1.44	–	72	90	84	–	79	104	97	–	57	70	66	–
	10d	3	1.54	1.78	1.62	2.30	87	105	101	–	101	121	115	–	68	83	80	–
	12d	3 1/4	1.78	1.78	1.62	2.30	87	105	101	138	101	121	115	157	68	83	80	111
	16d	3 1/2	1.62	1.94	1.78	2.48	94	121	112	147	108	138	128	166	74	96	89	119
3/4	20d	4	1.78	2.30	2.12	2.70	105	138	145	158	121	157	164	178	83	111	117	129
	30d	4 1/2	1.78	2.48	2.12	2.93	105	147	145	162	121	166	164	182	83	119	117	132
	40d	5	1.94	2.70	2.12	3.16	121	158	145	181	138	178	164	203	96	129	117	149
	50d	5 1/2	–	2.93	2.12	3.40	–	162	145	194	–	182	164	217	–	132	117	159
	60d	6	–	3.16	2.12	3.40	–	181	145	194	–	202	164	217	–	149	117	159
	6d	2	1.19	1.36	1.44	–	–	–	–	–	–	–	–	–	–	–	–	–
	8d	2 1/2	1.36	1.57	1.44	–	72	–	93	–	79	–	102	–	61	–	77	–
	10d	3	1.54	1.78	1.62	2.30	93	118	118	–	101	128	128	–	79	96	91	–
	12d	3 1/4	1.78	1.78	1.62	2.30	93	118	118	–	101	128	128	–	79	96	91	–
	16d	3 1/2	1.62	1.94	1.78	2.48	103	141	131	167	113	154	145	192	86	109	101	131
1	20d	4	1.78	2.30	2.12	2.70	118	159	164	177	128	183	188	202	96	124	129	140
	30d	4 1/2	1.78	2.48	2.12	2.95	118	167	164	181	128	192	188	207	96	131	129	143
	40d	5	1.94	2.70	2.12	3.16	141	177	164	199	154	202	188	227	109	140	129	159
	50d	5 1/2	–	2.93	2.12	3.40	–	181	164	214	–	207	188	243	–	143	129	171
	60d	6	–	3.16	2.12	3.40	–	199	164	214	–	227	188	243	–	159	129	171
	6d	2	1.19	1.36	1.44	–	–	–	–	–	–	–	–	–	–	–	–	–
	8d	2 1/2	1.36	1.57	1.44	–	–	–	–	–	–	–	–	–	–	–	–	–
	10d	3	1.54	1.78	1.62	2.30	–	–	–	–	–	–	–	–	–	–	–	–
	12d	3 1/4	1.78	1.78	1.62	2.30	93	–	118	–	101	–	128	–	79	–	100	–
	16d	3 1/2	1.62	1.94	1.78	2.48	103	141	133	–	113	154	145	–	88	120	113	–
1 1/2	20d	4	1.78	2.30	2.12	2.70	118	170	184	–	128	185	201	–	100	144	156	–
	30d	4 1/2	1.78	2.48	2.12	2.93	118	186	184	211	128	203	201	230	100	158	156	175
	40d	5	1.94	2.70	2.12	3.16	141	205	184	240	154	224	201	262	120	172	156	191
	50d	5 1/2	–	2.93	2.12	3.40	–	211	184	257	–	230	201	281	–	175	156	205
	60d	6	–	3.16	2.12	3.40	–	240	184	257	–	262	201	281	–	191	156	205

Table T-7 Lateral design loads for nails and spikes

[a]Source: National Design Specification for Wood Construction

Tabulated lateral design values for nailed connections shall be multiplied by all applicable adjustment factors [See discussions with Equation (30-2) of the text].

Tabulated lateral design values for box nails are for nails inserted in side grain with nail axis perpendicular to wood fibers, and with the following nail bending yield strengths (F_{yb}):
F_{yb} = 100,000 psi for 0.099", 0.113", 0.128" and 0.135" diameter box nails
F_{yb} = 90,000 psi for 0.148" and 0.162" diameter box nails

Tabulated lateral design values for common wire nails are for nails inserted in side grain with nail axis perpendicular to wood fibers, and with the following nail bending yield strengths (F_{yb}):
F_{yb} = 100,000 psi for 0.113" and 0.131" diameter common wire nails
F_{yb} = 90,000 psi for 0.148" and 0.162" diameter common wire nails
F_{yb} = 80,000 psi for 0.192", 0.207" and 0.225" diameter common wire nails
F_{yb} = 70,000 psi for 0.244" and 0.263" diameter common wire nails

Tabulated lateral design values for threaded hardened-steel nails are for nails inserted in side grain with nail axis perpendicular to wood fibers, and with the following nail bending yield strengths (F_{yb}):
F_{yb} = 130,000 psi for 0.120" and 0.135" diameter threaded hardened-steel nails
F_{yb} = 115,000 psi for 0.148" and 0.177" diameter threaded hardened-steel nails
F_{yb} = 100,000 psi for 0.207" diameter threaded hardened-steel nails
Tabulated lateral design values for threaded hardened-steel nails shall not apply for annularly threaded nails when threads occur at the shear plane. Refer to NDS 12.3.1 for computation of nail values when threads occur at the shear plane.

Tabulated lateral design values for common wire spikes are for spikes inserted in side grain with spike axis perpendicular to wood fibers, and with the following spike bending yield strengths (F_{yb}):
F_{yb} = 80,000 psi for 0.192", 0.207", and 0.225" diameter common wire spikes
F_{yb} = 70,000 psi for 0.244" and 0.263" diameter common wire spikes
F_{yb} = 60,000 psi for 0.283" and 5/16" diameter common wire spikes
F_{yb} = 45,000 psi for 3/8" diameter common wire spikes

Linear interpolation between sizes of side members is permitted.

Design values are not tabulated when the penetration in the main member is less than 12 diameters.

Design values are the same whether the loads are perpendicular to grain or parallel to grain.

For end nailed connections loaded laterally, design values are $2/3$ the tabular values.

For toenailed connections loaded laterally, design values are $5/6$ the tabular values.

For nails placed in predrilled holes in which the diameter is not larger than 75% of the nail diameter, the design values may be taken at 100% of the tabular values.

Required penetration for full design value is 12 diameters.

Actual penetration greater than the required penetration does not increase the design value.

Minimum penetration for half the design value is 6 diameters.

Actual penetration less than 6 diameters provides no allowable capacity.

For nails having a length of penetration between the minimum penetration and the required penetration for full design value, the design values may be interpolated linearly.

For withdrawal of nails in side plates, the withdrawal value per inch of penetration may be taken as the tabular value divided by the required length of penetration.

Table T-7 Lateral design loads for nails and spikes *(continued)*

Fastener Type		Condition of Wood[b]		Wet Service Factor C_M
		At Time of Fabrication	In service	
BOLTS		Dry Partially seasoned or wet Dry or wet Dry or wet	Dry Dry Exposed to weather Wet	1.00 see footnote c 0.75 0.67
COMMON NAILS, BOX NAILS, or COMMON SPIKES	Withdrawal Loads	Dry Partially seasoned or wet Partially seasoned or wet Dry	Dry Wet Dry Subject to wetting and drying	1.00 1.00 0.25 0.25
	Lateral Loads	Dry Partially seasoned or wet Dry	Dry Dry or wet Partially seasoned or wet	1.00 0.75 0.75
THREADED, HARDENED STEEL NAILS		Dry or wet	Dry or wet	1.00

[a] Source: National Design Specification for Wood Construction

[b] "Conditions of Wood" are defined as follows for determining wet service factors for connections:
"Dry" wood has a moisture content not greater than 19%.
"Wet" wood has a moisture content 30% or more (approximate fiber saturation point).
"Partially seasoned" wood has a moisture content between 19% and 30%.
"Exposed to weather" means that the wood will vary in moisture content from dry to partially seasoned, but is not expected to reach the fiber saturation point at times when the connection is supporting full design load.
"Subject to wetting and drying" means that the wood will vary in moisture content from dry to partially seasoned or wet, or vice versa, with consequent effects on the tightness of the connection.

[c] When bolts are installed in wood that is wet at the time of fabrication, but will be dry before full design load is applied, the following wet service factors C_M shall apply:

Arrangement of bolts	C_M
one fastener only	1.0
one or more fasteners placed in a single row parallel to grain	1.0
fasteners placed in two or more rows parallel to grain with separate splice plates for each row.	1.0
all other arrangements	0.4

When bolts are installed in wood that is partially seasoned at the time of fabrication, but that will be dry before full design load is applied, proportional intermediate wet service factors shall be permitted to be used.

Table T-8 Wet service factors C_M for bolts and nails[a]

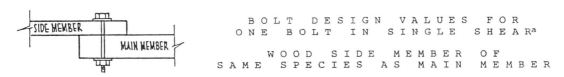

BOLT DESIGN VALUES FOR
ONE BOLT IN SINGLE SHEAR[a]

WOOD SIDE MEMBER OF
SAME SPECIES AS MAIN MEMBER

Member thickness			DOUGLAS FIR-LARCH			SOUTHERN PINE			SPRUCE-PINE-FIR		
main member in.	side member in.	Dia. of bolt in.	Parallel to grain P lbs	Perpendicular to grain side Q lbs.	main Q lbs.	Parallel to grain P lbs.	Perpendicular to grain side Q lbs.	main Q lbs.	Parallel to grain P lbs.	Perpendicular to grain side Q lbs.	main Q lbs.
1 1/2	1 1/2	1/2	480	300	300	530	330	330	410	240	240
		5/8	600	360	360	660	400	400	510	290	290
		3/4	720	420	420	800	460	460	610	340	340
		7/8	850	470	470	930	520	520	710	380	380
		1	970	530	530	1060	580	580	810	430	430
3	1 1/2	1/2	610	370	420	660	400	470	540	320	330
		5/8	880	520	480	940	560	550	780	410	390
		3/4	1190	590	550	1270	660	620	1000	450	440
		7/8	1390	630	610	1520	720	690	1160	490	490
		1	1590	680	670	1740	770	750	1330	530	540
3 1/2	1 1/2	1/2	610	370	430	660	400	470	540	320	370
		5/8	880	520	540	940	560	620	780	410	430
		3/4	1200	590	610	1270	660	690	1080	450	480
		7/8	1590	630	680	1680	720	770	1340	490	540
		1	1830	680	740	2010	770	830	1530	530	590
3 1/2	3 1/2	1/2	720	490	490	750	520	520	660	430	430
		5/8	1120	700	700	1170	780	780	1020	590	590
		3/4	1610	870	870	1690	960	960	1420	730	730
		7/8	1970	1060	1060	2170	1160	1160	1660	890	890
		1	2260	1230	1230	2480	1360	1360	1890	1000	1000
5 1/2	3 1/2	5/8	1120	700	730	1170	780	780	1020	590	650
		3/4	1610	870	1030	1690	960	1090	1480	730	900
		7/8	1970	1060	1260	2300	1160	1410	1920	910	1010
		1	2260	1290	1390	2870	1390	1550	2330	1120	1120
7 1/2	3 1/2	5/8	1120	700	730	1170	780	780	1020	590	650
		3/4	1610	870	1030	1690	960	1090	1480	730	910
		7/8	1970	1060	1360	2300	1160	1450	1920	910	1180
		1	2260	1290	1630	2870	1390	1830	2330	1120	1300

[a]Source: National Design Specification for Wood Construction

Tabulated lateral design values for bolted connections shall be multiplied by all applicable adjustment factors [see discussions with Equations (30-4a and b) in the text].

Tabulated values are for bolts having a bending yield strength of 45,000 psi or more.

Table T-9a Bolt design values

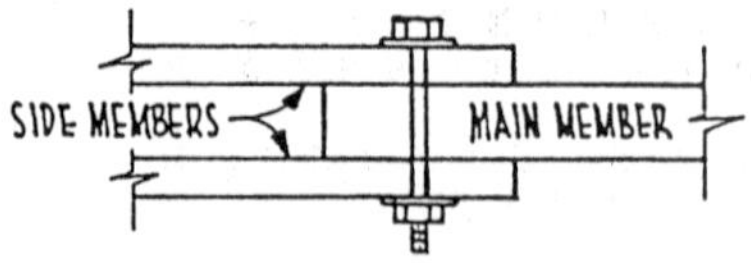

BOLT DESIGN VALUES FOR ONE BOLT IN DOUBLE SHEAR[a]

WOOD SIDE MEMBERS OF SAME SPECIES AS MAIN MEMBER

Member thickness main member in.	Member thickness side member in.	Dia. of bolt in.	DOUGLAS FIR-LARCH Parallel to grain P lbs	DOUGLAS FIR-LARCH Perpendicular to grain side Q lbs.	DOUGLAS FIR-LARCH Perpendicular to grain main Q lbs.	SOUTHERN PINE Parallel to grain P lbs.	SOUTHERN PINE Perpendicular to grain side Q lbs.	SOUTHERN PINE Perpendicular to grain main Q lbs.	SPRUCE-PINE-FIR Parallel to grain P lbs.	SPRUCE-PINE-FIR Perpendicular to grain side Q lbs.	SPRUCE-PINE-FIR Perpendicular to grain main Q lbs.
		1/2	1050	730	470	1150	800	550	880	640	370
		5/8	1310	1040	530	1440	1130	610	1100	830	410
1 1/2	1 1/2	3/4	1580	1170	590	1730	1330	660	1320	900	450
		7/8	1840	1260	630	2020	1440	720	1540	970	490
		1	2310	1350	680	2310	1530	770	1760	1050	530
		1/2	1230	730	860	1320	800	940	1080	640	740
		5/8	1760	1040	1050	1870	1130	1220	1570	830	830
3	1 1/2	3/4	2400	1170	1170	2550	1330	1330	2160	900	900
		7/8	3180	1260	1260	3360	1440	1440	2880	970	970
		1	4090	1350	1350	4310	1530	1530	3530	1050	1050
		1/2	1230	730	860	1320	800	940	1080	640	740
		5/8	1760	1040	1190	1870	1130	1290	1570	830	960
3 1/2	1 1/2	3/4	2400	1170	1370	2550	1330	1550	2160	900	1050
		7/8	3180	1260	1470	3360	1440	1680	2880	970	1130
		1	4090	1350	1580	4310	1530	1790	3530	1050	1230
		1/2	1430	970	970	1500	1040	1040	1310	870	860
		5/8	2240	1410	1230	2340	1560	1420	2050	1170	960
3 1/2	3 1/2	3/4	3220	1750	1370	3380	1910	1550	2950	1460	1050
		7/8	4290	2130	1470	4600	2330	1680	3600	1810	1130
		1	4900	2580	1580	5380	2780	1790	4110	2240	1230
		5/8	2240	1410	1460	2340	1560	1560	2050	1170	1310
5 1/2	3 1/2	3/4	3220	1750	2050	3380	1910	2180	2950	1460	1650
		7/8	4390	2130	2310	4600	2330	2650	3840	1810	1780
		1	5330	2580	2480	5740	2780	2810	4660	2240	1930
		5/8	2240	1410	1460	2340	1560	1560	2050	1170	1310
7 1/2	3 1/2	3/4	3220	1750	2050	3380	1910	2180	2950	1460	1820
		7/8	4390	2130	2720	4600	2330	2890	3840	1810	2420
		1	5330	2580	3380	5740	2780	3680	4660	2240	2630

[a]Source: National Design Specification for Wood Construction

Tabulated lateral design values for bolted connections shall be multiplied by all applicable adjustment factors [see discussions with Equations (30-4a and b) in the text).

Tabulated values are for bolts having a bending yield strength of 45,000 psi or more.

Table T-9b Bolt design values

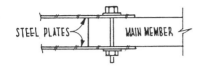

BOLT DESIGN VALUES FOR
ONE BOLT IN DOUBLE SHEAR[a]

ASTM A36 STEEL SIDE MEMBERS
$1/4$ INCH THICKNESS

Length of bolt in main member in.	Dia. of bolt in.	DOUGLAS FIR-LARCH		SOUTHERN PINE		SPRUCE-PINE-FIR	
		Parallel to grain	Perpendicular to grain	Parallel to grain	Perpendicular to grain	Parallel to grain	Perpendicular to grain
		P lbs	Q lbs	P lbs.	Q lbs	P lbs.	Q lbs
$1 1/2$	$1/2$	1050	470	1150	550	880	370
	$5/8$	1310	530	1440	610	1100	410
	$3/4$	1580	590	1730	660	1320	450
	$7/8$	1840	630	2020	720	1540	490
	1	2100	680	2310	770	1760	530
3	$1/2$	1510	940	1570	1000	1400	740
	$5/8$	2250	1050	2350	1220	2090	830
	$3/4$	3150	1170	3300	1330	2640	900
	$7/8$	3680	1260	4040	1440	3080	970
	1	4200	1350	4610	1530	3530	1050
$3 1/2$	$1/2$	1510	940	1570	1000	1400	840
	$5/8$	2250	1230	2350	1420	2090	960
	$3/4$	3170	1370	3300	1550	2940	1050
	$7/8$	4260	1470	4440	1680	3600	1130
	1	4900	1580	5380	1790	4110	1230
$5 1/2$	$5/8$	2250	1330	2350	1420	2090	1190
	$3/4$	3170	1800	3300	1910	2940	1590
	$7/8$	4260	2310	4440	2470	3940	1780
	1	5520	2480	5750	2810	5110	1930
$7 1/2$	$5/8$	2250	1330	2350	1420	2090	1190
	$3/4$	3170	1800	3300	1910	2940	1590
	$7/8$	4260	2320	4440	2470	3940	2060
	1	5520	2910	5750	3090	5110	2590
$9 1/2$	$3/4$	3170	1800	3300	1910	2940	1590
	$7/8$	4260	2320	4440	2470	3940	2060
	1	5520	2910	5750	3090	5110	2590
$11 1/2$	$7/8$	4260	2320	4440	2470	3940	2060
	1	5520	2910	5750	3090	5110	2590

[a]Source: National Design Specification for Wood Construction

Tabulated lateral design values for bolted connections shall be multiplied by all applicable adjustment factors [see discussions with Equations (30-4a and b) in the text].

Tabulated values are for bolts having a bending yield strength of 45,000 psi or more and a dowel bearing strength of 58,000 psi for ASTM A36 steel.

Table T-9c Bolt design values

Dimension	Parallel-to-Grain Loading	Perpendicular-to-Grain Loading
A, c–c spacing of bolts in a row	Four times bolt diameter for full design value. Minimum 3 times bolt diameter for 75% of full design value. Use straight-line interpolation for design values for intermediate spacing.	Design value limited by spacing requirements of attached member or members (whether of metal or of wood loaded parallel to grain).
Staggered bolts	Adjacent bolts are considered to be placed at critical section unless bolts in a row are spaced at a minimum of 8 times the bolt diameter (see sketch below).	Staggering of bolts is desirable for members loaded perpendicular to grain.
B, row spacing	Minimum of $1\frac{1}{2}$ times bolt diameter.	$2\frac{1}{2}$ times bolt diameter for l/D [b] ratio of 2; 5 times bolt diameter for l/D ratios of 6 or more; use straight-line interpolation for l/D between 2 and 6.

Spacing between rows paralleling a member may not exceed 5 in. unless separate splice plates are used for each row.

Dimension	Parallel-to-Grain Loading	Perpendicular-to-Grain Loading
C, end distance	In tension, 7 times bolt diameter for softwoods and 5 times bolt diameter for hardwoods for full design load. In compression, 4 times bolt diameter for full design load. Minimum end distance is $\frac{1}{2}$ that for full load for which the design value is 50% of that for full end distance. Interpolate for intermediate end distances.	4 times bolt diameter for full design load, 2 times bolt diameter for 50% of design load. Interpolate for intermediate loads. When members abut at a joint (not illustrated) the strength of the joint shall be evaluated also as a beam supported by fastenings
E, edge distance	$1\frac{1}{2}$ times the bolt diameter; except that for l/D ratios of more than 6, use $1\frac{1}{2}$ times bolt diameter or one-half the row spacing B, whichever is greater.	Minimum of 4 times bolt diameter at edge toward which load acts; minimum of $1\frac{1}{2}$ times bolt diameter at opposite edge.

[a] Source: National Design Specification for Wood Construction
[b] Ratio of length of bolt in main member l to diameter of bolt D.

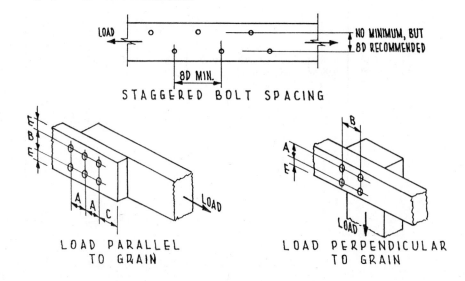

Table T-10 Spacing, edge and end distances for bolts[a]

GROUP ACTION FACTOR Cg FOR BOLTED CONNECTIONS[a]

WOOD SIDE PLATES

Bolt diameter 1"; spacing of bolts in a row, s = 4"; E = 1,400,000 psi

A_s/A_m [1]	A_s [1] in^2	Number of fasteners in a row										
		2	3	4	5	6	7	8	9	10	11	12
0.5	5	0.98	0.92	0.84	0.75	0.68	0.61	0.55	0.50	0.45	0.41	0.38
	12	0.99	0.96	0.92	0.87	0.81	0.76	0.70	0.65	0.61	0.57	0.53
	20	0.99	0.98	0.95	0.91	0.87	0.83	0.78	0.74	0.70	0.66	0.62
	28	1.00	0.98	0.96	0.93	0.90	0.87	0.83	0.79	0.76	0.72	0.69
	40	1.00	0.99	0.97	0.95	0.93	0.90	0.87	0.84	0.81	0.78	0.75
	64	1.00	0.99	0.98	0.97	0.95	0.93	0.91	0.89	0.87	0.84	0.82
1	5	1.00	0.97	0.91	0.85	0.78	0.71	0.64	0.59	0.54	0.49	0.45
	12	1.00	0.99	0.96	0.93	0.88	0.84	0.79	0.74	0.70	0.65	0.61
	20	1.00	0.99	0.98	0.95	0.92	0.89	0.86	0.82	0.78	0.75	0.71
	28	1.00	0.99	0.98	0.97	0.94	0.92	0.89	0.86	0.83	0.80	0.77
	40	1.00	1.00	0.99	0.98	0.96	0.94	0.92	0.90	0.87	0.85	0.82
	64	1.00	1.00	0.99	0.98	0.97	0.96	0.95	0.93	0.91	0.90	0.88

[a] Source: National Design Specification for Wood Construction.
NOTES: 1. When $A_s/A_m > 1.0$, use A_m/A_s and use A_m instead of A_s.
2. Tabulated group action factors (C_g) are conservative for D < 1", s < 4" or E > 1,400,000 psi.

Table T-11 Group action Cg for wood plates

GROUP ACTION FACTOR Cg FOR BOLTED CONNECTIONS[a]

STEEL SIDE PLATES

Bolt diameter 1"; spacing of bolts in a row, s = 4"; E = 1,400,000 psi

A_m/A_s	A_m in^2	Number of fasteners in a row										
		2	3	4	5	6	7	8	9	10	11	12
12	5	0.97	0.89	0.80	0.70	0.62	0.55	0.49	0.44	0.40	0.37	0.34
	8	0.98	0.93	0.85	0.77	0.70	0.63	0.57	0.52	0.47	0.43	0.40
	16	0.99	0.96	0.92	0.86	0.80	0.75	0.69	0.64	0.60	0.55	0.52
	24	0.99	0.97	0.94	0.90	0.85	0.81	0.76	0.71	0.67	0.63	0.59
	40	1.00	0.98	0.96	0.94	0.90	0.87	0.83	0.79	0.76	0.72	0.69
	64	1.00	0.99	0.98	0.96	0.94	0.91	0.88	0.86	0.83	0.80	0.77
	120	1.00	0.99	0.99	0.98	0.96	0.95	0.93	0.91	0.90	0.87	0.85
	200	1.00	1.00	0.99	0.99	0.98	0.97	0.96	0.95	0.93	0.92	0.90
18	5	0.99	0.93	0.85	0.76	0.68	0.61	0.54	0.49	0.44	0.41	0.37
	8	0.99	0.95	0.90	0.83	0.75	0.69	0.62	0.57	0.52	0.48	0.44
	16	1.00	0.98	0.94	0.90	0.85	0.79	0.74	0.69	0.65	0.60	0.56
	24	1.00	0.98	0.96	0.93	0.89	0.85	0.80	0.76	0.72	0.68	0.64
	40	1.00	0.99	0.97	0.95	0.93	0.90	0.87	0.83	0.80	0.77	0.73
	64	1.00	0.99	0.98	0.97	0.95	0.93	0.91	0.89	0.86	0.83	0.81
	120	1.00	1.00	0.99	0.98	0.97	0.96	0.95	0.93	0.92	0.90	0.88
	200	1.00	1.00	0.99	0.99	0.98	0.98	0.97	0.96	0.95	0.94	0.92
24	40	1.00	0.99	0.97	0.95	0.93	0.89	0.86	0.83	0.79	0.76	0.72
	64	1.00	0.99	0.98	0.97	0.95	0.93	0.91	0.88	0.85	0.83	0.80
	120	1.00	1.00	0.99	0.98	0.97	0.96	0.95	0.93	0.91	0.90	0.88
	200	1.00	1.00	0.99	0.99	0.98	0.98	0.97	0.96	0.95	0.93	0.92
30	40	1.00	0.98	0.96	0.93	0.89	0.85	0.81	0.77	0.73	0.69	0.65
	64	1.00	0.99	0.97	0.95	0.93	0.90	0.87	0.83	0.80	0.77	0.73
	120	1.00	0.99	0.99	0.97	0.96	0.94	0.92	0.90	0.88	0.85	0.83
	200	1.00	1.00	0.99	0.98	0.97	0.96	0.95	0.94	0.92	0.90	0.89

[a] Source: National Design Specification for Wood Construction.
NOTE: 1. Tabulated group action factors (C_g) are conservative for D < 1" or s < 4".

Table T-12 Group action factor Cg for steel plates

DESIGN VALUES F_g IN PSI FOR END GRAIN IN BEARING[a]

SAWN LUMBER AND TIMBER

Species	Wet service conditions[b]	Dry service conditions[c]	
		5 in. x 5 in. and larger	2 in. to 4 in. thick
DOUGLAS FIR-LARCH	1350	1480	2020
SOUTHERN PINE	1320	1450	1970
SPRUCE-PINE-FIR	940	1040	1410

[a] Source: National Design Specification for Wood Construction
[b] Wet service conditions are defined as exceeding 19% moisture content.

Table T-13 Design values F_g for end grain in bearing

MINIMUM LATERAL SUPPORT REQUIREMENTS FOR WOOD BEAMS[a]

(Compute depth-to thickness ratio d/t using nominal dimensions)

$d/t \leq 2$	No lateral supports are required
$d/t = 3$ or 4	Full-depth lateral supports are required at bearings (or ends) of the beam
$d/t = 5$	Continuous support is required at one edge over the entire length of the beam
$d/t = 6$	Full-depth lateral support is required at the bearings (or ends) and at intervals not to exceed 8 ft
	or, Continuous support is required at both edges
	or, Continuous support is required at the compression edge edge plus full-depth lateral support at the end bearings
$d/t = 7$	Continuous support is required at both edges

Following general practices are recommended for all d/t

1. Provide full-depth blocking or bridging at all bearing lines
2. Provide full-depth blocking or bridging at any line of concentrated loads above or below (such as a stud wall)
3. Provide full-depth blocking (or a fascia piece) along the free end of cantilevers

[a] Source: National Design Specification for Wood Construction

Table T-14 Minimum lateral support requirements for wood beams[a]

ALLOWABLE SPANS FOR FLOOR JOISTS — 40 LBS/FT² LIVE LOAD[a]

DESIGN CRITERIA:
Deflection — For 40 psf (1.92 kN/m²) live load.
Limited to span in inches (mm) divided by 360.
Strength — Live load of 40 psf (1.92 kN/m²) plus dead load of 10 psf (0.48 kN/m²) determines the required bending design value.

Joist Size (in)	Spacing (in)	Modulus of Elasticity, E, in 1,000,000 psi $\times$ 0.00689 for N/mm²																
$\times$ 25.4 for mm		0.8	0.9	1.0	1.1	1.2	1.3	1.4	1.5	1.6	1.7	1.8	1.9	2.0	2.1	2.2	2.3	2.4
2 x 6	12.0	8-6	8-10	9-2	9-6	9-9	10-0	10-3	10-6	10-9	10-11	11-2	11-4	11-7	11-9	11-11	12-1	12-3
	16.0	7-9	8-0	8-4	8-7	8-10	9-1	9-4	9-6	9-9	9-11	10-2	10-4	10-6	10-8	10-10	11-0	11-2
	19.2	7-3	7-7	7-10	8-1	8-4	8-7	8-9	9-0	9-2	9-4	9-6	9-8	9-10	10-0	10-2	10-4	10-6
	24.0	6-9	7-0	7-3	7-6	7-9	7-11	8-2	8-4	8-6	8-8	8-10	9-0	9-2	9-4	9-6	9-7	9-9
2 x 8	12.0	11-3	11-8	12-1	12-6	12-10	13-2	13-6	13-10	14-2	14-5	14-8	15-0	15-3	15-6	15-9	15-11	16-2
	16.0	10-2	10-7	11-0	11-4	11-8	12-0	12-3	12-7	12-10	13-1	13-4	13-7	13-10	14-1	14-3	14-6	14-8
	19.2	9-7	10-0	10-4	10-8	11-0	11-3	11-7	11-10	12-1	12-4	12-7	12-10	13-0	13-3	13-5	13-8	13-10
	24.0	8-11	9-3	9-7	9-11	10-2	10-6	10-9	11-0	11-3	11-5	11-8	11-11	12-1	12-3	12-6	12-8	12-10
2 x 10	12.0	14-4	14-11	15-5	15-11	16-5	16-10	17-3	17-8	18-0	18-5	18-9	19-1	19-5	19-9	20-1	20-4	20-8
	16.0	13-0	13-6	14-0	14-6	14-11	15-3	15-8	16-0	16-5	16-9	17-0	17-4	17-8	17-11	18-3	18-6	18-9
	19.2	12-3	12-9	13-2	13-7	14-0	14-5	14-9	15-1	15-5	15-9	16-0	16-4	16-7	16-11	17-2	17-5	17-8
	24.0	11-4	11-10	12-3	12-8	13-0	13-4	13-8	14-0	14-4	14-7	14-11	15-2	15-5	15-8	15-11	16-2	16-5
2 x 12	12.0	17-5	18-1	18-9	19-4	19-11	20-6	21-0	21-6	21-11	22-5	22-10	23-3	23-7	24-0	24-5	24-9	25-1
	16.0	15-10	16-5	17-0	17-7	18-1	18-7	19-1	19-6	19-11	20-4	20-9	21-1	21-6	21-10	22-2	22-6	22-10
	19.2	14-11	15-6	16-0	16-7	17-0	17-6	17-11	18-4	18-9	19-2	19-6	19-10	20-2	20-6	20-10	21-2	21-6
	24.0	13-10	14-4	14-11	15-4	15-10	16-3	16-8	17-0	17-5	17-9	18-1	18-5	18-9	19-1	19-4	19-8	19-11
F_b	12.0	718	777	833	888	941	993	1,043	1,092	1,140	1,187	1,233	1,278	1,323	1,367	1,410	1,452	1,494
	16.0	790	855	917	977	1,036	1,093	1,148	1,202	1,255	1,306	1,357	1,407	1,456	1,504	1,551	1,598	1,644
	19.2	840	909	975	1,039	1,101	1,161	1,220	1,277	1,333	1,388	1,442	1,495	1,547	1,598	1,649	1,698	1,747
	24.0	905	979	1,050	1,119	1,186	1,251	1,314	1,376	1,436	1,496	1,554	1,611	1,667	1,722	1,776	1,829	1,882

NOTE: The required bending design value, F_b, in pounds per square inch ($\times$ 0.00689 for N/mm²) is shown at the bottom of this table and is applicable to all lumber sizes shown. Spans are shown in feet-inches (1 foot = 304.8 mm, 1 inch = 25.4 mm) and are limited to 26 feet (7925 mm) and less.

[a]Source:Uniform Building Code, 1994

Table T-15 Allowable spans for floor joists

ALLOWABLE SPANS FOR CEILING JOISTS — 10 LBS/FT² LIVE LOAD

DESIGN CRITERIA:
Deflection — For 10 (0.48 kN/m²) psf live load.
Limited to span in inches (mm) divided by 240.
Strength — Live load of 10 psf (0.48 kN/mm²) plus dead load of 5 psf (0.24 kN/m²) determines the required fiber stress value.

Joist Size (in)	Spacing (in)	Modulus of Elasticity, E, in 1,000,000 psi $\times$ 0.00689 for N/mm²																
$\times$ 25.4 for mm		0.8	0.9	1.0	1.1	1.2	1.3	1.4	1.5	1.6	1.7	1.8	1.9	2.0	2.1	2.2	2.3	2.4
2 x 4	12.0	9-10	10-3	10-7	10-11	11-3	11-7	11-10	12-2	12-5	12-8	12-11	13-2	13-4	13-7	13-9	14-0	14-2
	16.0	8-11	9-4	9-8	9-11	10-3	10-6	10-9	11-0	11-3	11-6	11-9	11-11	12-2	12-4	12-6	12-9	12-11
	19.2	8-5	8-9	9-1	9-4	9-8	9-11	10-2	10-4	10-7	10-10	11-0	11-3	11-5	11-7	11-9	12-0	12-2
	24.0	7-10	8-1	8-5	8-8	8-11	9-2	9-5	9-8	9-10	10-0	10-3	10-5	10-7	10-9	10-11	11-1	11-3
2 x 6	12.0	15-6	16-1	16-8	17-2	17-8	18-2	18-8	19-1	19-6	19-11	20-3	20-8	21-0	21-4	21-8	22-0	22-4
	16.0	14-1	14-7	15-2	15-7	16-1	16-6	16-11	17-4	17-8	18-1	18-5	18-9	19-1	19-5	19-8	20-0	20-3
	19.2	13-3	13-9	14-3	14-8	15-2	15-7	15-11	16-4	16-8	17-0	17-4	17-8	17-11	18-3	18-6	18-10	19-1
	24.0	12-3	12-9	13-3	13-8	14-1	14-5	14-9	15-2	15-6	15-9	16-1	16-4	16-8	16-11	17-2	17-5	17-8
2 x 8	12.0	20-5	21-2	21-11	22-8	23-4	24-0	24-7	25-2	25-8								
	16.0	18-6	19-3	19-11	20-7	21-2	21-9	22-4	22-10	23-4	23-10	24-3	24-8	25-2	25-7	25-11		
	19.2	17-5	18-1	18-9	19-5	19-11	20-6	21-0	21-6	21-11	22-5	22-10	23-3	23-8	24-0	24-5	24-9	25-2
	24.0	16-2	16-10	17-5	18-0	18-6	19-0	19-6	19-11	20-5	20-10	21-2	21-7	21-11	22-4	22-8	23-0	23-4
2 x 10	12.0	26-0																
	16.0	23-8	24-7	25-5														
	19.2	22-3	23-1	23-11	24-9	25-5												
	24.0	20-8	21-6	22-3	22-11	23-8	24-3	24-10	25-5	26-0								
F_b	12.0	711	769	825	880	932	983	1,033	1,082	1,129	1,176	1,221	1,266	1,310	1,354	1,396	1,438	1,480
	16.0	783	847	909	968	1,026	1,082	1,137	1,191	1,243	1,294	1,344	1,394	1,442	1,490	1,537	1,583	1,629
	19.2	832	900	965	1,029	1,090	1,150	1,208	1,265	1,321	1,375	1,429	1,481	1,533	1,583	1,633	1,682	1,731
	24.0	896	969	1,040	1,108	1,174	1,239	1,302	1,363	1,423	1,481	1,539	1,595	1,651	1,706	1,759	1,812	1,864

NOTE: The required bending design value, F_b, in pounds per square inch ($\times$ 0.00689 for N/mm²) is shown at the bottom of this table and is applicable to all lumber sizes shown. Spans are shown in feet-inches (1 foot = 304.8 mm, 1 inch = 25.4 mm) and are limited to 26 feet (7925 mm) and less.

[a]Source: Uniform Building Code, 1994

Table T-16 Allowable spans for ceiling joists

DESIGN CRITERIA:
Strength—Live load of 20 psf (0.96 kN/m^2) plus dead load of 10 psf (0.48 kN/m^2) determines the required bending design value.
Deflection—For 20 psf (0.96 kN/m^2) live load.
Limited to span in inches (mm) divided by 240.

Bending Design Value, F_b (psi) ($\times$ 0.00689 for N/mm^2)

× 25.4 for mm

Size (in)	Spacing (in)	300	400	500	600	700	800	900	1000	1100	1200	1300	1400	1500	1600	1700	1800	1900	2000
2 × 6	12.0	7-1	8-2	9-2	10-0	10-10	11-7	12-4	13-0	13-7	14-2	14-9	15-4	15-11	16-5	17-5	17-10		
	16.0	6-2	7-1	7-11	8-8	9-5	10-0	10-8	11-3	11-9	12-4	12-10	13-3	13-9	14-2	14-8	15-1	15-6	15-11
	19.2	5-7	6-6	7-3	7-11	8-7	9-2	9-9	10-3	10-9	11-3	11-8	12-2	12-7	13-0	13-4	13-9	14-2	14-6
	24.0	5-0	5-10	6-6	7-1	7-8	8-2	8-8	9-2	9-7	10-0	10-5	10-10	11-3	11-7	11-11	12-4	12-8	13-0
2 × 8	12.0	10.10	12-1	13-3	14-4	15-3	16-3	17-1	17-11	18-9	19-6	20-3	21-7	22-3	22-11	23-7			
	16.0	9-1	9-4	10-6	11-6	12-5	13-3	14-0	14-10	15-6	16-3	16-10	17-6	18-1	18-9	19-4	19-10	20-5	20-11
	19.2	7-5	8-7	9-7	10-6	11-4	12-1	12-10	13-6	14-2	14-10	15-5	16-0	16-7	17-1	17-7	18-1	18-7	19-1
	24.0	6-7	7-8	8-7	9-4	10-1	10-10	11-6	12-1	12-8	13-3	13-9	14-4	14-10	15-3	15-9	16-3	16-8	17-1
2 × 10	12.0	11-11	13-9	15-5	16-11	18-3	19-6	20-8	21-10	23-11	24-10	25-10							
	16.0	10-4	11-11	13-4	14-8	15-10	16-11	17-11	18-11	19-10	20-8	21-8	22-4	23-1	23-11	24-7	25-4	26-0	
	19.2	9-5	10-11	12-2	13-4	14-5	15-5	16-4	17-3	18-1	18-11	19-8	20-5	21-1	21-10	22-6	23-1	23-9	24-5
	24.0	8-5	9-9	10-11	11-11	12-11	13-9	14-8	15-5	16-2	16-11	17-7	18-3	18-11	19-6	20-1	20-8	21-3	21-10
2 × 12	12.0	14-6	16-9	18-9	20-6	22-2	23-9	25-2											
	16.0	12-7	14-6	16-3	17-9	19-3	20-6	21-9	23-0	24-1	25-2								
	19.2	11-6	13-3	14-10	16-3	17-6	18-9	19-11	21-0	22-0	23-0	23-11	24-10	25-8					
	24.0	10-3	11-10	13-3	14-6	15-8	16-9	16-9	18-9	19-8	20-6	21-5	22-2	23-0	23-9	24-5	25-2	25-10	
E	12.0	0.15	0.24	0.33	0.44	0.55	0.67	0.80	0.94	1.09	1.24	1.40	1.56	1.73	1.91	2.09	2.28	2.47	
	16.0	0.13	0.21	0.29	0.38	0.48	0.58	0.70	0.82	0.94	1.07	1.21	1.35	1.50	1.65	1.81	1.97	2.14	2.31
	19.2	0.12	0.19	0.26	0.35	0.44	0.53	0.64	0.75	0.86	0.98	1.10	1.23	1.37	1.51	1.65	1.80	1.95	2.11
	24.0	0.11	0.17	0.24	0.31	0.39	0.48	0.57	0.67	0.77	0.88	0.99	1.10	1.22	1.35	1.48	1.61	1.75	1.83

NOTE: The required modulus of elasticity, E, in 1,000,000 pounds per square inch (psi) ($\times$ 0.00689 for N/mm^2) is shown at the bottom of this table, is limited to 2.6 million psi (17 914 N/mm^2) and less, and is applicable to all lumber sizes shown. Spans are shown in feet-inches (1 foot = 304.8 mm, 1 inch = 25.4 mm) and are limited to 26 feet (7925 mm) and less.

Table T-17 Allowable spans for low-slope rafters[a]

DESIGN CRITERIA:
Strength—Live load of 20 psf (0.96 kN/m^2) plus dead load of 10 psf (0.48 kN/m^2) determines the required bending design value.
Deflection—For 20 psf (0.96 kN/m^2) live load.
Limited to span in inches (mm) divided by 180.

Bending Design Value, F_b (psi) ($\times$ 0.00689 for N/mm^2)

× 25.4 for mm

Size (in)	Spacing (in)	300	400	500	600	700	800	900	1000	1100	1200	1300	1400	1500	1600	1700	1800	1900	2000
2 × 4	12.0	4-6	5-3	5-10	6-5	6-11	7-5	7-10	8-3	8-8	9-0	9-5	9-9	10-1	10-5	10-9	11-1	11-4	11-8
	16.0	3-11	4-6	5-1	5-6	6-0	6-5	6-9	7-2	7-6	7-10	8-2	8-5	8-9	9-0	9-4	9-7	9-10	10-1
	19.2	3-7	4-1	4-7	5-1	5-5	5-10	6-2	6-10	7-2	7-5	7-8	8-0	8-3	8-6	8-9	9-0	9-3	
	24.0	3-2	3-8	4-1	4-8	4-11	5-3	5-6	5-10	6-1	6-5	6-8	6-11	7-2	7-5	7-7	7-10	8-0	8-3
2 × 6	12.0	7-1	8-2	9-2	10-0	10-10	11-7	12-4	13-0	13-7	14-2	14-9	15-4	15-11	16-5	16-11	17-5	17-10	18-4
	16.0	6-2	7-1	7-11	8-8	9-5	10-0	10-8	11-3	11-9	12-4	12-10	13-3	13-9	14-2	14-8	15-1	15-6	15-11
	19.2	5-7	6-6	7-3	7-11	8-7	9-2	9-9	10-3	10-9	11-3	11-8	12-2	12-7	13-0	13-4	13-9	14-2	14-6
	24.0	5-0	5-10	6-6	7-1	7-8	8-2	8-8	9-2	9-7	10-0	10-5	10-10	11-3	11-7	11-11	12-4	12-8	13-0
2 × 8	12.0	9-4	10-10	12-1	13-3	14-4	15-3	16-3	17-1	17-11	18-9	19-6	20-3	20-11	21-7	22-3	22-11	23-7	24-2
	16.0	8-1	9-4	10-6	11-6	12-5	13-3	14-0	14-10	15-6	16-3	16-10	17-6	18-1	18-9	19-4	19-10	20-5	20-11
	19.2	7-5	8-7	9-7	10-6	11-4	12-1	12-10	13-6	14-2	14-10	15-5	16-0	16-7	17-1	17-7	18-1	18-7	19-1
	24.0	6-7	7-8	8-7	9-4	10-1	10-10	11-6	12-1	12-8	13-3	13-9	14-4	14-10	15-3	15-9	16-3	16-8	17-1
2 × 10	12.0	11-11	13-9	15-5	16-11	18-3	19-6	20-8	21-10	22-10	23-11	24-10	25-10						
	16.0	10-4	11-11	13-4	14-8	15-10	16-11	17-11	18-11	19-10	20-8	21-6	22-4	23-1	23-11	24-7	25-4	26-0	
	19.2	9-5	10-11	12-2	13-4	14-5	15-5	16-4	17-3	18-1	18-11	19-8	20-5	21-1	21-10	22-6	23-1	23-9	24-5
	24.0	8-5	9-9	10-11	11-11	12-11	13-9	14-8	15-5	16-2	16-11	17-7	18-3	18-11	19-6	20-1	20-8	21-3	21-10
2 × 12	12.0	14-6	16-9	18-9	20-6	22-2	23-9	25-2											
	16.0	12.7	14-6	16-3	17-9	19-3	20-6	21-9	23-0	24-1	25-2								
	19.2	11-6	13-3	14-10	16-3	17-6	18-9	19-11	21-0	22-0	23-0	23-11	24-10	25-8					
	24.0	10-3	11-10	13-3	14-6	15-8	16-9	16-9	18-9	19-8	20-6	21-5	22-2	23-0	23-9	24-5	25-2	25-10	
E	12.0	0.12	0.18	0.25	0.33	0.41	0.51	0.60	0.71	0.82	0.93	1.05	1.17	1.30	1.43	1.57	1.71	1.85	2.00
	16.0	0.10	0.15	0.22	0.28	0.36	0.44	0.52	0.61	0.71	0.80	0.91	1.01	1.13	1.24	1.36	1.40	1.60	1.73
	19.2	0.09	0.14	0.20	0.26	0.33	0.40	0.48	0.56	0.64	0.73	0.83	0.93	1.03	1.13	1.24	1.35	1.46	1.58
	24.0	0.03	0.13	0.18	0.23	0.29	0.35	0.43	0.50	0.58	0.66	0.74	0.83	0.92	1.01	1.11	1.21	1.31	1.41

NOTE: The required modulus of elasticity, E, in 1,000,000 pounds per square inch (psi) ($\times$ 0.00689 for N/mm^2) is shown at the bottom of this table, is limited to 2.6 million psi (17 914 N/mm^2) and less, and is applicable to all lumber sizes shown. Spans are shown in feet-inches (1 foot = 304.8 mm, 1 inch = 25.4 mm) and are limited to 26 feet (7925 mm) and less.
[a] Source: Uniform Building Code, 1994

Table T-18 Allowable spans for high-slope rafters

Recommended Uniform Floor Live Loads for APA RATED STURD-I-FLOOR and APA RATED SHEATHING with Long Dimension Perpendicular to Supports

Sturd-I-Floor Span Rating	Sheathing Span Rating	Maximum Span (in.)	Allowable Live Loads (psf)(a)						
			Joist Spacing (in.)						
			12	16	20	24	32	40	48
16 oc	24/16, 32/16	16	185	100					
20 oc	40/20	20	270	150	100				
24 oc	48/24	24	430	240	160	100			
32 oc	60/32	32		430	295	185	100		
48 oc		48			460	290	160	100	55

(a) 10 psf dead load assumed. Live load deflection limit is ℓ/360.

Note: Shaded joist spacings meet Code Plus recommendations.

PS 1 Plywood Recommendations for Uniformly Loaded Heavy Duty Floors(a)
(Deflection limited to 1/240 of span.) (Span Ratings apply to APA RATED SHEATHING and APA RATED STURD-I-FLOOR, respectively, marked PS 1.)

Uniform Live Load (psf)	Center-to-Center Support Spacing (inches) (Nominal 2-Inch-Wide Supports Unless Noted)					
	12(b)	16(b)	20(b)	24(b)	32	48(c)
50	32/16, 16 oc	32/16, 16 oc	40/20, 20 oc	48/24, 24 oc	48 oc	48 oc
100	32/16, 16 oc	32/16, 16 oc	40/20, 20 oc	48/24, 24 oc	48 oc	1-1/2(d)
150	32/16, 16 oc	32/16, 16 oc	40/20, 20 oc	48/24, 48 oc	48 oc	1-3/4(e), 2(d)
200	32/16, 16 oc	40/20, 20 oc	48/24, 24 oc	48 oc	1-1/8(e), 1-3/8(d)	2(e), 2-1/2(d)
250	32/16, 16 oc	40/20, 24 oc	48/24, 48 oc	48 oc	1-3/8(e), 1-1/2(d)	2-1/4(e)
300	32/16, 16 oc	48/24, 24 oc	48 oc	48 oc	1-1/2(e), 1-5/8(d)	2-1/4(e)
350	40/20, 20 oc	48/24, 48 oc	48 oc	1-1/8(e), 1-3/8(d)	1-1/2(e), 2(d)	
400	40/20, 20 oc	48 oc	48 oc	1-1/4(e), 1-3/8(d)	1-5/8(e), 2(d)	
450	40/20, 24 oc	48 oc	48 oc	1-3/8(e), 1-1/2(d)	2(e), 2-1/4(d)	
500	48/24, 24 oc	48 oc	48 oc	1-1/2(d)	2(e), 2-1/4(d)	

(a) Use plywood with T&G edges, or provide structural blocking at panel edges, or install a separate underlayment.

(b) A-C Group 1 sanded plywood panels may be substituted for span rated Sturd-I-Floor panels (1/2-inch for 16 oc; 5/8-inch for 20 oc; 3/4-inch for 24 oc).

(c) Nominal 4-inch wide supports.

(d) Group 1 face and back, any species inner plies, sanded or unsanded, single layer.

(e) All-Group 1 or Structural I plywood, sanded or unsanded, single layer.

a Source: American Plywood Association Design/Construction Guide for Residential and Commercial Construction

Table T-19 Allowable floor live loads on plywood floors[a]

Recommended Uniform Roof Live Loads for APA RATED SHEATHING(c) and APA RATED STURD-I-FLOOR With Long Dimension Perpendicular to Supports(e)

Panel Span Rating	Minimum Panel Thickness (in.)	Maximum Span (in.)		Allowable Live Loads (psf)(d)							
		With Edge Support(a)	Without Edge Support	Spacing of Supports Center-to-Center (in.)							
				12	16	20	24	32	40	48	60
APA RATED SHEATHING(c)											
12/0	5/16	12	12	30							
16/0	5/16	16	16	70	30						
20/0	5/16	20	20	120	50	30					
24/0	3/8	24	20(b)	190	100	60	30				
24/16	7/16	24	24	190	100	65	40				
32/16	15/32	32	28	325	180	120	70	30			
40/20	19/32	40	32	—	305	205	130	60	30		
48/24	23/32	48	36	—	—	280	175	95	45	35	
60/32	7/8	60	48	—	—	—	305	165	100	70	35
APA RATED STURD-I-FLOOR(f)											
16 oc	19/32	24	24	185	100	65	40				
20 oc	19/32	32	32	270	150	100	60	30			
24 oc	23/32	48	36	—	240	160	100	50	30	25	
32 oc	7/8	48	40	—	—	295	185	100	60	40	
48 oc	1-3/32	60	48	—	—	—	290	160	100	65	40

(a) Tongue-and-groove edges, panel edge clips (one midway between each support, except two equally spaced between supports 48 inches on center), lumber blocking, or other. For low slope roofs, see Table 22.

(b) 24 inches for 15/32-inch and 1/2-inch panels.

(c) Includes APA RATED SHEATHING/CEILING DECK.

(d) 10 psf dead load assumed.

(e) Applies to panels 24 inches or wider applied over two or more spans.

(f) Also applies to C-C Plugged grade plywood.

Note: Shaded support spacings meet Code Plus recommendations.

Recommended Maximum Spans for APA Panel Roof Decks for Low Slope Roofs(a)
(Long panel dimension perpendicular to supports and continuous over two or more spans)

Grade	Minimum Nominal Panel Thickness (in.)	Minimum Span Rating	Maximum Span (in.)	Panel Clips Per Span(b) (number)
APA RATED SHEATHING	15/32	32/16	24	1
	19/32	40/20	32	1
	23/32	48/24	48	2
	7/8	60/32	60	2

(a) Low slope roofs are applicable to built-up, single-ply and modified bitumen roofing systems. For guaranteed or warranted roofs contact membrane manufacturer for acceptable deck.
(b) Edge support may also be provided by tongue-and-groove edges or solid blocking.

a Source: American Plywood Association Design/Construction Guide for Residential and Commercial Construction

Table T-20 Allowable roof live loads on plywood roofsa

Recommended Shear (pounds per foot) for Horizontal APA Panel Diaphragms with Framing of Douglas-Fir:, Larch or Southern Pine[a] for Wind or Seismic Loading

Panel Grade	Common Nail Size	Minimum Nail Penetration in Framing (inches)	Minimum Nominal Panel Thickness (inch)	Minimum Nominal Width of Framing Member (inches)	Blocked Diaphragms				Unblocked Diaphragms	
					Nail Spacing (in.) at diaphragm boundaries (all cases), at continuous panel edges parallel to load (Cases 3 & 4), and at all panel edges (Cases 5 & 6)[b]				Nails Spaced 6″ max. at Supported Edges[b]	
					6	4	2½[c]	2[c]	Case 1 (No unblocked edges or continuous joints parallel to load)	All other configurations (Cases 2, 3, 4, 5 & 6)
					Nail Spacing (in.) at other panel edges (Cases 1, 2, 3, & 4)[b]					
					6	6	4	3		
APA STRUCTURAL I grades	6d[e]	1-1/4	5/16	2	185	250	375	420	165	125
				3	210	280	420	475	185	140
	8d	1-1/2	3/8	2	270	360	530	600	240	180
				3	300	400	600	675	265	200
	10d[d]	1-5/8	15/32	2	320	425	640	730	285	215
				3	360	480	720	820	320	240
APA RATED SHEATHING APA RATED STURD-I-FLOOR and other APA grades except Species Group 5	6d[e]	1-1/4	5/16	2	170	225	335	380	150	110
				3	190	250	380	430	170	125
			3/8	2	185	250	375	420	165	125
				3	210	280	420	475	185	140
	8d	1-1/2	3/8	2	240	320	480	545	215	160
				3	270	360	540	610	240	180
			7/16	2	255	340	505	575	230	170
				3	285	380	570	645	255	190
			15/32	2	270	360	530	600	240	180
				3	300	400	600	675	265	200
	10d[d]	1-5/8	15/32	2	290	385	575	655	255	190
				3	325	430	650	735	290	215
			19/32	2	320	425	640	730	285	215
				3	360	480	720	820	320	240

(a) For framing of other species: (1) Find specific gravity for species of lumber in AFPA National Design Specification. (2) Find shear value from table above for nail size for Structural I panels (regardless of actual grade). (3) Multiply value by 0.82 for species with specific gravity of 0.42 or greater, or 0.65 for all other species.

(b) Space nails maximum 12 in. o.c. along intermediate framing members (6 in. o.c. when supports are spaced 48 in. o.c.).

(c) Framing at adjoining panel edges shall be 3-in. nominal or wider, and nails shall be staggered where nails are spaced 2 inches o.c. or 2-1/2 inches o.c.

(d) Framing at adjoining panel edges shall be 3-in. nominal or wider, and nails shall be staggered where 10d nails having penetration into framing of more than 1-5/8 inches are spaced 3 inches o.c.

(e) 8d is recommended minimum for roofs due to negative pressures of high winds.

Notes: Design for diaphragm stresses depends on direction of continuous panel joints with reference to load, not on direction of long dimension of sheet. Continuous framing may be in either direction for blocked diaphragms.

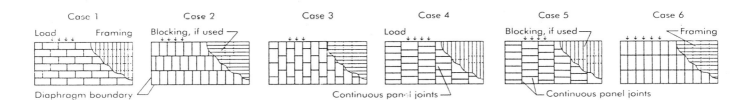

a Source: American Plywood Association Design/Construction Guide for Residential and Commercial Construction

Table T-21 Allowable shear load on horizontal plywood diaphragms

Recommended Shear (pounds per foot) for APA Panel Shear Walls with Framing of Douglas-Fir, Larch, or Southern Pine(a) for Wind or Seismic Loading(b)

Panel Grade	Minimum Nominal Panel Thickness (in.)	Minimum Nail Penetration in Framing (in.)	Panels Applied Direct to Framing — Nail Size (common or galvanized box)	Nail Spacing at Panel Edges (in.) 6	4	3	2(•)	Panels Applied Over 1/2" or 5/8" Gypsum Sheathing — Nail Size (common or galvanized box)	Nail Spacing at Panel Edges (in.) 6	4	3	2(•)
APA STRUCTURAL I grades	5/16	1-1/4	6d	200	300	390	510	8d	200	300	390	510
	3/8			230(d)	360(d)	460(d)	610(d)		280	430	550	730
	7/16	1-1/2	8d	255(d)	395(d)	505(d)	670(d)	10d(f)	—	—	—	—
	15/32			280	430	550	730		—	—	—	—
	15/32	1-5/8	10d(f)	340	510	665	870		—	—	—	—
APA RATED SHEATHING; APA RATED SIDING(g) and other APA grades except species Group 5	5/16 or 1/4(c)	1-1/4	6d	180	270	350	450	8d	180	270	350	450
	3/8	1-1/4	6d	200	300	390	510	8d	200	300	390	510
	3/8			220(d)	320(d)	410(d)	530(d)		260	380	490	640
	7/16	1-1/2	8d	240(d)	350(d)	450(d)	585(d)	10d(f)	—	—	—	—
	15/32			260	380	490	640		—	—	—	—
	15/32	1-5/8	10d(f)	310	460	600	770	—	—	—	—	—
	19/32			340	510	665	870		—	—	—	—
APA RATED SIDING(g) and other APA grades except species Group 5			Nail Size (galvanized casing)					Nail Size (galvanized casing)				
	5/16(c)	1-1/4	6d	140	210	275	360	8d	140	210	275	360
	3/8	1-1/2	8d	160	240	310	410	10d(f)	160	240	310	410(a)

(a) For framing of other species: (1) Find specific gravity for species of lumber in the AFPA National Design Specification.
  (2)(a) For common or galvanized box nails, find shear value from table above for nail size for STRUCTURAL I panels (regardless of actual grade).(b) For galvanized casing nails, take shear value directly from table above. (3) Multiply this value by 0.82 for species with specific gravity of 0.42 or greater, or 0.65 for all other species.

(b) All panel edges backed with 2-inch nominal or wider framing. Install panels either horizontally or vertically. Space nails maximum 6 inches o.c. along intermediate framing members for 3/8-inch and 7/16-inch panels installed on studs spaced 24 inches o.c. For other conditions and panel thicknesses, space nails maximum 12 inches o.c. on intermediate supports.

(c) 3/8-inch or APA RATED SIDING-16 oc is minimum recommended when applied direct to framing as exterior siding.

(d) Shears may be increased to values shown for 15/32-inch sheathing with same nailing provided (1) studs are spaced a maximum of 16 inches o.c., or (2) if panels are applied with long dimension across studs.

(e) Framing at adjoining panel edges shall be 3-inch nominal or wider, and nails shall be staggered where nails are spaced 2 inches o.c.

(f) Framing at adjoining panel edges shall be 3-inch nominal or wider, and nails shall be staggered where 10d nails having penetration into framing of more than 1-5/8 inches are spaced 3 inches o.c.

(g) Values apply to all-veneer plywood APA RATED SIDING panels only. Other APA RATED SIDING panels may also qualify on a proprietary basis. APA RATED SIDING-16 oc plywood may be 11/32-inch, 3/8-inch or thicker. Thickness at point of nailing on panel edges governs shear values.

Typical Layout for Shear Walls

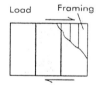
Load Framing

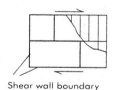
Shear wall boundary

Blocking

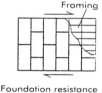
Framing

Foundation resistance

a Source: American Plywood Association Design/Construction Guide for Residential and Commercial Construction

Table T-22 Allowable shear loads on plywood shear panelsa

APA RATED SHEATHING

Thickness in.	Weight psf	Area A in²/ft	Moment of Inertia I in⁴/ft	Section Modulus S in³/ft	Design Stress	24/0 Parallel to strength axis psi	24/0 Perpendicular to strength axis psi	32/16 Parallel to strength axis psi	32/16 Perpendicular to strength axis psi	40/20 Parallel to strength axis psi	40/20 Perpendicular to strength axis psi	48/24 Parallel to strength axis psi	48/24 Perpendicular to strength axis psi
3/8	1.1	4.5	0.053	0.281	F_b	890	192	1320	327	2224	534		
					F_t	511	220	622	280	644	356		
					F_c	633	278	789	344	800	444		
					F_v	55	98	70	121	88	154		
					E_b	1,245,000	68,000	2,386,000	153,000	4,670,000	340,000		
					E_a	744,000	322,000	922,000	400,000	933,000	511,000		
1/2	1.5	6.0	0.125	0.500	F_b	600	194	888	331	1500	540	2030	810
					F_t	500	215	607	271	628	347	867	423
					F_c	475	208	592	258	600	333	833	400
					F_v	45	26	58	33	73	41	94	48
					E_b	528,000	89,000	1,012,000	201,000	1,980,000	446,000	3,520,000	732,000
					E_a	558,000	242,000	692,000	300,000	700,000	383,000	975,000	467,000
5/8	1.9	7.5	0.244	0.781	F_b	384	124	569	212	960	346	1300	518
					F_t	403	172	485	217	503	277	693	338
					F_c	380	167	473	207	480	267	667	320
					F_v	36	21	46	26	58	33	75	38
					E_b	271,000	45,700	518,000	103,000	1,014,000	229,000	1,803,000	375,000
					E_a	447,000	193,000	553,000	240,000	560,000	307,000	780,000	373,000
3/4	2.3	9.0	0.422	1.125	F_b	266	86	395	148	667	239	900	360
					F_t	338	143	404	181	419	231	578	282
					F_c	317	139	394	172	400	222	556	267
					F_v	30	18	39	22	49	28	62	32
					E_b	156,000	26,400	300,000	60,000	586,000	132,000	1,043,000	217,000
					E_a	372,000	161,000	461,000	200,000	467,000	256,000	650,000	311,000

APA RATED STURD-I-FLOOR

Thickness in.	Weight psf	Area A in²/ft	Moment of Inertia I in⁴/ft	Section Modulus S in³/ft	Design Stress	20 oc Parallel to strength axis psi	20 oc Perpendicular to strength axis psi	24 oc Parallel to strength axis psi	24 oc Perpendicular to strength axis psi	32 oc Parallel to strength axis psi	32 oc Perpendicular to strength axis psi	48 oc Parallel to strength axis psi	48 oc Perpendicular to strength axis psi
5/8	1.9	7.5	0.244	0.781	F_b	737	323	983	496	1337	876	2458	1567
					F_t	503	277	581	338	693	433	971	867
					F_c	480	267	560	320	660	413	933	833
					F_v	58	34	75	39	88	56	132	90
					E_b	947,000	165,000	1,352,000	330,000	2,480,000	877,000	5,184,000	2,033,000
					E_a	560,000	307,000	653,000	373,000	767,000	487,000	1,093,000	973,000
3/4	2.3	9.0	0.422	1.125	F_b	512	224	683	344	928	608	1707	1088
					F_t	419	231	484	282	578	361	809	722
					F_c	400	222	467	267	550	344	778	694
					F_v	49	28	62	33	73	47	110	75
					E_b	547,000	95,500	782,000	191,000	1,434,000	507,000	3,000,000	1,175,000
					E_a	467,000	256,000	544,000	311,000	639,000	406,000	911,000	811,000
7/8	2.6	10.5	0.670	1.531	F_b	376	165	502	253	682	447	1254	800
					F_t	359	198	415	241	495	310	693	619
					F_c	343	190	400	229	471	295	667	595
					F_v	42	24	53	28	63	40	94	64
					E_b	345,000	60,100	493,000	120,000	903,000	319,000	1,888,000	740,000
					E_a	400,000	219,000	467,000	267,000	548,000	348,000	781,000	695,000
1	3.0	12.0	1.000	2.000	F_b	288	126	384	194	522	342	960	612
					F_t	314	173	363	211	433	271	607	542
					F_c	300	167	350	200	413	258	583	521
					F_v	36	21	47	24	55	35	83	56
					E_b	231,000	40,300	330,000	80,600	605,000	214,000	1,265,000	496,000
					E_a	350,000	192,000	408,000	233,000	479,000	304,000	683,000	608,000

[a] Source: American Plywood Association Technical Note N-375, September, 1991
[b] The strength axis is the long panel dimension unless otherwise identified.
[c] E_b is the modulus of elasticity in bending; E_a is the modulus of elasticity under axial load. All other symbols are standard symbols defined in the text.

Table T-23 Allowable unit stress on APA rated plywood[a,b,c]